Carbohydrate-Active Enzymes (CAZymes)

This book provides a comprehensive exploration of Carbohydrate-Active Enzymes (CAZymes) in addressing the climate crisis where the biotechnological advances in utilizing this diverse enzyme group for climate change mitigation have been examined. An overview of the CAZymes from their sources, structure, diversity, biological significance, and potential for alleviating climate change have been discussed. It further examines the use of CAZymes in sustainable practices for their application in various sectors from food and agriculture to textile and pharmaceutical industries as well as medicine and combating climate change.

- Provides a detailed analysis of CAZymes for their structural and functional diversity
- Reviews the application of CAZymes in biorefineries for the generation of biofuel and value-added products
- Discusses other applications of CAZymes in agriculture and industrial sectors as biological replacements for harmful chemicals
- Addresses the role of the CAZymes in mitigating climate change
- Explains climate change with respect to the various causes and the evidence

This book is aimed at researchers in bioengineering, enzymology, and carbohydrates.

Anil Kumar Verma is the principal editor of the proposed book with expertise in the field of sustainable energy and biorefineries-based application of carbohydrate-active enzymes from lignocellulosic waste. He completed his PhD in May 2015 from the Indian Institute of Technology Guwahati, Assam. The PhD thesis work focused on the carbohydrate-active enzyme glucuronoxylan xylanohydrolase from *Clostridium thermocellum*. After completing his PhD, he received a prestigious DST-SERB National Post-Doctoral Fellowship (N-PDF) grant and joined The Energy and Resource Institute (TERI, New Delhi). He worked as Principal Investigator (PI) on a project, "Molecular cloning, biochemical characterization and structural and functional investigation of novel carbohydrate-active enzymes." During the N-PDF fellowship, he was selected as an assistant professor at Central University Sikkim, where he teaches industrial microbiology and instrumentation as a course subject. He has supervised 20 MSc dissertations and 3 PhDs (microbiology) are in progress. He has published around 51 publications with a cumulative impact factor of more than 60 in peer-reviewed journals and 435 citations with an h-index of 11.

Carbohydrate-Active Enzymes (CAZymes)

Biotechnological Applications for Mitigation of Climate Change

Edited by

Anil Kumar Verma

CRC Press
Taylor & Francis Group
Boca Raton London New York

CRC Press is an imprint of the
Taylor & Francis Group, an informa business

Designed cover image © 2026 Shutterstock

First edition published 2026
by CRC Press
2385 NW Executive Center Drive, Suite 320, Boca Raton FL 33431

and by CRC Press
4 Park Square, Milton Park, Abingdon, Oxon, OX14 4RN

CRC Press is an imprint of Taylor & Francis Group, LLC

ISBN: 978-1-032-88907-8 (hbk)
ISBN: 978-1-032-91166-3 (pbk)
ISBN: 978-1-003-56168-2 (ebk)

DOI: 10.1201/9781003561682

Typeset in Times
by Newgen Publishing UK

Contents

Preface

The Earth's climate is undergoing unprecedented change, primarily due to the accelerated release of greenhouse gases into the atmosphere. The consequences of this phenomenon, known as climate change, are far-reaching, affecting every facet of our planet's ecosystems and human societies. It is a global challenge that requires urgent and innovative solutions. One such key lies in the remarkable world of carbohydrate-active enzymes (CAZymes).

This book, *Carbohydrate-Active Enzymes (CAZymes): Biotechnological Applications for Mitigation of Climate Change*, delves into the pivotal role that CAZymes play in addressing the climate crisis. CAZymes, a diverse group of enzymes involved in carbohydrate metabolism, have emerged as powerful tools in biotechnology, with the potential to mitigate climate change through their application in various sectors, including agriculture, food production, bioenergy, and biomedicine.

This book explores the diverse aspects of CAZymes, including those involved in carbon sequestration, biomass utilization, biofuel production, and waste valorization, all of which contribute to reducing fossil fuel dependency and promoting the circular economy. The book also covers challenges and novel technologies in scaling up CAZyme production including technical barriers faced in CAZyme-based biorefineries, their economic, policy, and market impacts, with case studies of CAZyme-driven circular economy models.

This book equally highlights the vast structural diversity of CAZymes, examining their novel sources and various key functions, as well as their roles across different domains of life, including food and agriculture. One of the book's main themes covers CAZyme classification, their catalytic mechanisms, related databases dedicated to studying CAZymes, and bioinformatics tools used for analyzing CAZymes.

Additionally, this book presents recent advancements in nano-CAZymes, which are CAZymes enhanced with nanomaterials. It discusses nanoparticles used for targeted and controlled delivery of CAZymes, and how these carbohydrate-active nanoenzymes are further applied in bioenergy, biomedicine, and pollutant degradation.

In summary, this book serves as a comprehensive and current resource for researchers, students, professionals, and policymakers interested in the fascinating world of CAZymes and their crucial role in combating climate change, as well as in biorefineries and the circular economy. We hope that by exploring the functions of CAZymes and nano-CAZymes, along with their diverse applications, we can inspire new ideas and innovations that promote a sustainable and climate-resilient future.

Acknowledgments

First and foremost, we would like to express our gratitude to CRC Press and the series editors of the book series "Climate Change Resilience through Biorefinery Technologies" for giving us the opportunity to submit this book titled "*Carbohydrate-Active Enzymes (CAZymes): Biotechnological Applications for Mitigation of Climate Change*." Thank you for your assistance in formatting, coordination, and timely communication, which proved invaluable.

I sincerely thank all the fellow academicians, researchers, and faculty members who contributed and shared their expertise while writing the book chapters. Their knowledge and dedication have enriched this book, making it a truly comprehensive work. I am deeply grateful to the reviewers whose critical insights, constructive suggestions, and thoughtful feedback have significantly improved the quality and clarity of the content. Their collective efforts are evident in the high quality of writing on topics like nano-CAZymes and their applications in bioenergy, biomedicine, and pollutant degradation. The book highlights the role of carbohydrate-active enzymes (CAZymes) in addressing global environmental challenges. I hope this book inspires new ideas to combat pollution and reduce greenhouse gases, while promoting CAZyme-based technologies as tools to mitigate climate change.

I want to thank my PhD student, Ms. Dixita Chettri, for her consistent support since we first drafted the idea for this book. She has been essential at every stage of the project and has contributed greatly to this journey. I also wish to thank Department of Microbiology, Sikkim University, Department of Microbiology, for providing the necessary infrastructure and a supportive environment for the timely completion of this book.

I am deeply grateful to Almighty God, who gave me the strength and knowledge to finish this project. The successful completion of this work would not have been possible without my parents' blessings. As this is my first book as an editor, I am honored to dedicate it to my beloved father, the late Jagdish Verma.

Dr. Anil Kumar Verma

Contributors

Shadab Ahmed
Department of Biotechnology
Savitribai Phule Pune University
Pune, India

Shreya Biswas
Department of Biosciences and Bioengineering
Indian Institute of Technology
Guwahati, Assam, India

Manswama Boro
Department of Microbiology
Sikkim University
Gangtok, Sikkim, India

Di Cai
National Energy R&D Center for Biorefinery
Beijing University of Chemical Technology
Beijing, China

Changjing Chen
China Resources Double-Crane Pharmaceutical Co. Ltd.
Beijing, China

Dixita Chettri
Department of Microbiology
Sikkim University
Gangtok, Sikkim, India

Manisha Chirania
Department of Microbiology
Sikkim University
Gangtok, Sikkim, India

Sandip Das
Department of Molecular Biology and Bioinformatics
Tripura University
Suryamaninagar, Tripura, India

Satwika Das
Bioprocess Development Laboratory
Department of Biotechnology
National Institute of Technology
Warangal, Telangana, India

Yunxing Gao
National Energy R&D Center for Biorefinery
Beijing University of Chemical Technology
Beijing, China

Arabinda Ghosh
Department of Molecular Biology and Bioinformatics
Tripura University
Suryamaninagar, Tripura, India

Arun Goyal
Department of Biosciences and Bioengineering
Indian Institute of Technology
Guwahati, Assam, India

Nihar N. Jog
Department of Biotechnology
Savitribai Phule Pune University
Pune, India

Anbunithi K.
Bioprocess Development Laboratory
Department of Biotechnology
National Institute of Technology
Warangal, Telangana, India

Maryam Khan
Department of Biotechnology
American International University
Jahra, Kuwait

Zicheng Liao
National Energy R&D Center for Biorefinery
Beijing University of Chemical Technology
Beijing, China

Mohd Hasan Mujahid
Department of Polymer and Process Engineering
Indian Institute of Technology
Roorkee, Uttarakhand, India

Ashish A. Prabhu
Bioprocess Development Laboratory
Department of Biotechnology
National Institute of Technology
Warangal, Telangana, India

Rajendran Omana Rajesh
Bioprocess Development Laboratory
Department of Biotechnology
National Institute of Technology
Warangal, Telangana, India

Aruna Rani
Department of Biotechnology
IILM University
Greater Noida, India

Abhijit Reang
Department of Molecular Biology and Bioinformatics
Tripura University
Suryamaninagar, Tripura, India

Minal Richi
Department of Biological Sciences
Kuwait University
Shedadeyah, Kuwait

Shahin Mohammed S. K.
Department of Biotechnology
National Institute of Technology
Warangal, Telangana, India

Bhaskar Sharma
Department of Chemical and Environmental Engineering
Marlan and Rosemary Bourns College of Engineering
University of California
Riverside, CA, USA

Pallavi S. Shere
Department of Biotechnology
Savitribai Phule Pune University
Pune, India

Shuchi Singh
Department of Agricultural and Biological Engineering
University of Illinois Urbana-Champaign
Urbana, IL, USA

Divyanshi Srivastava
Department of Biotechnology
IILM University
Greater Noida, India

Anil Kumar Verma
Department of Microbiology
Sikkim University
Gangtok, Sikkim, India

Ashwani Kumar Verma
Department of Biosciences and Bioengineering
Indian Institute of Technology
Guwahati, Assam, India

Yankun Wang
National Energy R&D Center for Biorefinery
Beijing University of Chemical Technology
Beijing, China

Honggang Zhang
SDIC Advanced Biomass Fuel (Hailun) Co., Ltd.
Heilongjiang, China

Hongjia Zhang
National Energy R&D Center for Biorefinery
Beijing University of Chemical Technology
Beijing, China

1 Climate Change and Global Warming

Evidence, Causes, and Significance of Carbohydrate-Active Enzymes for Its Alleviation

Dixita Chettri and Anil Kumar Verma

ABBREVIATIONS

AAs	Auxiliary activities
CAZy	Carbohydrate-active enZYmes database
CAZymes	Carbohydrate-active enzymes
CBMs	Carbohydrate-binding modules
CD	Catalytic domains
CEs	Carbohydrate esterases
CH_4	Methane
CO_2	Carbon dioxide
GDP	Gross Domestic Product
GHG	Greenhouse gas
GHs	Glycoside hydrolases
GTs	Glycosyltransferases
IPCC	Intergovernmental Panel on Climate Change
N_2O	Nitrous oxide
NASA	National Aeronautics and Space Administration
NOAA	National Oceanic and Atmospheric Administration
OB fold	Oligonucleotide/oligosaccharide-binding fold
PLs	Polysaccharide lyases
SO_2	Sulfur dioxide
UV	Ultraviolet rays

1.1 INTRODUCTION: DEFINING GLOBAL WARMING AND CLIMATE CHANGE

The long-term increase in Earth's average temperature, on the surface and in the oceans, owing to the emission of greenhouse gases (GHGs) and their accumulation in

DOI: 10.1201/9781003561682-1

the atmosphere, is known as global warming. This phenomenon is caused due to the "greenhouse effect," of the gases such as carbon dioxide (CO_2), nitrous oxide (N_2O), and methane (CH_4), namely the GHGs, that act as an insulation blanket around the Earth, trapping heat that might otherwise radiate away into space. By trapping heat, global warming increases global temperatures with cascading adverse effects on natural ecosystems and human society (Zandalinas et al., 2021). Global warming is most commonly used interchangeably with climate change; however, climate change is a wider term that comprehends the long-term changes in climate patterns, precipitation, and seasonal cycles on Earth, with global warming being a key contributor to these changes (Penz, 2017). Since most of the changes occurred in the late 19th and early 20th centuries as a result of industrialization, increased deforestation, and burning of fossil fuels, which accelerated further into the 20th century, global warming represents an urgent environmental crisis with far-reaching consequences for the stability of the world's climate (Philander, 2008).

While the Earth's climate has been subject to natural fluctuations for millions of years, these changes have generally occurred over very long periods attributable to natural causes, for instance, volcanic eruptions, changes in solar radiation, or shifts in ocean circulation. Scientists note that current warming trends are occurring at an astonishing rate, which can be directly attributed to human activities since the beginning of the Industrial Revolution (Al-Ghussain, 2019). The burning of coal, oil, and natural gas for industrial, transportation, and residential purposes has led to a significant increase in the GHGs concentration in the atmosphere, especially carbon dioxide. During this period of rapid industrialization, a rise in global temperatures has been observed that correlates directly with increased emissions of these gases. This marks the beginning of human-induced warming, which scientists are now tracking and studying to understand its consequences. Studies by the Intergovernmental Panel on Climate Change (IPCC) and other organizations show that the continued warming of the Earth's atmosphere has set in motion a cascade of environmental changes, including stronger and more frequent weather events, melting glaciers, rising sea levels, and changes to ecosystems and biodiversity (Hansen et al., 2023).

At the core of global warming is the greenhouse effect: a natural process in which GHGs trap heat in the Earth's atmosphere. GHGs are naturally present in the atmosphere, helping to maintain a temperature on Earth that is suitable for life. However, human activities have disrupted this balance, causing the enhanced greenhouse effect and resulting in an increase in surface temperatures (Bargaoui & Nouri, 2021). The IPCC and other scientific bodies have analyzed this process using climate models, historical data, and temperature records to predict the potential impact of continued GHG emissions. The science is precise: as GHG concentrations rise, so does the Earth's temperature, triggering a cascade of environmental changes. Global warming has far-reaching consequences, affecting the dynamics of both natural ecosystems and human societies. Climate researchers warn that the global temperature could rise by more than 1.5–2°C above pre-industrial levels before the end of this century if no drastic measures are taken to reduce GHG emissions. This threshold is critical because, beyond this point, the most severe and irreversible impacts are expected, characterized by widespread droughts, flooding, and displacement of people's livelihoods (Masson-Delmotte et al., 2022). The global level is the most appropriate level for attempting to mitigate global warming, which includes activities like switching to renewable energy, increasing energy efficiency, reforestation, and developing carbon capture technologies (Abbass et al., 2022).

An important sustainable bio-based approach is the use of carbohydrate-active enzymes (CAZymes), a group of enzymes that play a crucial role in various bioprocesses, including the carbon cycle. They can be used to mitigate global warming via application in bioenergy production, carbon sequestration, and waste valorization (Monclaro et al., 2024). The promotion of CAZyme research and biotechnological implementation can significantly contribute to sustainable climate solutions.

1.2 CURRENT GLOBAL SCENARIO: EVIDENCE AND STATISTICS

The current global scenario of global warming is increasingly alarming, with average global temperatures rising at an unprecedented rate due to human activities. The past decade has been the warmest on record, with 2024 recognized as the hottest year globally with global temperatures surpassing the critical 1.5°C threshold above pre-industrial levels for the first time in a calendar year. This warming trend is driving more frequent and intense heatwaves, wildfires, floods, droughts, and sea-level rise. Polar ice caps and glaciers are rapidly melting, adding to rising ocean levels and threatening coastal communities. Biodiversity is under stress, and agricultural productivity is facing disruptions. Despite international efforts like the Paris Agreement aiming to limit temperature rise to below 2°C, current policies and emission trajectories suggest that the world is still on track for a 2.5–3°C increase by the end of the century, unless immediate and significant mitigation actions are taken.

The breach of the 1.5°C threshold, although temporary, serves as a stark warning. The Paris Agreement aims to constrain global warming to well below 2°C, preferably to 1.5°C, to avert the most catastrophic consequences of climate change. The events of 2024 highlight the urgency for immediate and substantial reductions in GHG emissions to prevent surpassing these critical limits permanently.

1.2.1 Factors Triggering Climate Change

Given the urgency to tackle the increasing risks of climate change, it is necessary to identify and understand the causes and effects of various factors leading to climate change (see Figure 1.1). Although climate change is generally associated with anthropogenic activities, some natural factors also contribute, albeit to a lesser extent, to the rise in temperature.

Natural factors: The Earth has experienced various climate change phenomena throughout geological time, even before the factors associated with human activities became significant. These natural factors, such as volcanic eruptions, natural GHG emissions, and solar radiation, which played an essential role in climate change in the pre-industrial era, now have a shorter influence or are generally limited to a region, with negligible effects on global warming.

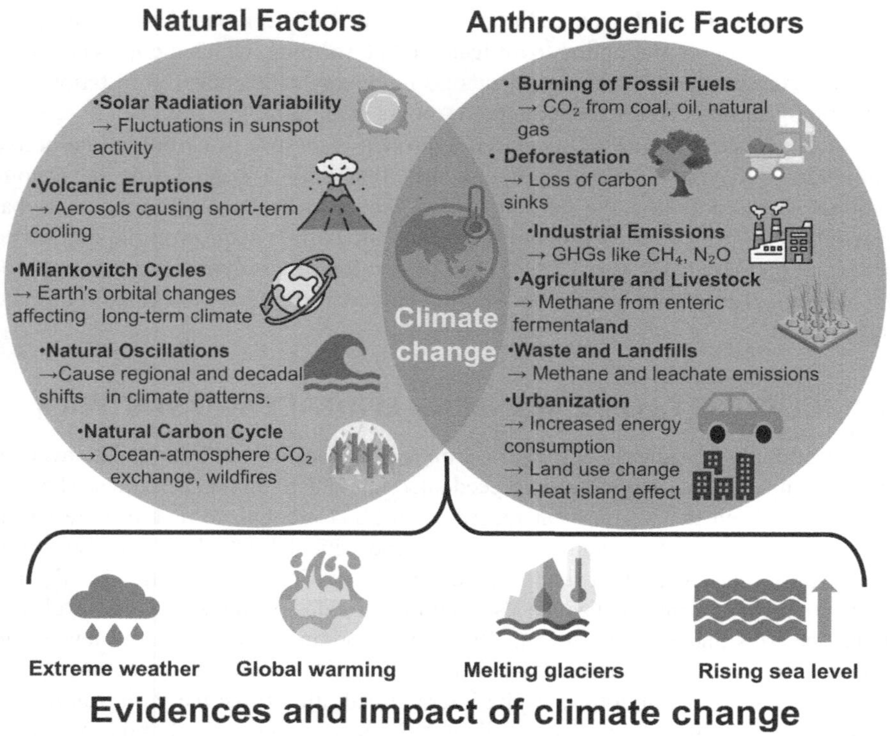

FIGURE 1.1 Human and natural factors influencing climate change and their global effects.

Solar radiation is the primary energy source on the Earth's surface, with its amount constantly fluctuating due to the Earth's wobbly motion and changes in its orbit and tilt, also known as Milankovitch cycles (Berger, 1988). These cycles are reportedly linked to solar irradiance and ice ages, with internal processes such as the size of the ice and the timing of its collapse being influenced by these external orbital cycles (Ganopolski, 2024). In addition, phenomena such as the 11-year sunspot cycle lead to fluctuations in total solar irradiance, which remarkably cause UV variability, resulting in changes in stratospheric chemistry that affect surface climate. UV variability is also linked to changes in the circulation pattern of the stratosphere-troposphere coupling, such as the jet stream. Although these fluctuations have been linked to climate change phenomena such as the Maunder Minimum, where low solar activity was associated with the ice age, the current temperature rise is inexplicable, as these factors play only a limited role (Gray et al., 2010).

Another factor is volcanic eruptions, during which large amounts of SO_2 are released into the atmosphere and converted into aerosols, which can cause temporary cooling of the Earth's surface by reflecting solar radiation. Depending on the altitude and latitude of the eruption site, tropical eruptions have been found to have a greater impact on global temperature. In addition, these aerosols also absorb the outgoing infrared radiation, which heats the stratosphere and changes the atmospheric circulation patterns. Furthermore, CO_2 is also released during volcanic eruptions, although its quantity is negligible compared to human-induced emissions (Marshall et al., 2022). Furthermore, natural oscillations such as the El Niño-Southern Oscillation, Pacific Decadal Oscillation, and North Atlantic Oscillation contribute to global climate change by redistributing heat and moisture, thereby altering patterns of precipitation, storms, and temperatures (Bharati et al., 2025; Raghuraman et al., 2024).

Anthropogenic factors: Although natural factors play a role in climate change over geological timescales, anthropogenic activities have been the leading cause of the changing global climate since the 20th century.

The main anthropogenic activity contributing to climate change is the emission of GHGs, which traps heat in the Earth's atmosphere that leads to an increase in global temperature. Industrialization and urbanization have exponentially increased energy demand, which is met by the extensive burning of fossil fuels, such as coal and oil, liberating large amounts of GHGs, including CO_2 and N_2O, into the atmosphere (Qureshi & Jamil, 2021; Saxena, 2025). The growth of industry is highly dependent on energy consumption, which is primarily driven by the high reliance on fossil fuels. In addition, modern infrastructure development relies on cement production, which releases CO_2 through both the combustion of fuels and chemical reactions during the calcination of limestone (Chaudhury et al., 2023). Industrial activities also generate large amounts of pollutants in the form of aerosols and particulate matter, which disrupt cloud formation and affect changes in local climate patterns. Modern transport systems, such as road vehicles, aviation, shipping, and railways, are also largely fueled by petroleum and diesel, which contribute significantly to global CO_2 emissions (Song et al., 2023).

Land cover change and extensive land use are also important factors contributing to climate change. Urbanization has led to large-scale deforestation as more land is used for the development of human settlements and for agricultural purposes to meet the increasing needs of a rapidly growing population. Overgrazing of land is another cause of deforestation, disrupting the natural process of carbon sequestration by plants and trees and leading to a reduction in the carbon sink and the release of sequestered carbon (Beillouin et al., 2022; Roy et al., 2022). Increased concrete and asphalt surfaces also have a heat island effect, as they retain the sun's heat, thereby increasing the local temperature (Chen et al., 2024).

The expansion of agriculture is altering ecosystems, impacting carbon and nitrogen cycles, and increasing emissions of CH_4 and N_2O. Enteric fermentation in livestock farming and the use of chemical fertilizers for crop production are also associated with the release of these GHGs into the atmosphere. The incineration of agricultural waste, livestock production and microbial digestion of waste also produce CH_4 and N_2O, which increase the amount of GHGs present in the atmosphere (Wu et al., 2024).

Waste management also contributes to the increased release of GHGs into the atmosphere. Large quantities of waste are disposed of in landfills, which also release considerable amounts of CH_4 into the atmosphere through the anaerobic fermentation of organic waste by microbial activity. The process of incineration also leads to the

TABLE 1.1
Contribution of the Global Top Five Economies (GDP-Wise) in Carbon Dioxide (CO_2) and Nitrous Oxide (N_2O) Emissions for the Year 2023

	CO_2 Emission		(N_2O) (GgN/yr)				
Countries	Fossil fuel (Mt/yr)	Land usage change (TgC/yr)	Direct emission form agriculture	Other direct emissions	Indirect emission	Perturbed fluxes	Natural emissions
USA	4911	14	33	170	99	–221.8	393
China	11,903	–63	575	266	236	–61.1	306
Germany	596	–5.5	44	22	15	–0.3	24
India	3062	–26	390	104	139	77	243
Japan	989	–1.8	12	21	6.6	–5.3	14

\# The data reproduced from the Global Carbon Atlas (www.globalcarbonatlas.org)
GgN/yr = Gigagrams of nitrogen per year
TgC/yr = Teragrams of carbon per year
Mt/yr = Metric tons per year

release of CO_2 and other particulates into the atmosphere, resulting in air pollution and subsequently a change in climate patterns (Blair & Mataraarachchi, 2021).

Thus, human activities are the leading cause of global climate change, with the emission of GHGs being the leading cause (see Table 1.1). Numerous events around the world underscore the seriousness of this problem, and urgent international action to mitigate it must be implemented.

1.2.2 Evidence of Climate Change

Direct evidence of climate change can be observed in the form of changing weather patterns and the associated phenomena. Data from various organizations such as NASA and NOAA show that the last 10 years have been the warmest on record (NASA, 2024a). The year 2023 was recorded as the warmest year in the previous 100,000 years with a record 1.45°C above the pre-industrial baseline (Romanello et al., 2024), which was immediately followed by the year 2024, which with a record 1.55°C ± 0.13°C was probably the first year to exceed the threshold of 1.5°C according to the Paris Agreement (Organization, 2025). The melting of ice sheets and glaciers has become an ongoing phenomenon, with billions of tonnes of ice melting from Greenland and Antarctica, reflecting rising temperatures in the polar regions (IPCC, 2021). In addition, recent events such as the report of the melting of the last glacier in Venezuela (NASA, 2024b), the absence of snowfall on Mount Fuji until early November, breaking a 130-year-old record (Reuters, 2024), and the first snowfall in the Al-Jawf region of the Saudi Arabian desert (Spanish, 2024) were some of the major highlights of 2024.

A direct consequence of the substantial melting of glaciers and ice is the rise in sea level. Sea levels have been observed to have risen by a total of 20 cm over the past 180 years, with the rate of increase significantly accelerating in recent decades due to the melting of glaciers and the thermal expansion of the oceans (NOAA, 2023). In addition, the oceans absorb a large amount of CO_2, accounting for almost 30% of total emissions, which lowers the pH of the oceans and leads to disruptions in the marine ecosystem, resulting in significant losses of coral reefs and shellfish populations (NOAA, 2022).

The frequency of extreme weather events, such as heatwaves, floods, cyclones, and erratic rainfall patterns, has also increased in recent years. For example, the 2022 heatwave in Europe, the bushfires in Australia, and the record-breaking 2020 Atlantic hurricane season (WMO, 2023). All of these events are attributed to rising global temperatures and changing weather patterns. According to a NASA report, the frequency of natural disasters, such as floods and droughts, has nearly doubled over the last five years (Guardian, 2025). These events have resulted in several lives lost, such as the recent 2025 heatwave in India and Pakistan, which saw an additional 8°C rise in temperature, deviating from the seasonal trend and reaching almost 48°C (Sophia Saifi, 2025). This event was immediately followed by severe monsoon flooding, with around 250 victims reported from Pakistan and 120 from India (Jilani, 2025).

All these events emphasize the urgency of the situation and the need for the world government to prioritize the search for alternatives to the factors that trigger the release of GHGs and the rise in global temperatures.

1.3 CARBOHYDRATE-ACTIVE ENZYMES IN CLIMATE CHANGE ALLEVIATION

In light of the threats and challenges posed by climate change, CAZymes play a crucial role in mitigating its effects, particularly in the context of biomass conversion and the carbon cycle (see Figure 1.2). These enzymes are responsible for biomass conversion and carbon cycling, efficiently catalyzing the degradation of lignocellulosic biomass into fermentable sugars that can be used for the production of biofuels, thus reducing dependence on fossil fuels (Chettri & Verma, 2024). In carbon sequestration, they mediate the conversion of plant biomass into simple organic matter, e.g. biochar. They also contribute to sustainable agriculture by breaking down organic matter into bioavailable nutrients, which improves soil fertility. CAZymes also play a crucial role in waste utilization and bioremediation while offsetting GHG emissions and environmental pollutants, making them extremely important from both ecological and industrial perspectives (Xiong et al., 2023). Their use is part of bioeconomy initiatives that endeavor to tackle climate challenges sustainably.

1.3.1 CAZymes: Definition and Classification

CAZyme refers to a diverse group of enzymes that play a central role in the synthesis, modification, and degradation of carbohydrates. Since carbohydrates are important biomolecules that function as both structural and storage polysaccharides in all areas of life, the study of CAZymes is of fundamental importance (Chettri, Chirania, et al., 2024). These enzymes have co-evolved with their hosts to carry out a wide range of biochemical reactions and have a central role in various biological processes and industrial applications (Zhang et al., 2023). CAZymes exhibit a variety of structural

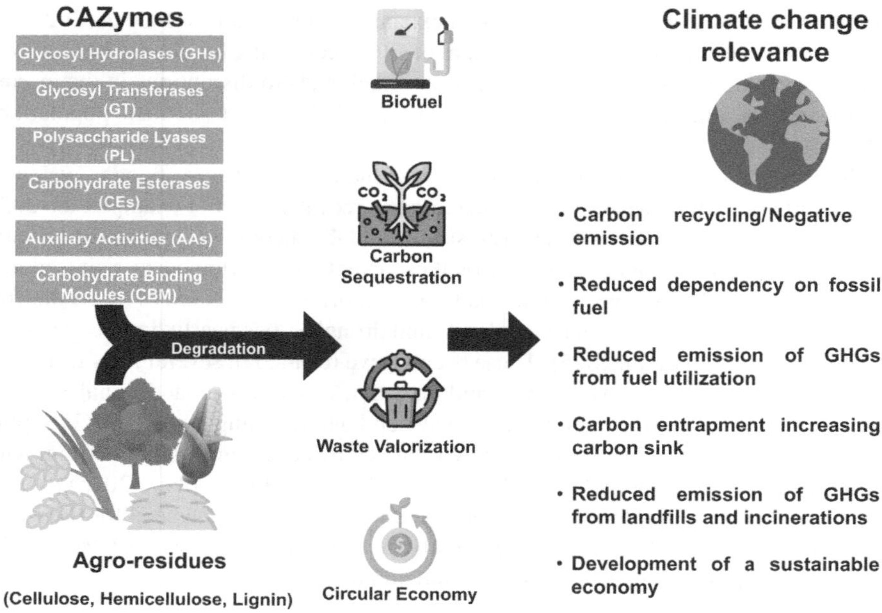

FIGURE 1.2 Role of CAZymes in mitigation of climate change.

and functional properties, which have been used for their systematic classification, and a database for the Carbohydrate-Active enZYmes Database (CAZy) has been developed (www.cazy.org/) (Terrapon et al., 2017).

The CAZy classification system categorizes these enzymes into 5 primary catalytic families with a non-catalytic module, namely: glycoside hydrolases (GHs), which hydrolyze glycosidic bonds in complex carbohydrates, releasing smaller oligosaccharides or monosaccharides; glycosyltransferases (GTs) are the enzymes involved in the biosynthesis of glycosidic bonds by transferring sugar units from activated donor molecules to acceptors; carbohydrate esterases (CEs) remove ester-bound functional groups from carbohydrate molecules and thus support carbohydrate degradation; polysaccharide lyases (PLs) cleave polysaccharides via β-elimination mechanisms, resulting in unsaturated products; auxiliary activities (AAs) include enzymes that help the degradation of recalcitrant polysaccharides such as lignin by removing reactive intermediates (e.g., lytic polysaccharide monooxygenases) and the non-catalytic carbohydrate-binding modules (CBMs), which increase the activity of catalytic CAZymes by targeting specific carbohydrate substrates (Gavande et al., 2023). As of June 2025, approximately 37,706 CAZymes from different domains of life have been discovered, with bacteria making the most considerable contribution (www.cazy.org/).

1.3.2 Structural and Functional Diversity of CAZymes

CAZymes exhibit outstanding functional and structural diversity, which enables them to act on a broad spectrum of polysaccharides with high specificity and efficiency. Their structural modularity, substrate specificity, catalytic mechanism, and synergy with auxiliary domains make them versatile in many biological and industrial processes (Shanmugam & Yin, 2025). The modular structure of CAZymes often consists of three prominent domains, i.e., catalytic domains (CD) and the corresponding CBMs, which are connected by a linker region (Masson-Delmotte et al., 2022). The CD is the central region involved in enzymatic activity. For each class of CAZymes, there are different architectures of the CDs that determine the mechanism of action. Non-catalytic domains, or CBMs, facilitate enzyme-substrate interactions by increasing the enzyme's affinity for specific polysaccharides. The linker region, a flexible peptide fragment, connects the CD and CBMs, enabling dynamic substrate interactions. The association of these domains differs from one CAZyme family to another and determines their substrate specificity, stability, and efficiency, thereby facilitating a targeted enzymatic response to complex carbohydrate substrates in nature.

The CDs of CAZymes adopt different three-dimensional folds, which determine their reaction mechanisms. The general structural folds are:

1. **$(\beta/\alpha)_8$-Barrel (TIM barrel) Fold**: In most GHs and GTs, they have alternating β-strands and α-helices in a barrel-shaped conformation. Example: cellulases (GH5) (Sanjaya et al., 2021) and α-amylases (GH13) (Mareček et al., 2024), which cleave β-1,4-glycosidic bonds in the degradation of cellulose and starch, respectively.
2. **β-Sandwich Fold:** It consists of two stacked β-sheets that have a sandwich-like structure. This fold is often found in CBMs, which are responsible for recognizing and binding substrates. Examples: CBM3, which binds crystalline cellulose and improves the catalytic activity of related enzymes (Cai et al., 2011).
3. **β-Jelly Roll Fold:** A particular arrangement of the β-sheet in which the β-strands are folded in a barrel-shaped or curved structure like a jelly roll. It is particularly observed in enzymes that act on complex carbohydrates, such as CEs and PLs. Example: pectin lyases (PL1) that degrade pectin in plant cell walls and alginate lyases (PL7) that degrade alginate (Chernysheva et al., 2021; Yadav et al., 2023).
4. **β-Helix Fold:** Possesses a helical β-sheet topology, confers stability when attacking crystalline substrates and is found in enzymes that target rigid polysaccharides, e.g., chitinases and certain hemicellulases. Example: chitin-binding CBM family 54 (Itoh, 2021) & polygalacturonases (GH28) that hydrolyze pectin (Tu et al., 2021).
5. **β-Propeller Fold:** Features multiple β-folds arranged radially around a central axis with a propeller-like appearance. This fold is capable of accommodating large polysaccharide substrates. Example: Contains β-xylosidases and arabinofuranosidases (GH43) for the degradation of hemicellulose (Gavande et al., 2024; Pisalwar et al., 2023).
6. **$(\alpha/\alpha)_6$ Toroid Fold:** Consists of six α-helices that are present in a toroidal (doughnut-like) arrangement and is found in certain GTs. Example: endo-β-1,2-glucanase (GH162), which hydrolyzes β-1,2-glucans (Tanaka et al., 2019).

7. **Rossmann-like Fold:** The Rossmann fold, often found in association with nucleotide binding, consists of alternating β-strands and α-helices in a three-layer structure. It contains enzymes such as lipopolysaccharide galactosyltransferase (GT4) (Chen et al., 2023) & UDP-glucuronosyltransferases (GT1) (Xue et al., 2025).
8. **OB-Fold:** The OB fold is a small β-barrel structure responsible for nucleic acid and carbohydrate binding and is present in the CBM class of CAZymes—example: CBM10 (Oliveira et al., 2015).

It is essential to comprehend these structural folds, as they have a direct impact on the catalytic mechanisms and substrate specificities of CAZymes, and thus on their functional role in biological processes and industrial applications. Due to the diversity and functional importance of CAZymes, a database for their structure has also been developed (Shanmugam & Yin, 2025).

1.3.3 Biological, Environmental, and Industrial Significance

Carbohydrates, along with DNA and RNA, are the fundamental units of biological systems that exhibit great diversity and complexity and fulfill a multitude of functions, including cell signaling, receptor molecules, storage molecules, and maintaining structural integrity. Since CAZymes are directly involved in the biotransformation of these carbohydrate molecules, they play an important role in biological and ecological systems (Chettri & Verma, 2023). The primary function of CAZymes is carbohydrate metabolism, in which these enzymes break down complex carbohydrates into simpler forms to be used for energy production or other metabolic activities (Lombard et al., 2014). CAZymes are also crucial for maintaining plant cell integrity, as they are involved in the synthesis, modification, and degradation of cell wall components, including cellulose, hemicellulose, and lignin (Alshareef, 2024). CAZymes are crucial for the carbon cycle, as they degrade complex carbohydrates in plant biomass into simpler and more stable organic compounds, thereby supporting carbon utilization by other organisms and carbon sequestration (Yu et al., 2024; Zhao et al., 2020). An essential application of CAZymes is observed in host-pathogen interactions and pathogenesis, where the pathogen releases the CAZyme to degrade the plant cell wall. At the same time, the host can modify the carbohydrate components as a defence mechanism against the pathogens (Liu et al., 2025).

CAZymes have also gained importance in industry, where they can be used in biotechnological research as commercial enzymes for the bioconversion of plant biomass in biorefineries to produce value-added products such as platform chemicals, biochar, and biofertilizers for sustainable agricultural practices with a focus on biofuels (bioethanol, biogas, etc.) (F. Contesini et al., 2021). In the medical industry, CAZymes are utilized for the development of drugs, therapeutics, and inhibitors, as well as for diagnostics. In the textile industry, CAZymes are utilized for bioleaching, biostone removal, decolorization, dyeing, and other applications, while in the paper industry, they are employed for biopulping. In the detergent industry, CAZymes play a crucial role in detergent composition, as well as in the addition of animal feed to enhance the digestibility of nutrients. They are also found in the food industry, where they are used for food preservation and fermentation, as well as in fruit and vegetable processing, baking, and the production of prebiotics and nutraceuticals. They are now also used extensively in bioremediation processes to remove organic waste from soil and water (F. J. Contesini et al., 2021).

1.3.4 Importance in Climate Change Mitigation

The burning of fossil fuels is a primary factor contributing to climate change, alongside large-scale deforestation and waste generation, which leads to the release of trapped carbon into the atmosphere (Omemen & Aldbbah, 2025). CAZymes, with their ability to convert biomass into biofuels, sequester carbon, and reduce GHG emissions, hold immense potential to address the challenges of climate change. The production of alternative renewable energy, the utilization of waste, and the promotion of sustainable agricultural practices are the approaches with which CAZymes help to reduce the carbon footprint and, at the same time, contribute to the circular bioeconomy (Monclaro et al., 2024). CAZymes have thus become an indispensable tool in tackling climate change and reducing pollution.

1.3.4.1 Carbon Sequestration and Biomass Utilization

Plant biomass is the most abundant carbon source on Earth, with the process of photosynthesis playing an essential role in the carbon cycle, converting atmospheric CO_2 into complex carbohydrates, which act as a neutral carbon sink (Green & Keenan, 2022). The microbes present in the soil produce various CAZymes, such as cellulases, hemicellulases, and lignin-modifying enzymes, which convert complex carbohydrates mainly composed of cellulose, hemicellulose, and lignin into simpler and more stable soil organic matter (humus). This organic matter is then utilized by the soil microbes for their metabolic activities, sequestering carbon in the soil ecosystem. Thus, CAZymes play a crucial role in enriching the soil with chemically stable organic residues for long-term carbon storage, and therefore, play a significant role in carbon sequestration. In addition, CAZymes are crucial for the restoration of vegetation on terrestrial soils and contribute to carbon sequestration by creating new natural carbon sinks (Yu et al., 2024).

1.3.4.2 Biofuel Production and Reduction of Fossil Fuel Dependence

In view of the GHG emissions from the usage of fossil fuels, which are one of the principal causes of global climate change, the search for alternative energy sources for rapid industrialization and urbanization is the order of the day. As plant biomass is abundant in nature, the use of CAZymes to break it down into simple fermentable sugars for biofuel production represents an alternative carbon-neutral energy source. In biorefineries, the hydrolysis of biomass is the bottleneck step, where various CAZymes, such as cellulase and hemicellulase, are used for enzymatic treatment to increase the efficiency of producing simple sugars, including glucose, fructose, and xylose. These released sugars are then used in the production of biofuels. These released sugars are then utilized in bioethanol production through microbial fermentation, while the biochar produced as a by-product can be used for long-term carbon sequestration and to improve soil fertility (Chettri, Boro, et al., 2024; Chettri & Verma, 2024). In addition, CAZymes are also actively involved in biogas production by assisting in anaerobic digestion by converting complex polysaccharides into simple sugars that are utilized by the methanogenic microbes for biomethane production (Song et al., 2024).

1.3.4.3 Waste Valorization and Circular Economy

Agricultural and industrial processes generate large amounts of waste, which are often disposed of directly in landfills or water bodies, resulting in water and soil pollution. As these wastes are a rich source of complex carbohydrates, CAZymes can be utilized for their valorization into biofuels and value-added products, while promoting the principles of waste management and circular bioeconomy. The biological conversion of organic waste into value-added products through the application of CAZymes reduces the amount of waste sent to landfills, thereby supporting waste management and mitigating methane gas emissions from these landfills, a major GHG (Deng et al., 2022). In addition, agricultural waste is widely utilized in biorefineries for enzymatic conversion into biofuels and other value-added products, such as organic acids, various metabolites, platform chemicals, and biofertilizers (Joshi et al., 2024). The use of CAZymes in biorefineries has the advantage that more than one product is produced, with little waste, and the whole process is more environmentally friendly and sustainable. CAZymes are also used in soil improvement, where they degrade various organic pollutants that alter the physicochemical properties of the soil, thereby promoting soil microflora, increasing soil fertility, and promoting sustainable agriculture, which further contributes to climate change mitigation (Kozjek et al., 2023).

1.4 CONCLUSION AND FUTURE PROSPECT

Industrial development, growing population, and the unsustainable use of resources, with the release of GHGs into the environment, have led to global climate change, with impacts on both the environment and human health. Therefore, there is an urgent need for sustainable measures to mitigate climate change and enhance resilience, driven by changes in national and international policies, as well as scientific advancements. As their role in carbon sequestration, biomass degradation, and biofuel production aligns with the global sustainable development goals, CAZymes have become an essential biotechnological tool for mitigating climate change. CAZymes also contribute to a circular bioeconomy by facilitating effective bioconversion processes for biofuel production and offering substitutes for fossil fuels, while also creating value-added products.

The future of climate change mitigation depends on interdisciplinary approaches with scalable solutions. The development of technologies to sequester carbon and replace GHG-emitting fuels with renewable sources, including biofuels, will be crucial. The integration of new technologies, such as bioinformatics, artificial intelligence, enzyme engineering, and synthetic biology, is a future prospect for CAZymes in mitigating climate change. In addition, international agreements and the development of a national policy framework to promote alternatives to CO_2-releasing processes at the local level, such as incentivizing biofuels, investing in renewable energy infrastructure, and taxing carbon, are a necessary step. While CAZyme may not provide a complete solution to climate change, it is an important biological approach that will be important to integrate with other climate change mitigation strategies. Collaboration between scientists, industry, and policymakers will be a crucial step in this regard, where the combination of technological advances, behavioral changes, and policy incentives will ensure that climate resilience is both achievable and sustainable for future generations.

ACKNOWLEDGMENT

The authors thank the Department of Microbiology, Sikkim University, for providing the computational infrastructure and central library facilities for procuring references and plagiarism analysis (Drillbit Plagiarism Detection Software).

AUTHOR CONTRIBUTION STATEMENT

Dixita Chettri: Writing, original draft preparation, reviewing, and editing. Anil Kumar Verma: Reviewing and editing.

FUNDING

This research did not receive any specific grant from funding agencies in the public, commercial, or not-for-profit sectors.

CONFLICT OF INTEREST STATEMENT

The author declares that the research was conducted without any commercial or financial relationships that could be construed as potential conflicts of interest.

REFERENCES

Abbass, K., Qasim, M. Z., Song, H., Murshed, M., Mahmood, H., & Younis, I. (2022). A review of the global climate change impacts, adaptation, and sustainable mitigation measures. *Environmental Science and Pollution Research*, *29*(28), 42539–42559.

Al-Ghussain, L. (2019). Global warming: review on driving forces and mitigation. *Environmental Progress & Sustainable Energy*, *38*(1), 13–21.

Alshareef, S. A. (2024). Metabolic analysis of the CAZy class glycosyltransferases in rhizospheric soil fungiome of the plant species *Moringa oleifera*. *Saudi Journal of Biological Sciences*, *31*(4), 103956.

Bargaoui, S. A., & Nouri, F. Z. (2021). History of the human and nature relationship, discovery of greenhouse effect and awareness of the environmental problem. *Journal of Economic Science Research*, *4*(3), 23–24.

Beillouin, D., Cardinael, R., Berre, D., Boyer, A., Corbeels, M., Fallot, A., ... Demenois, J. (2022). A global overview of studies about land management, land-use change, and climate change effects on soil organic carbon. *Global Change Biology*, *28*(4), 1690–1702.

Berger, A. (1988). Milankovitch theory and climate. *Reviews of Geophysics*, *26*(4), 624–657.

Bharati, P., Deb, P., & Hunt, K. M. R. (2025). Pacific decadal oscillation-driven interdecadal variability of snowfall over the Karakoram and the Western Himalayas. *Weather and Climate Dynamics*, *6*(1), 197–210.

Blair, J., & Mataraarachchi, S. (2021). A review of landfills, waste and the nearly forgotten nexus with climate change. *Environments*, *8*(8), 73.

Cai, S., Zheng, X., & Dong, X. (2011). CBM3d, a novel subfamily of family 3 carbohydrate-binding modules identified in Cel48A exoglucanase of *Cellulosilyticum ruminicola*. *Journal of Bacteriology*, *193*(19), 5199–5206.

Chaudhury, R., Sharma, U., Thapliyal, P., & Singh, L. (2023). Low-CO_2 emission strategies to achieve net zero target in cement sector. *Journal of Cleaner Production*, *417*, 137466.

Chen, Y., Gu, J., Ashworth, G., Wang, Z., Zhang, Z., & Dong, C. (2023). Crystal structure of the lipopolysaccharide outer core galactosyltransferase WaaB involved in pathogenic bacterial invasion of host cells. *Frontiers in Microbiology*, *14*, 1239537.

Chen, Y., Sha, A., Jiang, W., Du, P., Lu, Q., & Hu, K. (2024). Development and evaluation of STRC coating for cooling asphalt pavement and mitigating urban heat island effects. *Construction and Building Materials*, *449*, 138494.

Chernysheva, N., Bystritskaya, E., Likhatskaya, G., Nedashkovskaya, O., & Isaeva, M. (2021). Genome-wide analysis of PL7 alginate lyases in the genus *Zobellia*. *Molecules*, *26*(8), 2387.

Chettri, D., Boro, D., Chirania, M., & Verma, A. K. (2024). Integrating biochar production in biorefineries: towards a sustainable future and circular economy. *Biofuels, Bioproducts and Biorefining*, *18*(6), 2156–2176.

Chettri, D., Chirania, M., Boro, D., & Verma, A. K. (2024). Glycoconjugates: advances in modern medicines and human health. *Life Sciences*, *348*, 122689.

Chettri, D., & Verma, A. K. (2023). Biological significance of carbohydrate active enzymes and searching their inhibitors for therapeutic applications. *Carbohydrate Research*, *529*, 108853.

Chettri, D., & Verma, A. K. (2024). Statistical optimization of cellulase production from *Bacillus* sp. YE16 isolated from yak dung of the Sikkim Himalayas for its application in bioethanol production using pretreated sugarcane bagasse. *Microbiological research*, *281*, 127623.

Contesini, F., Frandsen, R., & Damasio, A. (2021). Editorial: CAZymes in biorefinery: from genes to application. *Frontiers in Bioengineering and Biotechnology*, *9*, 622817. doi: 10.3389/fbioe.

Contesini, F. J., Frandsen, R. J. N., & Damasio, A. (2021). CAZymes in biorefinery: from genes to application. *Frontiers in Bioengineering and Biotechnology*, *9*, 622817.

Deng, C., Zhao, R., Qiu, Z., Li, B., Zhang, T., Guo, F., ... Zhang, L. (2022). Genome-centric metagenomics provides new insights into the microbial community and metabolic potential of landfill leachate microbiota. *Science of the Total Environment*, *816*, 151635.

Ganopolski, A. (2024). Toward generalized Milankovitch theory (GMT). *Climate of the Past*, *20*(1), 151–185.

Gavande, P. V., Goyal, A., & Fontes, C. M. (2023). Carbohydrates and carbohydrate-active enzymes (CAZyme): an overview. *Glycoside Hydrolases*, *1*, 1–23.

Gavande, P. V., Ji, S., Cardoso, V., Fontes, C. M., & Goyal, A. (2024). Reassigning the role of a mesophilic xylan hydrolysing family GH43 β-xylosidase from *Bacteroides ovatus*, BoExXyl43A as exo-β-1, 4-xylosidase. *Current Research in Biotechnology*, *7*, 100191.

Gray, L. J., Beer, J., Geller, M., Haigh, J. D., Lockwood, M., Matthes, K., ... Hood, L. (2010). Solar influences on climate. *Reviews of Geophysics*, *48*(4), 1–53.

Green, J. K., & Keenan, T. F. (2022). The limits of forest carbon sequestration. *Science*, *376*(6594), 692–693.

Guardian, T. (2025). Nasa data reveals dramatic rise in intensity of weather events. www.theguardian.com/world/2025/jun/17/nasa-data-reveals-dramatic-rise-in-intensity-of-weather-events

Hansen, J. E., Sato, M., Simons, L., Nazarenko, L. S., Sangha, I., Kharecha, P., ... Osman, M. B. (2023). Global warming in the pipeline. *Oxford Open Climate Change*, *3*(1), kgad008.

IPCC (2021). *Sixth Assessment Report – Summary for Policymakers.* www.ipcc.ch/report/ar6/wg1/

Itoh, T. (2021). Structures and functions of carbohydrate-active enzymes of chitinolytic bacteria *Paenibacillus* sp. str. FPU-7. *Bioscience, Biotechnology, and Biochemistry*, *85*(6), 1314–1323.

Jilani, H. (2025). Monsoons claim hundreds of lives across South Asia. *Financial Times.* / www.ft.com/content/e4eb48a3-a178-4858-b11f-98e704db4dce

Joshi, N., Grewal, J., Drewniak, L., & Pranaw, K. (2024). Bioprospecting CAZymes repertoire of aspergillus fumigatus for eco-friendly value-added transformations of agro-forest biomass. *Biotechnology for Biofuels and Bioproducts*, *17*(1), 3.

Kozjek, K., Manoharan, L., Urich, T., Ahrén, D., & Hedlund, K. (2023). Microbial gene activity in straw residue amendments reveals carbon sequestration mechanisms in agricultural soils. *Soil Biology and Biochemistry*, *179*, 108994.

Liu, F., Xiao, J., Wang, X.-F., Wang, Y.-X., Yang, H.-H., Cai, Y.-B., ... Wan, P.-J. (2025). Role of carbohydrate-active enzymes in brown planthopper virulence and adaptability. *Frontiers in Plant Science, 16*, 1554498.

Lombard, V., Golaconda Ramulu, H., Drula, E., Coutinho, P. M., & Henrissat, B. (2014). The carbohydrate-active enzymes database (CAZy) in 2013. *Nucleic Acids Research*, *42*(D1), D490–D495.

Mareček, F., Terrapon, N., & Janeček, Š. (2024). Two newly established and mutually related subfamilies GH13_48 and GH13_49 of the α-amylase family GH13. *Applied Microbiology and Biotechnology*, *108*(1), 415.

Marshall, L. R., Maters, E. C., Schmidt, A., Timmreck, C., Robock, A., & Toohey, M. (2022). Volcanic effects on climate: recent advances and future avenues. *Bulletin of Volcanology*, *84*(5), 54.

Masson-Delmotte, V., Zhai, P., Pörtner, H.-O., Roberts, D., Skea, J., & Shukla, P. R. (2022). *Global Warming of 1.5 C: IPCC special report on impacts of global warming of 1.5 C above pre-industrial levels in context of strengthening response to climate change, sustainable development, and efforts to eradicate poverty.* Cambridge University Press.

Monclaro, A. V., Gomes, H. A. R., Duarte, G. C., de Souza Moreira, L. R., & Filho, E. X. F. (2024). Unveiling the biomass valorization: the microbial diversity in promoting a sustainable socio-economy. *BioEnergy Research, 17*(3), 1355–1374.

NASA (2024a). *Global Climate Change: Vital Signs of the Planet.* Earth Observatory coverage, Issue. https://climate.nasa.gov/vital-signs/global-temperature/

NASA (2024b). *Humboldt Glacier's Demise.* https://earthobservatory.nasa.gov/images/152893/humboldt-glaciers-demise

NOAA (2022). *Ocean Acidification.* www.noaa.gov/education/resource-collections/ocean-coasts/ocean-acidification

NOAA (2023). *Sea Level Rise.* https://oceanservice.noaa.gov/facts/sealevel.html

Oliveira, C., Carvalho, V., Domingues, L., & Gama, F. M. (2015). Recombinant CBM-fusion technology—applications overview. *Biotechnology Advances*, *33*(3–4), 358–369.

Omemen, K. M., & Aldbbah, M. O. (2025). Climate change: key contributors and sustainable solutions. *International Journal of Electrical Engineering and Sustainability, 3*(1), 10–27.

Organization, W. M. (2025). *State of the global climate 2024 (WMO-No. 1368).* World Meteorological Organization. https://wmo.int/publication-series/state-of-global-climate-2024

Penz, H. (2017). 'Global warming'or 'climate change'? In A. F. Fill & H. Penz (Eds.), *The Routledge handbook of ecolinguistics* (pp. 277–292). Routledge.

Philander, S. G. (2008). *Encyclopedia of global warming and climate change: AE* (Vol. 1). Sage Publications Ltd.

Pisalwar, P., Fernandes, A., Tribhuvan, D., Gite, S., & Ahmed, S. (2023). Arabinofuranosidases. In Arun Goyal & Kedar Sharma (Eds.), *Glycoside hydrolases* (pp. 187–211). Elsevier.

Qureshi, A., & Jamil, M. (2021). The footprint of industrialization on climate change. *Journal of Business and Economics, 13*(1), 95–112.

Raghuraman, S. P., Soden, B., Clement, A., Vecchi, G., Menemenlis, S., & Yang, W. (2024). The 2023 global warming spike was driven by the El Niño–Southern Oscillation. *Atmospheric Chemistry and Physics, 24*(19), 11275–11283.

Reuters. (2024, November 7, 2024). Snow cap on Japan's Mt Fuji this year is latest spotted in 130 years. www.reuters.com/world/asia-pacific/snowcap-japans-mt-fuji-this-year-is-latest-spotted-130-years-2024-11-07/

Romanello, M., Walawender, M., Hsu, S.-C., Moskeland, A., Palmeiro-Silva, Y., Scamman, D., ... Ayeb-Karlsson, S. (2024). The 2024 report of the Lancet Countdown on health and climate change: facing record-breaking threats from delayed action. *The Lancet, 404*, 1847–1896..

Roy, P. S., Ramachandran, R. M., Paul, O., Thakur, P. K., Ravan, S., Behera, M. D., ... Kanawade, V. P. (2022). Anthropogenic land use and land cover changes—A review on its environmental consequences and climate change. *Journal of the Indian Society of Remote Sensing, 50*(8), 1615–1640.

Sanjaya, R. E., Putri, K. D. A., Kurniati, A., Rohman, A., & Puspaningsih, N. N. T. (2021). In silico characterization of the GH5-cellulase family from uncultured microorganisms: physicochemical and structural studies. *Journal of Genetic Engineering and Biotechnology, 19*(1), 143.

Saxena, V. (2025). Water quality, air pollution, and climate change: investigating the environmental impacts of industrialization and urbanization. *Water, Air, & Soil Pollution, 236*(2), 73.

Shanmugam, N. S., & Yin, Y. (2025). CAZyme3D: a database of 3D structures for carbohydrate-active enzymes. *Journal of Molecular Biology, 437*, 169001.

Song, B., Guo, H., Deng, Z., Chen, L., Ren, H., & Xu, X. (2024). Metagenomic characterization of anaerobic fermentation in weathered coal for biomethane production enhanced by landfill leachate. *Process Safety and Environmental Protection, 191*, 275–286.

Song, M.-J., Seo, Y.-J., & Lee, H.-Y. (2023). The dynamic relationship between industrialization, urbanization, CO_2 emissions, and transportation modes in Korea: empirical evidence from maritime and air transport. *Transportation, 50*(6), 2111–2137.

Sophia Saifi, R. M. A. S. I. (2025). Extended heatwave in India, Pakistan to test survivability limits, with temperatures reaching Death Valley levels. https://edition.cnn.com/2025/04/15/asia/india-pakistan-heatwave-climate-crisis-intl-hnk#:~:text=When%20the%20laborers%20or%20farmers,leading%20cause%20of%20maternal%20mortality.%E2%80%9D

Spanish, A. (2024). *Nieva por primera vez en el desierto de Al Nafūd (Al-Jawf).* https://as.com/deportes_accion/snow/nieva-por-primera-vez-en-el-desierto-de-al-nafud-n/

Tanaka, N., Nakajima, M., Narukawa-Nara, M., Matsunaga, H., Kamisuki, S., Aramasa, H., ... Terada, T. (2019). Identification, characterization, and structural analyses of a fungal endo-β-1, 2-glucanase reveal a new glycoside hydrolase family. *Journal of Biological Chemistry*, *294*(19), 7942–7965.

Terrapon, N., Lombard, V., Drula, E., Coutinho, P. M., & Henrissat, B. (2017). The CAZy database/the carbohydrate-active enzyme (CAZy) database: principles and usage guidelines. In Kiyoko F. Aoki-Kinoshita (Eds.) *A practical guide to using glycomics databases* (pp. 117–131). Springer.

Tu, T., Wang, Z., Luo, Y., Li, Y., Su, X., Wang, Y., . . . Hakulinen, N. (2021). Structural insights into the mechanisms underlying the kinetic stability of GH28 endo-polygalacturonase. *Journal of Agricultural and Food Chemistry*, *69*(2), 815–823.

WMO (2023). *State of the Global Climate 2023*. https://public.wmo.int/en/our-mandate/climate/wmo-statement-state-of-global-climate

Wu, X., Zhang, Y., Han, Y., Zhang, Y., Zhang, Y., Cheng, X., . . . Li, Z. (2024). Advances in methane emissions from agricultural sources: Part I. Accounting and mitigation. *Journal of Environmental Sciences*, *140*, 279–291.

Xiong, M., Jiang, W., Zou, S., Kang, D., & Yan, X. (2023). Microbial carbohydrate-active enzymes influence soil carbon by regulating the of plant-and fungal-derived biomass decomposition in plateau peat wetlands under differing water conditions. *Frontiers in Microbiology*, *14*, 1266016.

Xue, J., Li, Q., Wang, Y., Yin, R., & Zhang, J. (2025). Insight into the structure, oligomerization, and the role in drug resistance of human UDP-glucuronosyltransferases. *Archives of Toxicology*, *99*, 1–13.

Yadav, K., Yadav, S., Anand, G., Yadav, P. K., & Yadav, D. (2023). Hydrolysis of complex pectin structures: biocatalysis and bioproducts. In Rosana Goldbeck & Patrícia Poletto (Eds.) *Polysaccharide-degrading biocatalysts* (pp. 205–225). Elsevier.

Yu, Z., Zhang, W., He, H., Li, Y., Xie, Z., Sailike, A., ... Liang, Y. (2024). The CAZyme family regulates the changes in soil organic carbon composition during vegetation restoration in the Mu Us desert. *Geoderma*, *452*, 117109.

Zandalinas, S. I., Fritschi, F. B., & Mittler, R. (2021). Global warming, climate change, and environmental pollution: recipe for a multifactorial stress combination disaster. *Trends in Plant Science*, *26*(6), 588–599.

Zhang, L., Dong, C., Wang, J., Liu, M., Wang, J., Hu, J., ... Zhong, L. (2023). Predation of oomycetes by myxobacteria via a specialized CAZyme system arising from adaptive evolution. *The ISME Journal*, *17*(7), 1089–1103.

Zhao, Z., Baltar, F., & Herndl, G. J. (2020). Linking extracellular enzymes to phylogeny indicates a predominantly particle-associated lifestyle of deep-sea prokaryotes. *Science Advances*, *6*(16), eaaz4354.

2 Analyzing the Structural Diversity of Carbohydrate-Active Enzymes and Their Functional Role in Various Domains of Life

Divyanshi Srivastava and Aruna Rani

ABBREVIATIONS

CAZymes	Carbohydrate-active enzymes
GHs	Glycoside hydrolases
GTs	Glycosyl transferases
PLs	Polysaccharide lyases
CEs	Carbohydrate esterases
AAs	Auxiliary activities
CBMs	Carbohydrate-binding modules
PMEs	Pectin methylesterases
NSP	Non-starch polysaccharide
LPMOs	Lytic polysaccharide monooxygenases
XTHs	Xyloglucan endotransglucosylases/hydrolases
ROS	Reactive oxygen species
ABA	Abscisic acid
ATP	Adenosine triphosphate
Ara4N	4-Deoxy-4-amino-L-arabinose
AMF	Arbuscular mycorrhizal fungi
EMF	Ectomycorrhizal fungi
HGT	Horizontal gene transfer
SCFAs	Short-chain fatty acids
GH5	Glycoside hydrolase family 5
GH10	Glycoside hydrolase family 10

DOI: 10.1201/9781003561682-2

GH11	Glycoside hydrolase family 11
GH13	Glycoside hydrolase family 13
GH16	Glycoside hydrolase family 16
GH20	Glycoside hydrolase family 20
GH28	Glycoside hydrolase family 28
GH32	Glycoside hydrolase family 32
GH33	Glycoside hydrolase family 33
GH61	Glycoside hydrolase family 61
GH68	Glycoside hydrolase family 68
GH79	Glycoside hydrolase family 79
PL1	Polysaccharide lyase family 1
PL8	Polysaccharide lyase family 8
AA1	Auxiliary activity family 1
AA9	Auxiliary activity family 9
CE4	Carbohydrate esterase family 4
CE8	Carbohydrate esterase family 8
CE10	Carbohydrate esterase family 10
dbCAN-seq	Database for CAZyme Annotation Using Sequencing
UDP	Uridine diphosphate

2.1 INTRODUCTION

2.1.1 Overview of Carbohydrate-Active Enzymes

The carbon cycle relies on enzymatic processes to break down diverse polysaccharides into simple sugars. This recycling of organic matter is facilitated by CAZymes, which play a crucial role in polysaccharide deconstruction. These enzymes function through glycosidic bond synthesis and hydrolysis, β-elimination, and ester bond cleavage. They are modular proteins that contain non-catalytic carbohydrate-binding modules (CBMs), which enhance substrate recognition and catalytic efficiency. They often function as independent enzymatic systems or assemble into high-molecular-weight multi-enzyme complexes known as cellulosomes. They are categorized into distinct families based on sequence homology (Davies et al., 2005).

The main CAZymes are hydrolases, glycosyltransferases, lyases, carbohydrate esterases (CEs), and other auxiliary activities (AAs). Their intriguing reactions have attracted researchers from different domains. The growing inquisitiveness is due to CAZymes' extraordinary abilities, comprising building and degrading biopolymers, formation of bacterial biofilm, and glycosylation of proteins and lipids (Benini et al., 2020; Davies et al., 2005).

2.1.2 Importance of CAZymes in Biological Systems

CAZymes are crucial in metabolic pathways, disease pathogenesis, and immune defense mechanisms. Given their biological significance, studying CAZyme inhibitors is vital, as these molecules serve as potential therapeutic agents for glycan-related and CAZyme-associated diseases (Gavande et al., 2023; Garron et al., 2019).

2.1.2.1 Digestive System

CAZymes are critical in breaking complex dietary carbohydrates into simpler absorbable forms for the intestinal epithelium in humans. The microbiota and diet interact in the human gastrointestinal tract, representing distinct niches (Chettri et al., 2023). Carbohydrates act as the source of carbon for these bacteria, a medium to attach to the host, and a barrier to host infection. These enzymes have been used to genetically modify the human gut microbiota for improved health and treatment of diseases, in transforming the glycans in biopharmaceuticals, and in the potential production of universal O-type donor blood (Onyango et al., 2021).

2.1.2.2 Plant Metabolism

CAZymes such as glycoside hydrolases (GHs) and glycosyl transferases (GTs) synthesize and modify plant cell walls, which are composed of polysaccharides like cellulose, hemicellulose, and pectin (Wardman et al., 2022). The starch synthases and amylases biosynthesize and degrade starch, the primary form of energy storage in plants, and aid in the conversion of starch to glucose, which is imperative for energy production during phases of low photosynthetic activity (Cantarel et al., 2009). CAZymes modify oligosaccharides, the signaling molecules, and generate or degrade molecules like callose and β-glucans, involved in the plant's defense to pathogen attack, thereby providing a defense mechanism in plants (Smith et al., 1997).

2.1.2.3 Microbial Systems

The degrading, modifying, and synthesizing capabilities of CAZymes make them an integral part of diverse biological processes such as nutrient acquisition, energy metabolism, and ecological interactions within microbial communities. CAZymes like GHs, polysaccharide lyase (PLs), and CEs work symbiotically to depolymerize plant cell wall components like chitin and other complex carbohydrates. Microbial CAZymes break down lignocellulosic biomass, vital for carbon cycling in ecosystems and biofuel production, thereby degrading plant biomass (Strasser et al., 2016). Pathogenic microbes produce CAZymes, which contribute to the virulence of these organisms by degrading host tissues and evading immune responses. Understanding the pathogenicity of these enzymes can lead to developing new therapeutic strategies, e.g., *Streptococcus pneumoniae* utilizes CAZymes to degrade host glycoproteins, promoting colonization and infection (Flint et al., 2012).

2.1.2.4 Glycosylation Processes

Glycosylation is a post-translational modification where GTs and GHs catalyze the attachment of carbohydrate moieties to proteins and lipids, an essential part of protein folding, stability, and cellular communication. These enzymes break down complex carbohydrates, emphasizing their importance in constructing and degrading biological polymers. Their study provides insights into molecular mechanisms and potential therapeutic interventions (W. Yi et al., 2024).

2.2 CLASSIFICATION OF CARBOHYDRATE-ACTIVE ENZYMES

CAZymes have been classified into GHs, GTs, PLs, CEs, and AAs based on the mechanism of action. These enzymes are further classified into families and subfamilies based on their sequences and fold similarities (Bhattacharya et al., 2015).

2.2.1 Glycoside Hydrolases (GHs)

GHs catalyze the hydrolysis of the glycosidic bonds in O-, N- and S-linked glycosides, forming a sugar hemiacetal or hemiketal and the corresponding free aglycon. These enzymes are also known as glycosidases and glycosyl hydrolases. The catalysis of polysaccharides by GHs involves hydrolysis, where adding a water molecule breaks the glycosidic bond between the two monosaccharide units (Cantarel et al., 2009). Their catalytic mechanism involves catalyzation by retaining enzymes (net retention of anomeric configuration), inverting enzymes (net inversion of anomeric configuration), and NADH Co-factoring (Henrissat et al., 1997).

2.2.2 Glycosyl Transferases (GTs)

GTs are CAZymes that catalyze the formation of a glycosidic linkage to form a glycoside. These enzymes use the "activated" sugar phosphates as glycosyl donors and catalyze transfer of the glycosyl group to a nucleophile, usually an alcohol. The product of glycosyl transfer may be an O-, N-, S-, or C-glycoside belonging to a monosaccharide, oligosaccharide, or polysaccharide (Lairson et al., 2008; Coutinho et al., 2003; Campbell et al., 1997; Coutinho et al., 2009). The transfer of glycosyl groups to nucleophilic acceptor proceeds with either retention or inversion of configuration at the anomeric center (CAZypedia.org). Glycosylation reactions that are catalyzed by GTs occur everywhere in the biosphere and play crucial roles in all vital processes (Liang et al., 2015).

2.2.3 Polysaccharide Lyases (PLs)

PLs are enzymes that cleave activated glycosidic linkages in acidic polysaccharides. These enzymes act via an elimination mechanism, producing an unsaturated oligosaccharide product (Linhardt et al., 1987). PLs cleave polysaccharide chains containing uronic acid through a β-elimination mechanism that generates an unsaturated hexenuronic acid residue and a new reducing end (Lombard et al., 2010; Beeson et al.,2012). The sequence-based families of these enzymes are poly-specific, meaning they contain enzymes that act on different substrates and produce unique products (Garron et al., 2010). These lyases of microbial origin come from a wide variety of pathogenic and non-pathogenic bacteria and fungi (Linhardt et al., 1987).

2.2.4 Carbohydrate Esterases (CEs)

CEs release acyl or alkyl groups attached by ester linkage to carbohydrates (Armendáriz-Ruiz et al., 2018). This class of carbohydrate-active enzymes (CAZymes) catalyzes the de-O or de-N-acylation of substituted saccharides. Since an ester results from an acid-alcohol reaction, two classes of substrates are considered for CEs: those in which the sugar behaves as an "acid", such as pectin methyl esters, and the other in which the sugar acts as an alcohol, such as in acetylated xylan. The most common reaction mechanism of CEs involves a Ser-His-Asp catalytic triad catalyzed by deacetylation homologous to the action of classical lipase and serine proteases (CAZy database).

2.2.5 Auxiliary Activities (AAs)

This group of enzymes includes redox-active proteins that function alongside CAZymes. The identification of specific members within the CBM33 and GH61 families as lytic polysaccharide monooxygenases (LPMOs) led to their reclassification under a distinct category known as "Auxiliary Activities". AA enzymes play a crucial role in the breakdown of plant cell components, enhancing the efficiency of GHs, PLs, and CEs in accessing and processing sugars embedded in the plant cell wall (Levasseur et al., 2013).

2.3 STRUCTURAL DIVERSITY OF CARBOHYDRATE-ACTIVE ENZYMES

Structural studies of CAZymes have shown that these enzymes typically possess a modular architecture with multiple domains, primarily a Catalytic Domain (CD) and one or more CBMs (Davies et al., 2005). The CD is responsible for enzymatic activity, while CBMs enhance substrate binding and enzyme stability by specifically interacting with polysaccharides like cellulose. This modular design improves catalytic efficiency and allows precise recognition of complex carbohydrates as shown in Figure 2.1.

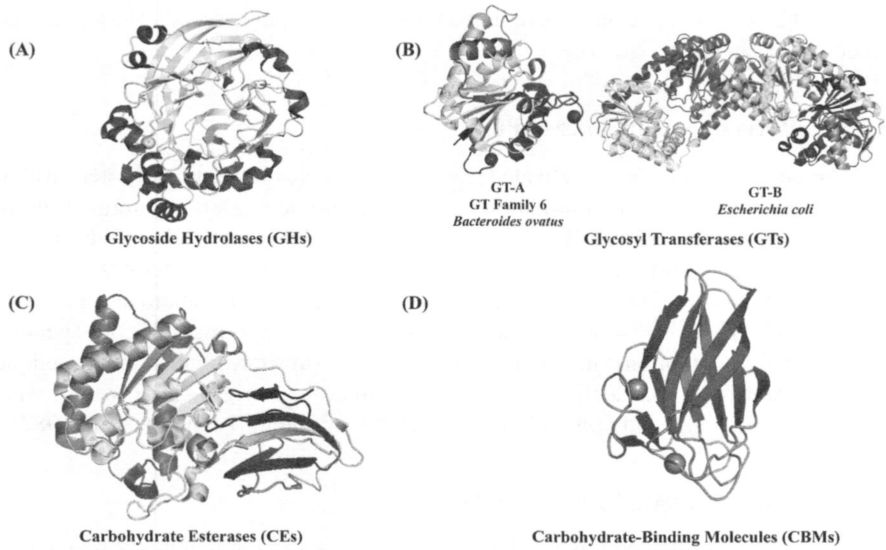

FIGURE 2.1 Structural modifications of Carbohydrate active enzymes.

CBMs are structurally distinct sequences that adopt specific folds to facilitate binding, making CAZymes highly effective in both biological and industrial contexts (Singh et al., 2021).

2.3.1 Structural Features and Fold Types

CAZymes have distinct structural features and fold types, which enhance their role in digesting complex carbohydrates and give them functional versatility. The specificity in the fold type of these enzymes aids in hydrolysis, modification, and binding of carbohydrates through varied biological contexts (Hashimoto et al., 2006).

The structural features and fold types of different classes of CAZymes are as follows:

2.3.1.1 TIM-Barrel or (β/α)$_8$ Fold

The TIM-barrel fold, also known as the (β/α)$_8$ barrel, is a predominant structural motif in many GHs. This fold provides a highly stable and versatile framework for positioning catalytic residues, enabling efficient cleavage of glycosidic bonds. The enclosed barrel structure forms a sheltered environment for catalysis, contributing to enhanced enzymatic reactivity and resistance to denaturation (Hashimoto et al., 2006; Terrapon et al., 2017).

2.3.1.2 GT-A and GT-B Folds

GT-A fold: Comprises a mixed α/β structure with a single-domain architecture. It facilitates binding of nucleotide sugar donors and acceptor substrates, essential for biosynthetic functions like glycoprotein assembly and cell wall biosynthesis (Terrapon et al., 2017).

GT-B fold: Consists of two α/β domains that form a deep cleft. The N-terminal domain generally binds the nucleotide donor, while the C-terminal domain interacts with the acceptor. This bi-domain structure provides high catalytic efficiency in glycosyl group transfer (Albesa-Jové et al., 2013).

2.3.1.3 β-Sandwich and β-Jelly Roll Folds

CBMs frequently adopt β-sandwich architectures, such as the β-jelly roll fold. These structures play a non-catalytic yet essential role by enhancing substrate affinity and targeting enzymes to carbohydrate surfaces. This interaction is crucial for enzymes involved in degrading recalcitrant polysaccharides like cellulose and chitin, as it helps align the catalytic domain with the substrate (Terrapon et al., 2017).

2.3.1.4 β-Helix Fold

PLs utilize a β-helix fold, composed of repetitive β-sheet motifs that form a helical structure. This conformation provides a large surface area suitable for interacting with linear polysaccharide chains such as pectins and alginates. PLs perform β-elimination reactions rather than hydrolysis, aiding in bacterial and plant cell wall degradation (Terrapon et al., 2017).

2.3.1.5 β-Sheet-Rich Folds with Metal-Binding Sites

AA enzymes, particularly LPMOs, are characterized by β-sheet-rich structures incorporating metal-binding centers, often copper (Paradisi et al., 2019). These structural elements are integral to their oxidative mechanism of cleaving polysaccharides like cellulose. The metal-binding site plays a critical role in redox reactions that enhance the accessibility of recalcitrant biomass (Terrapon et al., 2017; Albesa-Jové et al., 2013).

2.3.2 Active Site Architecture and Catalytic Mechanisms

It involves unique structural folds and conserved residues, contributing to their specific activity in carbohydrate metabolism. Many of these enzymes utilize an active site that has specific pockets and binding subsites, known as subsites-1, -2, etc., each of which interacts with distinct carbohydrate residues. These subsites allow CAZymes to engage the sugar substrate at varied positions, thereby optimizing binding and ensuring accurate cleavage of glycosidic bonds. For instance, in GHs, amino acid residues present in the active site engage with carbohydrates through hydrogen bonds, whereas catalytic residues aid in the hydrolysis reaction. GH family 11, xylanases, have a compact active site architecture optimized for hydrolysis of xylan backbones. These introduce aspartate or glutamate residues that behave as acid/base catalysts and nucleophiles (Wu et al., 2010; Fredslund et al., 2011).

The catalytic mechanisms in CAZymes vary according to the enzyme family and substrate. GHs mostly use two primary mechanisms: retaining and inverting mechanisms. For example, GH11 xylanases comprise an acid/base residue that donates a proton (H) to the leaving group while the other residue performs stabilization of the substrate transition state (Wu et al., 2020).

2.3.2.1 Active Site Architecture in CAZymes

The active site architecture of an enzyme is a specialized area within the enzyme that facilitates the binding and transformation of carbohydrate substrates. This region contains residues that behave as catalytic acids, bases, or nucleophiles to assist in the cleavage of glycosidic bonds. The enzyme architecture positions these residues in the right place, supported by structural motifs like loops or helices, stabilizing the substrate in the active site.

Glycoside Hydrolases (GHs): These enzymes hydrolyze glycosidic bonds present between carbohydrate units in complex polysaccharides and oligosaccharides. GHs have a "pocket" or a "cleft" -like structure in their active sites, which binds the substrate. The active sites contain catalytic residues of acidic amino acids such as aspartate and glutamate, which act as proton donors and nucleophiles. GH families like GH12 possess conserved aromatic residues within the active site, usually tryptophan or tyrosine. They are vital in the positioning and binding of the substrates. Any sort of minor variation in these residues can create an adverse impact on enzyme efficiency by altering thermodynamic properties (Davies et al., 1997; Gilbert et al., 2008).

Glycosyltransferases (GTs): These enzymes catalyze the transfer of sugar moieties from nucleotide sugars to acceptor molecules, thus forming glycosidic bonds. These enzymes either use a retaining or an inverting mechanism, determined by the active site structure of the enzyme. For example, the GT-B fold found in many glycosyltransferase enzymes enables a flexible, open-to-closed conformational change, thereby attracting the substrates into proximity for catalysis. This fold promotes different orientations for the donor and acceptor molecules, enabling the enzyme to retain or invert the stereochemistry of the sugar (Biely et al., 2014).

Polysaccharide Lyase (PLs): These enzymes catalyze the β-elimination of uronic acid-containing polysaccharides and forms unsaturated sugar moieties. The active site architecture of PLs is intricately designed to facilitate this reaction, involving specific amino acid residues and structural motifs. The catalytic mechanism of PLs typically involves a base that abstracts a proton from the C5 position of the sugar ring, leading to the formation of a carbanion intermediate, which is followed by the β-elimination of the glycosidic bond (Beeson et al., 2015). The key residues involved in this mechanism are histidine (His), lysine (Lys), or tyrosine (Tyr) residues for abstraction of the protons and glutamate (Glu) or aspartate (Asp) residues that donate a proton to the leaving group. The docking and mutations study for CS lyase of PL8 family inferred that Asn153, His203, Tyr212, Arg266, Glu349 were the key catalytic residues (Rani et al., 2018). Based on the crystal structures and sequence analyses of several alginate lyase of PL7 family it was reported that residues Y195, H119, Q117, and R72 are likely involved in the formation of the catalytic site (Zhu et al., 2015).

Carbohydrate Esterases (CEs): These enzymes exhibit active site architecture highly specialized for hydrolyzing ester bonds within complex sugars. A two-domain structure, including a catalytic α/β-hydrolase, NodB fold domain, and carbohydrate-binding molecules, is projected by these CEs, which facilitates substrate affinity and orientation. For instance, CE2 esterases contain a catalytic α/β-hydrolase core having residues like serine and histidine. This domain has a characteristic central β-sheet flanked by α-helices, forming a hydrophobic binding pocket that interacts with the sugar backbone of the substrate. During deacetylation reactions, aromatic residues such as tryptophan and tyrosine create a hydrophobic platform for stabilizing the polysaccharide chains and substrate binding (Harris et al., 2014; Lombard et al., 2014).

Auxiliary Activity Enzymes (AA): These enzymes play a crucial role in the oxidative degradation of polysaccharides. LPMOs are well-studied AA enzymes with distinct active site architectures for their peculiar biomass conversion. LPMOs contain a flat substrate-binding surface instead of deep channels or clefts in GHs. This flat architecture enhances the interaction between the enzyme and the crystalline surfaces of polysaccharides, such as cellulose. A mononuclear copper ion coordinated by the nitrogen of the N-terminal histidine and side chains of residues is present at the core of the active site of the enzyme. This copper center, present in a distorted square-planar geometry, is necessary for catalyzing oxygen insertion reactions that cleave polysaccharide chains (Armendáriz-Ruiz et al., 2018; Paradisi et al., 2019).

2.3.2.2 Catalytic Mechanisms in CAZymes

1. **Glycoside Hydrolases (GHs):** These enzymes exhibit retaining and inverting mechanisms, the two primary catalytic processes, each dependent on specific active site configurations and amino acid residues crucial for hydrolysis of the substrate.

   In retaining GHs, a two-step double-displacement reaction takes place. The catalytic site includes two essential amino acids, a nucleophile (e,g., glutamate or aspartate) and an acid/base catalyst. The first step involves a nucleophilic attack on the anomeric carbon, forming a covalent glycosyl-enzyme intermediate, and the next step involves the activation of water by the base, which hydrolyzes the glycosyl-enzyme intermediate, retaining the anomeric configuration. This mechanism commonly occurs in families such as GH1 and GH20, which process β-glucosides and N-acetylglucosaminides substrates, respectively.

   The inverting mechanism involves a single displacement reaction, where two catalytic residues are on opposite sides of the substrate. One of these residues acts as an acid to donate a proton, while the other functions as a base to deprotonate a water molecule that attacks the anomeric carbon. This causes the inversion of anomeric configuration. GH families like GH5 involved in cellulose degradation use this mechanism.

2. **Carbohydrate Esterases (CEs):** These enzymes, known for catalyzing the deacetylation of carbohydrates and their derivatives by breaking ester bonds, play a pivotal role in the metabolism of complex polysaccharides. The general mechanism of CE metabolism comprises nucleophilic attack on the ester bond and stabilization of transition states.

   Acid/Base Catalysis- The hydrolysis is facilitated by two amino acid residues that reside in the active site of the enzyme. They have a catalytic triad or dyad, where basic residues like histidine act as a general base, and acidic residues like aspartate or glutamate act as an acid. A water molecule is activated by deprotonation and hydroxyl ion formation by the base present in the active site of the enzyme. The hydroxyl ion, thus formed, attacks the carbonyl carbon of the ester linkage in the carbohydrate. This leads to bond breakage and the formation of a tetrahedral intermediate. This mechanism is common in acetyl xylan esterases (CE1 family), known for removing acetyl groups from xylan. This family of CEs supports this mechanism due to a classical α/β-hydrolase fold (Biely et al., 2016).

   Nucleophilic Attack Mechanism- In this mechanism, an activated nucleophile, residing in the enzyme's active site, directly attacks the ester bond. An acyl-enzyme intermediate is formed when a nucleophile, a serine or cysteine residue, attacks the carbonyl carbon of the ester bond. Simultaneously, a histidine residue in proximity activates a water molecule, which further attacks the acyl-enzyme intermediate to release the deacylated carbohydrate, freeing the enzyme. This type of mechanism is commonly found in feruloyl esterases (CE1, CE6), which are integral in breaking down ferulic acid linkages in hemicelluloses.

3. **Glycosyltransferases (GTs):** These enzymes catalyze the transfer of sugar moieties from activated donor molecules to acceptor molecules, leading to the formation of glycosidic bonds. The catalytic mechanisms of GTs primarily involve a retaining or inverting mechanism of glycosidic bond formation (CAZypedia).

   Retaining Mechanism – This mechanism is often called a "double displacement" reaction where the anomeric configuration of the sugar is maintained (e.g., α remains α) even after the transfer. The active site of glycosyltransferases has a nucleophilic amino acid residue, usually glutamate or aspartate, that attacks the anomeric carbon of the sugar donor (e.g., UDP-sugar). This results in a covalent glycosyl-enzyme intermediate by departing a nucleotide diphosphate. One of the prominent examples of a retaining mechanism is seen in UDP-Glycosyltransferases (UGTs), prevailing in plants and fungi. These enzymes utilize UDP-linked sugars to transfer glycosyl units to a spectrum of acceptors, influencing processes such as secondary metabolite biosynthesis and cellular defense.

   Inverting Mechanism – This mechanism proceeds through a direct SN2-like or bimolecular nucleophilic substitution reaction, involving a single displacement reaction, which results in the inversion of the stereochemistry (anomeric configuration) of the donor sugar. The alpha becomes beta or vice versa. Two critical residues are present in the enzyme's active site, the general acid and the general base (often glutamate or aspartate), where the latter activates the acceptor molecule for nucleophilic attack by deprotonating it. Concurrently, the leaving group (for example, UDP) undergoes protonation by the general acid, which stabilizes the leaving group and facilitates its departure. This causes the formation of a glycosidic bond with inverted stereochemistry. The GT2 family of these enzymes, comprising cellulose synthase and chitin synthase, typically goes through an inverting mechanism (Hemsworth et al., 2013).
4. **Polysaccharide Lyases (PLs):** These lyases catalyze the cleavage of glycosidic bonds via a β-elimination mechanism. The conserved arginine or lysine residues, present proximal to the acidic residues such as aspartate or glutamate, facilitate proton abstraction. The basic residues (mostly a lysine or histidine) in the enzyme's active site extract a proton from the C5 carbon of the polysaccharide's uronic acid. This forms a carbanion intermediate. A β-elimination reaction is then initiated by the negatively charged carbanion, which breaks the existing glycosidic bond and forms a double bond between C4 and C5 carbons (Filandr et al.,2020). Pectate lyases (PL1, PL3) have a parallel β-helix structure that forms an active site resembling a groove for binding to the polygalacturonic acid backbone of pectin (Horn et al., 2012).
5. **Auxiliary Activity Enzymes (AA):** Auxiliary Activities in CAZymes are primarily oxidative enzymes that assist GHs by making recalcitrant polysaccharide bonds more accessible via oxidative cleavage and generating or reacting with reactive oxygen species (ROS) and redox mediators (Agger et al., 2014; Rieder et al., 2023). LPMOs are well-studied AA enzymes that

carry out oxidative cleavage of glycoside bonds in chitin and cellulose. The active site of LPMO consists of a mononuclear copper center with three conserved coordinating groups. This mononuclear Cu site performs the key redox chemistry that activates O_2 or H_2O_2 to enable hydrogen-abstraction and hydroxylation of strong C–H bonds in polysaccharide sugar rings (Bouchey et al., 2022). An external reductant reduces the resting Cu(II)-LPMO to Cu(I). In the presence of carbohydrates, oxidative cleavage of the polymer occurs (peroxygenase activity). Cu(I) reacts with H_2O_2 to produce a short-lived activated copper-oxygen species that abstracts a hydrogen atom from C1/C4. The hydroxylation yields an oxidized sugar that undergoes spontaneous cleavage of the glycosidic bond (Bouchey et al., 2022; Rieder et al 2023).

2.3.3 Evolutionary Adaptations of CAZymes

CAZymes have undergone various evolutionary adaptations that have enhanced their functionality and adaptability to several ecological and host surroundings as depicted in Figure 2.2.

For instance, genomic studies have found that herbivorous arthropods possess expanded CAZyme repertoires, precisely in GHs and carbohydrate-binding molecules (CBMs), for breaking down plant polysaccharides. This adaptation eases nutrient access from rigid plant cell walls, allowing such insects to act on plant-based diets more efficaciously.

2.3.3.1 Evolutionary Adaptations in Glycoside Hydrolases (GHs)

These enzymes that break down complex sugars have evolved through diverse mechanisms such as horizontal gene transfer (HGT), modifications in the structural domain, and adaptive changes to niches of the residing environment. HGT has caused the acquisition of GH genes, prominently observed in the GH32 family, that

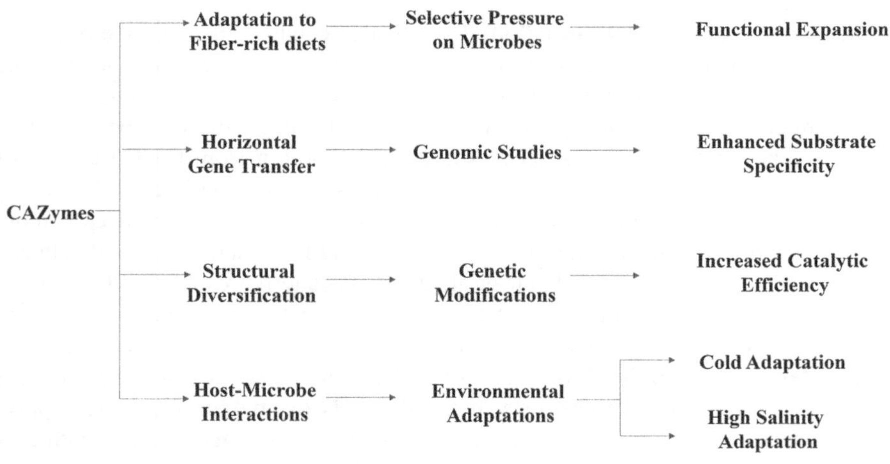

FIGURE 2.2 Overview of Evolutionary Adaptations of CAZymes.

are known for hydrolyzing fructans. Evolution in these enzymes makes them a vital part of digestion as well as in enhancing survival in different niches by manipulation of host plants (Hemsworth et al., 2014). GHs exhibit adaptability in their structure that enables them to function seamlessly across various substrates. GH16 can act on diverse terrestrial and marine polysaccharides because of the β-jelly-roll fold present in it. This structural versatility helps these enzymes maintain functionality under changing temperatures, pH levels, and ecological conditions (Harris et al., 2010). GH32 and GH68 have specific residues that have evolved for better interaction with fructans derived from plant extracts and other polysaccharides. Microbes and invertebrates that rely on contrasting carbohydrate sources in altering environments require this enzymatic flexibility. Such adaptations exhibit how GHs work to diversify, making them a crucial part of all biological processes and potential tools in biotechnology (Hemsworth et al., 2014; Harris et al., 2010).

2.3.3.2 Evolutionary Adaptations in Carbohydrate Esterases (CEs)

These enzymes have evolved to support diverse biological roles in microbial, plant, and animal systems. The NodB domain is a conserved catalytic module in certain CE families, particularly CE4 and CE18. It was initially characterized in the *Rhizobium* NodB protein, which plays a role in modifying carbohydrate substrates through deacetylation. Enzymes containing this domain are critical for bacterial nodulation, polysaccharide degradation, and cell wall remodeling. Enzymes in the CE4 family have undergone structural rearrangements in the NodB domain, mainly through circular permutation. This rearrangement alters the order of β-strands and α-helices, leading to changes in the active site geometry. The CE18 family exhibits deviations in the conserved catalytic motifs of the NodB domain. Some key residues are repositioned or replaced, leading to differences in substrate binding and enzyme function (Aragunde et al., 2018). Due to such evolutionary adaptations, these enzymes are important in biofilm formation and virulence. For instance, *Aspergillus fumigatus* Agd3, a CE18 member, has a peculiar catalytic mechanism for exopolysaccharide deacetylation (Li et al., 2012; Vu et al., 2014). Diversities in CE domain topology (e.g., NodB1 vs. NodB2) have significantly expressed distinct enzymatic functions, including cell wall remodeling and resistance to surrounding stresses (Vu et al., 2014).

Host-Specific Enzyme Induction – CE families such as CE1 and CE10 display gene expression patterns unique to infection stages. Comparative transcriptomic analysis indicates that clusters containing CE1 or CE10 sequences exhibit elevated expression levels, regardless of species composition. Genes encoding CE1 and CE10 are predominantly upregulated in plant pathogens, while their homologs display distinct expression patterns depending on the organism's lifestyle during infection. Notably, hemibiotrophic pathogens demonstrate the highest levels of gene expression among different lifestyles (Li et al., 2012).

Adaptations to Adverse Environments – CEs have evolved to adapt better to changing conditions to survive inside the host tissue. CEs derived from marine sources and soil show adaptations for stability and activity under extreme conditions of high salinity, alkaline pH, or low temperatures. This is primitively seen in halophilic microbes, where structural elasticity is critical for survival in saline environments (Li et al., 2012).

Antibiotic Resistance Mechanisms – Salmonella typhimurium possesses the ArnD protein, which plays a crucial role in lipid A modification and participates in the biosynthetic pathway of 4-deoxy-4-amino-L-arabinose (Ara4N). This modification enhances bacterial resistance to antimicrobial peptides and polymyxin antibiotics by covalently altering lipid A.

2.3.3.3 Evolutionary Adaptations in Polysaccharide Lyases (PLs)

The evolutionary adaptations of PLs are the result of structural innovation, catalytic optimization, gene duplication, and HGT. These adaptations have enabled PLs to degrade a wide array of polysaccharides across diverse habitats, contributing significantly to carbon cycling and biotechnological applications (Zhu et al., 2015; Garron et al. 2019). PLs have evolved multiple fold types, such as jelly-roll β-sandwiches (PL7), β-helical structures (PL1), and barrel-like folds (PL17). Evolutionary divergence in fold types reflects adaptation to various substrate complexities and environmental constraints. The Michaelis complex formed between an inactive mutant of polygalacturonate lyase of (PL10) and its substrate, trigalacturonate/Ca^{2+}, reveals the catalytic machinery employed by the enzyme. Remarkably, this architecture closely resembles, likely as a result of convergent evolution, the tetragalacturonic acid complex seen in a structurally unrelated polygalacturonate lyase from the PL-1 family (Charnock et al., 2002).

2.3.3.4 Evolutionary Adaptations in Glycosyl Transferases (GTs)

Structural Divergence and Functional Specialization – The categorization of glycosyltransferases is primarily based on their structural folds: GT-A, GT-B, and GT-C. Enzymes with the GT-A fold feature a Rossmann-like nucleotide-binding domain, while GT-B enzymes possess two distinct domains that are separated by a substrate-binding cleft. These structural features make the GTs adaptable to diverse donor and acceptor substrates, enhancing their substrate specificity and catalytic efficiency (Walton et al., 2016; Bisaaro et al., 2017).

Substrate Specificity and Convergent Evolution – The ability of glycosyltransferases to recognize different types of sugar donors and acceptors exhibits convergent evolution. GT6 and GT3 use similar sets of conserved residues to stabilize sugar nucleotide donors, such as UDP-glucose, even after having structural variations in their protein scaffolds. These models discovered structural and evolutionary features that can predict donor substrate specificity with approximately 90 percent accuracy (Bissaro et al., 2017).

Adaptation in Host-Microbe Interactions- In the microbiome, GTs enable modification of glycans by the microbes, evading the immune system and nutrient acquisition. They modify cell-surface molecules for virulence, an adaptation led by pressures of the host immune system (Walton et al., 2016).

In GTs, HGT supports symbiotic relationships with host organisms by optimizing nutrient extraction from plant fibers. This improves the ability of organisms to exploit new ecological niches and endure environmental stresses (Walton et al., 2016).

2.4 FUNCTIONAL ROLES OF CAZYMES IN VARIOUS DOMAINS OF LIFE

CAZymes are critical across varied domains in life. They break down dietary fibers into short-chain fatty acids (SCFAs) in microbes, as this process is essential for host metabolism and immunity.

2.4.1 CAZymes in BACTERIA: Degradation of Plant Biomass and Pathogenesis

2.4.1.1 Degradation of Plant Biomass

Clostridium phytofermentans codes for 171 types of CAZymes, of which 116 belong to 44 families of GHs. These enzymes are synergistic in behavior, targeting specific polysaccharides such as xylan and galactomannan (Tandrup et al., 2018;). *Caldicoprobacter algeriensis* is a thermophilic bacterium that utilizes CE4 and CE1 enzymes to act on acetylated glucuronoxylan in high-temperature environments and various hemicelluloses (Müller et al., 2018). Xylanases and cellulases exhibit promising industrial applications in biofuel production due to their stable and highly efficient capability to degrade lignocellulosic materials (Tandrup et al., 2018).

Endoglucanases (GH5 family) are found in *Clostridium thermocellum,* which work within cellulosomes and multi-enzyme complexes to enhance cellulose degradation efficiency. It is useful in the paper, textile, and bioethanol industries for converting plant biomass into fermentable sugars (Tandrup et al., 2018; Müller et al., 2018). Xylanases crush down xylan by cleaving the β-1,4 backbone into xylooligosaccharides. GH10 enzymes tend to have broad specificity, while GH11 enzymes are precisely specialized to operate on linear xylan. *Bacillus subtilis* produces GH10 xylanases to decompose agricultural residues for soil nutrient cycling (Tandrup et al., 2018; Müller et al., 2018).

Plant pathogens like *Erwinia chrysanthemi* have PL1 enzymes to facilitate colonization by pectin degradation (Müller et al., 2018). Acetylxylan esterases (CE4 family) remove acetyl groups from hemicellulose; deacetylation improves the accessibility of substrate and enhances enzymatic hydrolysis rates. *β*-Glucosidases (GH3 Family) hydrolyze cellobiose and short cello-oligosaccharides into glucose, finishing cellulose breakdown. The GH3 *β*-glucosidase from *Thermotoga maritima* is thermostable, making it suitable for industrial processes requiring high temperatures (Tandrup et al., 2018; Müller et al., 2018).

2.4.1.2 Pathogenesis

CAZymes in bacteria aid in pathogenesis by breaking down host glycans and facilitating colonization and infection. Sialidases (GH33 family) found in *Clostridium perfringens* cleave sialic acids from host glycoproteins and glycolipids. Hyaluronidases (GH79 family) present in *Streptococcus pyogenes* degrade hyaluronic acid and promote bacterial dissemination and invasion of tissues (Tandrup et al., 2018). β-N-Acetylglucosaminidases (GH20 family) found in *Helicobacter pylori* degrade mucins in the gastric mucosal layer, allowing the bacterium to colonize and evade host defenses (Frandsen et al., 2016; Tandrup et al., 2018). Chondroitin lyases (PL8 Family) retrieved from *Vibrio cholerae* degrade chondroitin sulfate, and promote bacterial colonization and toxin delivery (Tandrup et al., 2018; Müller et al., 2018). Alginate lyase of PL1 family of *Pseudomonas aeruginosa* degrades alginate, a component of biofilms, enabling bacteria to persist in host lungs in cystic fibrosis patients (Müller et al., 2018).

2.4.2 CAZymes in Archaea: Adaptations to Extreme Environments

CAZymes in archaebacteria show remarkable features that enable them to function even in extreme conditions of high temperature, salinity, or acidic conditions. Archaeal CAZymes from *the Pyrococcus* and *Thermococcus* genera are thermostable and help them to thrive in hydrothermal vents. Haloarchaea bacteria such as *Haloferax* and *Halorubrum* possess CAZymes that are adapted to high saline conditions, maintaining activity in extremely ionic environments. *Sulfolobus acidocaldarius,* an acidophilic archaeabacterium, produces enzymes that function optimally at very low pH, necessary for survival in acidic hot springs. Their distinct protein structures include increased ionic interactions and core hydrophobic regions, which improve stability under extreme conditions (Forsberg et al., 2016; Hemsworth et al., 2018).

α-Amylase of GH13 family from *Pyrococcus furiosus* is highly active at temperatures up to 110°C due to its compact protein structures, greater ionic interactions, and a higher proportion of α-helices. It hydrolyzes α-1,4-glycosidic bonds in starch and is important for carbohydrate metabolism in hydrothermal vents. It is also used in the processing of starch and sugar syrup production due to its high-temperature activity, making it perfect for bioethanol production and heated laundering (Forsberg et al., 2016; Hemsworth et al., 2018). Xylanases (GH10 Family) extracted from *Thermococcus zilligii*, a hyperthermophilic archaeon, operate against both linear and branched xylans. Their structural adaptations comprise disulfide bonds and salt-bridging interactions to attain stability. They degrade hemicellulose by cleaving β-1,4-glycosidic bonds in xylan and facilitate biomass decomposition in extreme ecosystems. These serve as key enzymes in the paper and pulp industries for eco-friendly bleaching. They also play a crucial role in biorefining for converting plant biomass into fermentable sugars (Forsberg et al., 2016; Hemsworth et al., 2018).

β-Glucosidase (GH1 Family) found in *Sulfolobus solfataricus*, an acidophilic (pH ~ 2–3) and thermophilic archaeon, has a robust structure possessing high surface charge density for acid resistance. It hydrolyzes cellobiose and various cello-oligosaccharides into glucose, and serves as a model enzyme for researching extremophile carbohydrate metabolism and used for biofuel production (Forsberg et al., 2016; Hemsworth et al., 2018). Pullulanase (GH57) Family of *Thermococcus kodakarensis*, a hyperthermophile, resistant to EDTA, making it operational in high-salt and alkaline environments. It hydrolyzes α-1,6-glycosidic bonds in pullulan, amylopectin, and glycogen, and is useful in industries that process starch to produce glucose syrups and maltodextrins. Pullulanases have also been explored for their role in brewing and baking industries (Forsberg et al., 2016; Hemsworth et al., 2018). Chitinase (GH18 Family) of *Halobacterium salinarum* has salt tolerance due to more acidic amino acids on the enzyme's surface, maintaining solubility and activity in ionic conditions. It digests chitin, a polysaccharide found in marine invertebrates and fungal cell walls, into N-acetylglucosamine (GlcNAc). Agricultural industries are widely using the chitinase enzyme to control fungal pathogens (Forsberg et al., 2016; Hemsworth et al., 2018).

2.4.3 CAZymes in Fungi: Decomposition and Symbiosis

Fungi play an important role in ecosystems through their CAZymes, facilitating decomposition and symbiosis. Research has elucidated their diversity and functional adaptations, implicating their applications in biotechnology, agriculture, and environmental sustainability (Courtade et al., 2018; Bissaro et al., 2017).

2.4.3.1 Decomposition of Plant Biomass

Fungi play a crucial role in breaking down lignocellulosic biomass, efficiently degrading components such as cellulose, hemicellulose, and lignin. A key group of enzymes in this process is cellulases (GH7), also known as cellobiohydrolases, which are found in saprophytic fungi like *Trichoderma reesei*. These enzymes specifically target β-1,4-glycosidic bonds in cellulose, breaking them down into cellobiose and eventually glucose. Structurally, cellulases consist of a catalytic domain responsible for hydrolysis and a CBM that enhances enzyme attachment to crystalline cellulose, improving catalytic efficiency. Advances in enzyme engineering have further optimized their efficiency, enabling higher yields in biofuel production and promoting sustainable energy solutions (Courtade et al., 2018; Bissaro et al., 2017).

Laccases (AA1 Family) are produced by white-rot fungi such as *Phanerochaete chrysosporium*, and digest lignin into smaller, more accessible components. Multicopper oxidases have a unique trinuclear copper center, which is responsible for electron transfer (Paradisi et al., 2019). The terminal electron acceptor is oxygen, producing water as a by-product. Laccases facilitate nutrient cycling in forest ecosystems by lignin decomposition, providing access to cellulose and hemicellulose (Courtade et al., 2018; Bissaro et al., 2017). Xylanases (GH10 Family) are found in fungi like *Aspergillus niger* and *Penicillium spp.* These enzymes hydrolyze β-1,4-xylosidic linkages in xylan, one of the main components of hemicellulose, releasing xylose and smaller oligosaccharides, which fungi use as a carbon source. GH10 xylanases are endo-acting enzymes having broad substrate specificity. Their active site can accommodate both linear and branched xylan chains, which enhances the degradation of hemicellulose (Courtade et al., 2018; Bissaro et al., 2017).

2.4.3.2 Symbiosis

Mycorrhizal fungi are known for their symbiotic roles, utilizing CAZymes to modify plant cell walls during colonization and nutrient exchange. Arbuscular Mycorrhizal Fungi (AMF) are a group of fungi belonging to the Glomeromycota phylum that form symbiotic relationships with over 80% of terrestrial plants. They form intracellular connections (arbuscules) within the cortical cells of the root, facilitating the exchange of nutrients and carbohydrates. AMF secretes enzymes like cellulases, xylanases, and glucanases to hydrolyze plant-derived polysaccharides. These enzymes enhance fungal penetration and colonization by degrading cell wall components. Enzymes released by AMF help in phosphate mobilization, nutrient uptake, carbon source exchange, and root extension beyond zones to access soil regions (Courtade et al., 2018; Bissaro et al., 2017).

Ectomycorrhizal Fungi (EMF) establish symbiotic relationships with woody plants by covering root tips and forming a Hartig net, which is a nutrient exchange interface in the root cortex. Chitinases digest chitin and release nitrogen for plant use, while ligninolytic and cellulolytic enzymes decompose organic matter, including lignocellulosic compounds (Chylenski et al., 2019). Proteases hydrolyze proteins to release amino acids and ammonium. Some of the remarkable fungal CAZymes include AA9 lytic polysaccharide monooxygenases (LPMOs) sourced from *Neurospora crassa*, GH28 polygalacturonases found in *Aspergillus flavus,* AA1 laccases present in *Phanerochaete chrysosporium,* and CE8 pectin methylesterases possessed by *Penicillium expansum* (Courtade et al., 2018; Bissaro et al., 2017).

2.4.4 CAZymes in Eukaryotes: Plant Cell Wall Biosynthesis and Remodeling

Plant CAZymes play vital roles in the synthesis and remodification of cell walls, critical for growth, defense, and environmental adaptation. The cell wall, composed of cellulose, hemicellulose, pectin, and lignin, is a dynamic structure that requires precise enzymatic regulation to balance rigidity and flexibility.

2.4.4.1 Biosynthesis of Polysaccharides

Cellulose synthases (CESA, GT2 family), polymerize residues of glucose into β-1,4-linked glucan chains that assemble into cellulose microfibrils. Microfibrils provide structural integrity and lay the foundation of plant morphology and defense. CESA has a cytoplasmic glycosyltransferase domain for polymerization of glucose, a transmembrane domain that anchors it into the plasma membrane, facilitates glucan extrusion into the cell wall. CESAs function as multi-protein cellulose synthase complexes (CSCs), often arranged as rosettes of six units, at the plasma membrane synthesizing cellulose fiber. Specific CESA isoforms (e.g., CESA1, CESA3, and CESA6 in primary walls) are designed for special developmental roles. These enzymes are regulated by mechanosensory pathways that detect wall integrity and external signals like light, hormones (e.g., auxin), or environmental stresses (e.g., salinity). Cell wall mechanosensors sense cellulose microfibril tension and send inhibitory feedback signals, adjusting CESA activity as required (Frommhagen et al., 2016; Agger et al., 2014). *Hemicellulose synthases (GT47 family)* such as xyloglucans and arabinoxylans, are synthesized by GT47 enzymes. *Pectin synthases (GT8 family)* are branched polysaccharides primarily composed of homogalacturonan, rhamnogalacturonan, and arabinogalactan. The enzymes of this family, such as *galacturonosyl transferases (GalATs),* catalyze the formation of the homogalacturonan backbone. Their activity controls cell wall adhesion, rigidity, and defense responses against pathogens by regulating wall permeability (Frommhagen et al., 2016; Agger et al., 2014).

Both of these enzymes are present in the Golgi apparatus, assembling polysaccharides before secretion into the extracellular matrix. They work in groups with pectin methylesterases (PMEs) and xylanases for modifications required post-synthesis. Hemicellulose and pectin biosynthesis modification leads to crop resilience and promotes the production of digestible bio-feedstocks for animal feed and industrial applications (Frommhagen et al., 2016; Agger et al., 2014).

2.4.4.2 Remodeling during Growth and Stress

Expansins are unique proteins, non-enzymatic in nature, involved in loosening the cellulose-hemicellulose network in plant cell walls for cell elongation and growth. They act by disturbing hydrogen bonds between cellulose microfibrils and matrix polysaccharides, initiating wall extensibility devoid of polymer hydrolysis. These enzymes are critically required for root and shoot growth, fruit ripening, enhancing softening, and responding to abiotic stresses (Frommhagen et al., 2016; Agger et al., 2014).

Xyloglucan endotransglucosylases/hydrolases (XTHs) belong to the GH16 family, which modify xyloglucan, a major hemicellulose that crosslinks cellulose microfibrils, maintaining the integrity and flexibility of the cell wall. *Hydrolase activity* cleaves xyloglucan polymers to facilitate cell wall remodeling. XTHs enable controlled loosening and reassembly of cell walls during rapid growth and modulate wall architecture in response to mechanical stress, pathogen attacks, or drought salinity (Frommhagen et al., 2016; Agger et al., 2014).

PMEs are enzymes that de-esterify methyl groups on homogalacturonan, a pectin polysaccharide, and convert them into a more acidic form. During salt or drought stress, PMEs adjust the rigidity and permeability of the wall, enhancing cell survival. These enzymes alter pectin properties to make walls more resilient to microbial degradation (Frommhagen et al., 2016).

2.4.4.3 Signaling and Response to Abiotic Stress

Peroxidases (class III plant peroxidases) and *laccases* (multicopper oxidases) are enzymes that catalyze the oxidative polymerization of monolignols (e.g., p-coumaryl alcohol, coniferyl alcohol, and sinapyl alcohol) into lignin. Lignin, an aromatic polymer, fortifies the plant cell wall and provides structural rigidity and resistance against biotic and abiotic stresses. *Peroxidases* utilize H_2O_2 as a co-substrate for the oxidation of monolignols, initiating their polymerization. They are expressed in specific cell types, particularly during secondary wall thickening in xylem and sclerenchyma cells. Laccases oxidize monolignols and phenolics using molecular oxygen as an electron acceptor. Peroxidases and laccase activities are upregulated under stress conditions, often modulated by signaling molecules like abscisic acid (ABA) (Frommhagen et al., 2016; Agger et al., 2014). ROS are produced in the apoplast by *NADPH oxidases,* which signal the activation of CAZymes at low concentrations, whereas, at higher concentrations, ROS introduce oxidative stress. They regulate the expression of CAZyme genes coding for peroxidases, laccases, and xyloglucan-modifying enzymes (Frommhagen et al., 2016; Agger et al., 2014).

2.4.5 CAZymes in Animals: Digestion and Gut Microbiota Interactions

Microbes in the gut help animals break down polysaccharides, such as cellulose and hemicellulose, into monosaccharides or simple sugars. This digestive process produces SCFAs like butyrate, acetate, and propionate, which serve as energy sources for the host and thereby modulate immune responses. Acetate and propionate are absorbed into the bloodstream, serving as a gluconeogenesis or fatty acid synthesis precursor. SCFAs govern both innate and adaptive immune systems, reducing inflammatory cytokine production and enhancing immune homeostasis. Butyrate helps restore gut barrier integrity and alleviate inflammation in diseases like colitis. It promotes regulatory T-cell differentiation, critical for maintaining immune homeostasis (Forsberg et al., 2023; Kuusk et al., 2020).

GHs are a central part of carbohydrate digestion in animals and microbial ecosystems like the gut microbiome. Cellulases (GH5 and GH9) hydrolyze β-1,4 bonds in cellulose to release glucose. Amylases (GH13) break down starch into maltose and maltotriose. Lactase (GH2) cleaves lactose into glucose and galactose in mammals. In the gut microbiome, GHs from bacteria like *Bacteroides* and *Firmicutes* degrade dietary fibers such as cellulose and hemicellulose, producing SCFAs (Forsberg et al., 2023; Kuusk et al., 2020). PLs cleave uronic acid-containing polysaccharides, contributing to the metabolism of plant-derived carbohydrates (Forsberg et al., 2023). GTs catalyze glycosylation reactions, crucial for the modification of carbohydrate structures to perform diverse metabolic roles.

Cellulases (GH5 Family) hydrolyze β-1,4-glycosidic bonds in cellulose to give oligosaccharides and glucose. These enzymes are vital for herbivores as cellulose makes up a major portion of plant biomass. These are found predominantly in the gut microbiota of *Fibrobacter* and *Ruminococcus* species. These enzymes improve livestock nutrition by enhancing feed digestibility (Kuusk et al., 2020).

Xylanases (GH10 Family) produces xylo-oligosaccharides and xylose, which are successively fermented by microbes. These are endo-acting enzymes having broad substrate specificity, potentially enough to degrade highly substituted xylans. These are present in *Bacteroides* species, performing fiber digestion in monogastric and herbivorous animals. They are ruling the prebiotic industry owing to their inherent capability to produce functional oligosaccharides (Forsberg et al., 2023; Kuusk et al., 2020).

α-Amylases (GH13 family) produces maltose, maltotriose, and dextrins. Their active site containing conserved catalytic residues, such as aspartate and glutamine, mediates hydrolysis. Evolutionary pressures in omnivores, carnivores, and herbivores have modified the efficiency and expression of α-amylases to comply with dietary patterns. α-Amylases are used in feed formulations for monogastric animals like poultry and pigs to bring about improvement in starch digestibility, hence reducing feed costs and enhancing growth rates. Gut bacteria, including *Bacteroides* and *Prevotella*, complement the activities of host amylase by breaking down resistant starch, leading to a better energy harvest and metabolic health (Forsberg et al., 2023; Kuusk et al., 2020).

β-Mannanases (GH26 family) hydrolyze β-1,4-glycosidic bonds in mannans, glucomannans, and galactomannans found in plant cell walls. They are found in the gut microbes of herbivorous animals, such as *Prevotella* and *Clostridium* species, which promote gut health by enhancing fiber digestibility and energy yield (Kuusk et al., 2020). By degrading non-starch polysaccharides (NSP), β-mannanases refine nutrient absorption and feed efficiency in livestock. They reduce the antinutritional effects of mannans, which can hinder protein digestion in animals. The study of β-mannanases in the gut microbiota of *Prevotella* species has been insightful for carbohydrate utilization and microbial adaptation to fiber-rich diets (Forsberg et al., 2023).

2.5 ROLE OF CAZYMES IN INDUSTRY

CAZymes are crucial in various industries because they can degrade, modify, and synthesize complex carbohydrates. These enzymes are employed in biorefineries, food and beverage production, textiles, pharmaceuticals, and bioenergy (Wang et al., 2019).

2.5.1 Applications in Biofuel Production and Biomass Conversion

The fermentable sugars produced by complex plant biomass breakdown are processed into biofuels like ethanol, biodiesel, and biogas. Lignocellulose is complex and recalcitrant due to its intricate matrix of cellulose microfibrils embedded in hemicellulose and lignin. In industrial processes, pretreating the fermentable sugars often modifies lignocellulose to improve enzyme accessibility. CEs remove acetyl and feruloyl groups from hemicelluloses and optimize substrate accessibility for downstream enzymatic hydrolysis. LPMOs use oxidative mechanisms to cleave recalcitrant polysaccharides and enhance cellulose depolymerization efficiency (Kracher et al., 2016; Hangasky et al., 2018). Industries utilize enzyme cocktails designed for specific biomass sources and engineer CAZymes for higher thermal stability and substrate specificity. This enzymatic hydrolysis of lignocellulose releases monosaccharides such as glucose and xylose, which further serve as fermentation substrates (Hangasky et al., 2018).

Fermentation of glucose by *Saccharomyces cerevisiae* produces ethanol. Anaerobic digestion of sugars by microbial consortia results in methane-rich biogas. CAZymes indirectly assist in biodiesel production by digesting feedstock into usable components. Lipases are used in transesterification reactions to synthesize biodiesel. Biofuel production driven by CAZymes supports sustainable energy goals. Lignocellulosic biofuels are carbon-neutral, reducing greenhouse gas emissions compared to fossil fuels. Synthetic biology and microbial fermentation are innovations in the CAZyme industry that aim to cut costs and improve enzyme yields (Kracher et al., 2016; Hangasky et al., 2018).

2.5.2 Industrial Use in Food, Feed, and Pharmaceuticals

Amylases, xylanases, and pectinases improve processing efficiency and quality of food by modifying food texture and enhancing flavor. Amylases hydrolyze starch in bread-making, which handles the dough much better and gives it a longer shelf life. Invertases and cellulases aid in the production of sweeteners, such as glucose and fructose syrups. Pectinases and all other cell wall-degrading enzymes are used in beverage clarification to clarify juices and wines by disintegrating polysaccharides that cause turbidity (Wang et al., 2019; Paradisi et al., 2019; Scott et al., 2020). NSP degrading enzymes, such as β-glucanases and xylanases highly improve the digestibility of feed by breaking down indigestible fibers, enhancing energy yield, reducing anti-nutritional factors, and amplifying nutrient absorption. (Wang et al., 2019; Paradisi et al., 2019; Scott et al., 2020).

Glycosyltransferases and glycosidases play a vital role in drug synthesis by altering glycosylation patterns. These enzymes are therapeutically used to improve the efficacy and stability of drugs in the pharmaceutical industry. CAZymes help in the modification of polysaccharides in bioactive compounds to enhance their absorption capacity and bioavailability (Paradisi et al., 2019; Scott et al., 2020).

2.6 EMERGING TRENDS AND FUTURE DIRECTIONS

CAZymes are undergoing innovations with their applications spreading across environmental, industrial, and biomedical domains. Research in sequencing technologies and bioinformatics tools like the CAZyme database and dbCAN-seq has identified and classified new CAZyme families. This advancement exhibits the diversity of microbial genomes and environmental metagenomes (Borisova et al., 2015; Beeson et al., 2015). Synthetic biology combines CAZymes with designer microorganisms that allow novel biosynthetic pathways for biopolymers, biofuels, and value-added chemicals (Borisova et al., 2015). CAZymes have revolutionized eco-friendly alternatives to fossil fuels by converting lignocellulosic biomass into biofuels and biochemicals (Beeson et al., 2015). CAZymes have been leveraged in soil microbial communities to bring about carbon sequestration and mitigate climate change (Borisova et al., 2015). Artificial intelligence-driven modeling and machine learning are being used for CAZyme structure prediction, active site architecture, and catalytic mechanisms, which boost enzyme discovery and functional annotation (Borisova et al., 2015; Beeson et al., 2015).

The study of Kinetic Isotope Effects (KIEs) in CAZymes reveals reaction mechanisms, specifically in enzymatic hydrogen transfer and bond cleavage processes. KIEs are influenced by isotopic substitution, such as replacing hydrogen with deuterium, as it alters the reaction rate due to differences in mass and quantum tunneling effects. In CAZymes, these effects can show whether bond cleavage involving a specific atom is part of the rate-determining step or if quantum mechanical tunneling is involved in the enzymatic mechanism (Hangasky et al., 2019).

2.7 CONCLUSION

CAZymes play a crucial role in the synthesis and degradation of carbohydrates and glycoconjugates, contributing to diverse metabolic pathways. These pathways encompass both anabolic and catabolic processes, facilitating the formation and breakdown of complex biomolecules, including bacterial exopolysaccharides, starch, cellulose, and lignin. CAZymes also perform adenosine triphosphate (ATP)-dependent glycosylation of proteins and lipids, covalently attaching the carbohydrate to a macromolecule. The classification and structural diversity of these enzymes exhibit their adaptation to unique biological roles, including the degradation, modification, and synthesis of complex carbohydrates. The organization of these enzymes into families, such as GHs, glycosyltransferases, and CEs, reveals diverse catalytic mechanisms. CAZymes aid in nutrient cycling, structural carbohydrate metabolism, and ecological interactions across all domains of life. Industrial applications include their diverse functions in biomass conversion, food processing, and pharmaceutical synthesis, signifying sustainable biotechnological advancements and innovations. Therefore, these enzymes are essential for survival, driving the synthesis, modification, and degradation of complex carbohydrates, which govern all crucial processes at macro and micro levels across extensive domains of life.

ACKNOWLEDGMENT

The authors thank the Department of Microbiology, Sikkim University, for providing the computational infrastructure and central library facilities for procuring references and plagiarism analysis (Original Plagiarism Detection Software).

AUTHOR CONTRIBUTION STATEMENT

Divyanshi Srivastava: Writing, original draft preparation; Aruna Rani: Writing, original draft preparation, reviewing and editing, Supervision.

FUNDING

This research did not receive any specific grant from funding agencies in the public, commercial, or not-for-profit sectors.

CONFLICT OF INTEREST STATEMENT

The author declares that the research was conducted without any commercial or financial relationships that could be construed as potential conflicts of interest.

REFERENCES

Agger, J. W., Isaksen, T., Varnai, A., Vidal-Melgosa, S., Willats, W. G. T., Ludwig, R., & Horn, S. J. (2014). Discovery of LPMO activity on hemicelluloses shows the importance of oxidative processes in plant cell wall degradation. *Proceedings of the National Academy of Sciences*, 111(17), 6287–6292. https://doi.org/10.1073/pnas.1323629111

Albesa-Jové, D., Giganti, D., Jackson, M., Alzari, P. M., & Guerin, M. E. (2014). Structure–function relationships of membrane-associated GT-B glycosyltransferases. Glycobiology, 24(2), 108–124. https://doi.org/10.1093/glycob/cwt101

Aragunde, H., Biarnés, X., & Planas, A. (2018). Substrate Rrecognition and Sspecificity of Cchitin Ddeacetylases and Rrelated Ffamily 4 Ccarbohydrate Eesterases. *International Journal of Molecular Sciences*, 19(2), 412. https://doi.org/10.3390/ijms19020412

Armendáriz-Ruiz, M., Rodríguez-González, J. A., Camacho-Ruíz, R. M., & Mateos-Díaz, J. C. (2018). Carbohydrate esterases: An overview. In G. Sandoval (Ed.), *Lipases and*

Phospholipases (Vol. 1835, pp. 23–45). Humana Press. https://doi.org/10.1007/978-1-4939-8672-9_2

Armendáriz-Ruiz, M., Rodríguez-González, J. A., Camacho-Ruíz, R. M., & Mateos-Díaz, J. C. (2018). Carbohydrate esterases: An overview. *In Methods in Molecular Biology (Vol. 1835, pp. 39–68)*. Humana Press. https://doi.org/10.1007/978-1-4939-8672-9_2

Beeson, W. T., Phillips, C. M., Cate, J. H. D., & Marletta, M. A. (2012). Oxidative cleavage of cellulose by fungal copper-dependent polysaccharide monooxygenases. *Journal of the American Chemical Society*, 134(2), 890–892. https://doi.org/10.1021/ja210657t

Beeson, W. T., Vu, V. V., Span, E. A., Phillips, C. M., & Marletta, M. A. (2015). Cellulose degradation by polysaccharide monooxygenases. *Annual Review of Biochemistry*, 84, 923–946. https://doi.org/10.1146/annurev-biochem-060614-034439

Benini, S. (2020). Carbohydrate-Aactive Eenzymes: Structure, Aactivity, and Rreaction Pproducts. *International Journal of Molecular Sciences*, 21(8), 2727. https://doi.org/10.3390/ijms21082727

Bhattacharya, T., Ghosh, T. S., & Mande, S. S. (2015). Global profiling of carbohydrate-active enzymes in the human gut microbiome. *PLOS ONE*, 10(11), e0142038. https://doi.org/10.1371/journal.pone.0142038

Biely, P., Cziszárová, M., Uhliariková, I., Agger, J. W., & Westereng, B. (2014). Fluorimetric detection of oxidative dehydrogenation of reducing carbohydrates by lytic polysaccharide monooxygenase. *Biotechnology for Biofuels*, 7(1), 98. https://doi.org/10.1186/1754-6834-7-98

Biely, P., Singh, S., & Puchart, V. (2016). Towards enzymatic breakdown of complex plant xylan structures: State of the art. *Biotechnology Advances*, 34(7), 1260–1274. https://doi.org/10.1016/j.biotechadv.2016.09.001

Bissaro, B., Rohr, A. K., Muller, G., Chylenski, P., Skaugen, M., Forsberg, Z., ... & Eijsink, V. G. H. (2017). Oxidative cleavage of polysaccharides by monocopper enzymes depends on H2O2. *Nature Chemical Biology*, 13(10), 1123–1128. https://doi.org/10.1038/nchembio.2470

Borisova, A. S., Isaksen, T., Dimarogona, M., Kognole, A. A., Mathiesen, G., Várnai, A., ... & Eijsink, V. G. H. (2015). Structural and functional characterization of a lytic polysaccharide monooxygenase with broad substrate specificity. *Journal of Biological Chemistry*, 290(38), 22955–22969. https://doi.org/10.1074/jbc.M115.660183

Bouchey, C. J., Shopov, D. Y., Gruen, A. D., & Tolman, W. B. (2022). Mimicking the Cu active site of lytic polysaccharide monooxygenase using monoanionic tridentate N-donor ligands. *ACS Omega*, 7(39), 35217–35232. https://doi.org/10.1021/acsomega.2c04432

Campbell, J. A., Davies, G. J., Bulone, V., & Henrissat, B. (1997). A classification of nucleotide-diphospho-sugar glycosyltransferases based on amino acid sequence similarities. *Biochemical Journal*, 326(Pt 3), 929–939. https://doi.org/10.1042/bj3260929u

Cantarel, B. L., Coutinho, P. M., Rancurel, C., Bernard, T., Lombard, V., & Henrissat, B. (2009). The Carbohydrate-Active Enzymes database (CAZy): An expert resource for glycogenomics. *Nucleic Acids Research*, 37(Database issue), D233–8. https://doi.org/10.1093/nar/gkn663

CAZy Database. (2024). Retrieved from www.cazy.org

CAZypedia. (2021, January 7). Glycosyltransferases. CAZypedia: A Lliving Eencyclopedia of Ccarbohydrate-Aactive Eenzymes. Retrieved from http://www.cazypedia.org/index.php?title=Glycosyltransferases&oldid=16153

Charnock, S. J., Brown, I. E., Turkenburg, J. P., Black, G. W., & Davies, G. J. (2002). Convergent evolution sheds light on the anti-β-elimination mechanism common to family 1 and 10 polysaccharide lyases. *Proceedings of the National Academy of Sciences*, 99(19), 12067–12072. https://doi.org/10.1073/pnas.182431199

Chettri, D., & Verma, A. K. (2023). Biological significance of carbohydrate-active enzymes and searching their inhibitors for therapeutic applications. *Carbohydrate Research*, 529, 108853. https://doi.org/10.1016/j.carres.2023.108853

Chylenski, P., Bissaro, B., Sørlie, M., Røhr, Å. K., Várnai, A., Horn, S. J., & Eijsink, V. G. H. (2019). Lytic polysaccharide monooxygenases in enzymatic processing of lignocellulosic biomass. ACS Catalysis, 9(6), 4970–4991. https://doi.org/10.1021/acscatal.9b00246

Courtade, G., Forsberg, Z., Heggset, E. B., Eijsink, V. G. H., & Aachmann, F. L. (2018). The carbohydrate-binding module and linkers of a modular lytic polysaccharide monooxygenase promote localized substrate oxidation. *Journal of Biological Chemistry*, 293(4), 1306–1314. https://doi.org/10.1074/jbc.RA117.000326

Coutinho, P. M., Deleury, E., Davies, G. J., & Henrissat, B. (2003). An evolving hierarchical family classification for glycosyltransferases. *Journal of Molecular Biology*, 328(2), 307–317. https://doi.org/10.1016/s0022-2836(03)00307-3

Davies, G. J., Gloster, T. M., & Henrissat, B. (2005). Recent structural insights into carbohydrate-active enzymes. *Current Opinion in Structural Biology*, 15(6), 637–645. https://doi.org/10.1016/j.sbi.2005.10.008

Davies, G. J., Wilson, K. S., & Henrissat, B. (1997). Nomenclature for sugar-binding subsites in glycosyl hydrolases. *Biochemical Journal*, 321(2), 557–559. https://doi.org/10.1042/bj3210557

Filandr, F., Man, P., Halada, P., Chang, H., Eijsink, V. G. H., & Väljamäe, P. (2020). The H_2O_2-dependent activity of a fungal lytic polysaccharide monooxygenase investigated with a turbidimetric assay. Biotechnology for Biofuels, 13(1), Article 37. https://doi.org/10.1186/s13068-020-01673-4

Flint, H. J., Scott, K. P., Duncan, S. H., Louis, P., & Forano, E. (2012). Microbial degradation of complex carbohydrates in the gut. *Gut Microbes*, 3(4), 289–306. https://doi.org/10.4161/gmic.19897

Forsberg, Z., & Courtade, G. (2023). On the impact of carbohydrate-binding modules (CBMs) in lytic polysaccharide monooxygenases (LPMOs). Essays in Biochemistry, 67(3), 561–574. https://doi.org/10.1042/EBC20220162

Forsberg, Z., Mackenzie, A. K., Sørlie, M., Røhr, Å. K., Helland, R., Arvai, A. S., ... & Vaaje-Kolstad, G. (2016). Structural and functional characterization of a lytic polysaccharide monooxygenase with broad substrate specificity. *Journal of Biological Chemistry*, 291(13), 6553–6562. https://doi.org/10.1074/jbc.M115.705202

Frandsen, K. E. H., Simmons, T. J., Dupree, P., Poulsen, J. C. N., Hemsworth, G. R., Ciano, L., ... & Lo Leggio, L. (2016). The molecular basis of polysaccharide cleavage by lytic polysaccharide monooxygenases. *Nature Chemical Biology*, 12(4), 298–303. https://doi.org/10.1038/nchembio.2029

Fredslund, F., Katayama, T., Jensen, M. T., Nielsen, B. W., Brøndsted, L., Vogensen, F. K., & Lo Leggio, L. (2011). Structural and biochemical characterization of a novel type of galactose 6-phosphate isomerase. *The Journal of Biological Chemistry*, 286(45), 40550–40559. https://doi.org/10.1074/jbc.M111.286088

Frommhagen, M., Koetsier, M. J., Westphal, A. H., Visser, J., Hinz, S. W. A., Vincken, J. P., van Berkel, W. J. H., Kabel, M. A., & Gruppen, H. (2016). Lytic polysaccharide monooxygenases from Myceliophthora thermophila C1 differ in substrate preference and reducing agent specificity. Biotechnology for Biofuels, 9(1), Article 186. https://doi.org/10.1186/s13068-016-0594-y

Frommhagen, M., Koetsier, M. J., Westphal, A. H., Visser, J., Hinz, S. W. A., Vincken, J.-P., ... & Kabel, M. A. (2016). Lytic polysaccharide monooxygenases assist in cellulose depolymerization by oxidative cleavage—an in-depth study on the effect of oxygen, electron donors, and interactions with cellobiose dehydrogenase. *Biotechnology for Biofuels*, 9(1), 186. https://doi.org/10.1186/s13068-016-0619-8

Garron, M. L., & Cygler, M. (2010). Structural and mechanistic classification of uronic acid-containing polysaccharide lyases. *Glycobiology*, 20, 1547–1573. https://doi.org/10.1093/glycob/cwq083

Garron, M. L., & Henrissat, B. (2019). The continuing expansion of CAZymes and their families. *Current Opinion in Chemical Biology*, 53, 82–87. https://doi.org/10.1016/j.cbpa.2019.08.004

Gavande, P. V., Goyal, A., & Fontes, C. M. G. A. (2023). Carbohydrates and carbohydrate-active enzymes (CAZymes): An overview. In A. Goyal & K. Sharma (Eds.), *Foundations and Frontiers in Enzymology, Glycoside Hydrolases* (pp. 1–23). Academic Press. https://doi.org/10.1016/B978-0-323-91805-3.00012-5

Gilbert, H. J., Stålbrand, H., & Brumer, H. (2008). How the walls come crumbling down: Recent structural and functional insights into bacterial cellulase and xylanase systems. *Current Opinion in Plant Biology*, 11(3), 338–348. https://doi.org/10.1016/j.pbi.2008.02.008

Hangasky, J. A., Detomasi, T. C., & Marletta, M. A. (2019). Glycosidic bond oxidation by polysaccharide monooxygenases. *Current Opinion in Structural Biology*, 59, 54–64. https://doi.org/10.1016/j.sbi.2019.02.002

Hangasky, J. A., Iavarone, A. T., & Marletta, M. A. (2018). Reactivity of O2 versus H2O2 with polysaccharide monooxygenases. *Proceedings of the National Academy of Sciences of the United States of America*, 115(19), 4915–4920. https://doi.org/10.1073/pnas.187021153115606115

Harris, P. V., Welner, D., McFarland, K. C., Re, E., Navarro Poulsen, J. C., Brown, K., ... & Larsen, S. (2010). Stimulation of lignocellulosic biomass hydrolysis by proteins of glycoside hydrolase family 61: Structure and function of a large, enigmatic family. *Biochemistry*, 49(15), 3305–3316. https://doi.org/10.1021/bi100009p

Harris, P. V., Xu, F., Kreel, N. E., Kang, C., & Fukuyama, S. (2014). New enzyme insights drive advances in commercial ethanol production. *Current Opinion in Chemical Biology*, 19, 162–170. https://doi.org/10.1016/j.cbpa.2014.02.015

Hashimoto, H. (2006). Recent structural studies of carbohydrate-binding modules. *Cellular and Molecular Life Sciences*, 63, 2954–2967. https://doi.org/10.1007/s00018-006-6195-3

Hemsworth, G. R., Ciano, L., Davies, G. J., & Walton, P. H. (2018). Production and spectroscopic characterization of lytic polysaccharide monooxygenases. *In Methods in Enzymology* (Vol. 613, pp. 63–90). Academic Press. https://doi.org/10.1016/bs.mie.2018.10.014

Hemsworth, G. R., Davies, G. J., & Walton, P. H. (2013). Recent insights into copper-containing lytic polysaccharide monooxygenases. *Current Opinion in Structural Biology*, 23(5), 660–668. https://doi.org/10.1016/j.sbi.2013.05.006

Hemsworth, G. R., Henrissat, B., Davies, G. J., & Walton, P. H. (2014). Discovery and characterization of a new family of lytic polysaccharide monooxygenases. *Nature Chemical Biology*, 10(2), 122–126. https://doi.org/10.1038/nchembio.1417

Henrissat, B., & Davies, G. (1997). Structural and sequence-based classification of glycoside hydrolases. *Current Opinion in Structural Biology*, 7(5), 637–644. https://doi.org/10.1016/s0959-440x(97)80072-3

Horn, S. J., Forsberg, Z., Vaaje-Kolstad, G., Westereng, B., & Eijsink, V. G. H. (2012). Analysing lytic polysaccharide monooxygenases using different polysaccharide substrates. *Carbohydrate Research*, 448, 1–5. https://doi.org/10.1016/j.carres.2012.09.016

Kracher, D., Scheiblbrandner, S., Felice, A. K. G., Breslmayr, E., Preims, M., Ludwicka, K., ... & Ludwig, R. (2016). Extracellular electron transfer systems fuel cellulose oxidative degradation. *Science*, 352(6289), 1098–1101. https://doi.org/10.1126/science.aaf3165

Kuusk, S., Bissaro, B., Kuusk, P., Forsberg, Z., Eijsink, V. G. H., Sørlie, M., ... & Vaaje-Kolstad, G. (2020). Kinetics of H2O2-driven degradation of chitin by a bacterial lytic

polysaccharide monooxygenase. *Journal of Biological Chemistry*, 295(10), 2981–2992. https://doi.org/10.1074/jbc.RA119.012284
Lairson, L. L., Henrissat, B., Davies, G. J., & Withers, S. G. (2008). Glycosyltransferases: Structures, functions, and mechanisms. *Annual Review of Biochemistry*, 77, 521–555. https://doi.org/10.1146/annurev.biochem.76.061005.092322
Levasseur, A., Drula, E., Lombard, V., Coutinho, P. M., & Henrissat, B. (2013). Expansion of the enzymatic repertoire of the CAZy database to integrate auxiliary redox enzymes. *Biotechnology for Biofuels*, 6(1), 41. https://doi.org/10.1186/1754-6834-6-41
Li, X., Beeson, W. T., Phillips, C. M., Marletta, M. A., & Cate, J. H. D. (2012). Structural basis for substrate targeting and catalysis by fungal polysaccharide monooxygenases. *Structure*, 20(6), 1051–1061. https://doi.org/10.1016/j.str.2012.04.002
Liang, D.-M., Liu, J.-H., Wu, H., Wang, B.-B., Zhu, H.-J., & Qiao, J.-J. (2015). Glycosyltransferases: Structures, functions, and applications. *Chemical Society Reviews*, 44, 8350–8374. https://doi.org/10.1039/C5CS00600G
Linhardt, R. J., Galliher, P. M., & Cooney, C. L. (1987). Polysaccharide lyases. *Applied Biochemistry and Biotechnology*, 12, 135–176. https://doi.org/10.1007/BF02798420
Lombard, V., Bernard, T., Rancurel, C., Brumer, H., Coutinho, P. M., & Henrissat, B. (2010). A hierarchical classification of polysaccharide lyases for glycogenomics. *Biochemical Journal*, 432, 437–444. https://doi.org/10.1042/BJ20101185
Lombard, V., Golaconda Ramulu, H., Drula, E., Coutinho, P. M., & Henrissat, B. (2014). The carbohydrate-active enzymes database (CAZy) in 2013. *Nucleic Acids Research*, 42(D1), D490–D495. https://doi.org/10.1093/nar/gkt1178
Müller, G., Chylenski, P., Bissaro, B., Eijsink, V. G. H., & Horn, S. J. (2018). The impact of hydrogen peroxide supply on lytic polysaccharide monooxygenase action. *FEBS Letters*, 592(22), 4051–4066. https://doi.org/10.1002/1873-3468.13174
Onyango, S. O., Juma, J., De Paepe, K., & Van de Wiele, T. (2021). Oral and gut microbial carbohydrate-active enzymes landscape in health and disease. *Frontiers in Microbiology*, 12, 653448. https://doi.org/10.3389/fmicb.2021.653448
Paradisi, A., Hakala, T. K., Kallio, P., & Bailey, M. J. (2019). Lytic polysaccharide monooxygenases facilitate plant polysaccharide degradation in the presence of oxygen and cellobiose dehydrogenase. *ACS Sustainable Chemistry & Engineering*, 7(2), 1394–1400.
Paradisi, A., Johnston, E. M., Tovborg, M., Nicoll, C. R., Ciano, L., Dowle, A., McMaster, J., Hancock, Y., Davies, G. J., & Walton, P. H. (2019). Formation of a copper(II)-tyrosyl complex at the active site of lytic polysaccharide monooxygenases following oxidation by H_2O_2. Journal of the American Chemical Society, 141(46), 18585–18599. https://doi.org/10.1021/jacs.9b09833
Rani, A., Dhillon, A., Sharma, K., & Goyal, A. (2018). Insights into the structural characteristics and substrate binding analysis of chondroitin AC lyase (PsPL8A) from Pedobacter saltans. *International Journal of Biological Macromolecules*, 109, 980–991. https://doi.org/10.1016/j.ijbiomac.2017.11.087
Rieder, L., & Sørlie, M. (2023). Recent advances in understanding LPMO catalysis. *Biochemistry*, 62(22), 3170–3172. https://doi.org/10.1021/acs.biochem.3c00458.
Scott, B. R., Huang, H. Y., Frandsen, K. E. H., Lo Leggio, L., & Johansen, K. S. (2020). The role of lytic polysaccharide monooxygenases in biomass degradation. *Biotechnology Advances*, 42, 107580. https://doi.org/10.1016/j.biotechadv.2020.107580
Singh, A., Bajar, S., Devi, A., & Pant, D. (2021). Bioresource technology reports. *Bioresource Technology Reports*, 14, 100652. https://doi.org/10.1016/j.biteb.2021.100652
Smith, A. M., Denyer, K., & Martin, C. (1997). The synthesis of the starch granule. *Annual Review of Plant Physiology and Plant Molecular Biology*, 48, 67–87. https://doi.org/10.1146/annurev.arplant.48.1.67

Strasser, R. (2016). Plant protein glycosylation. *Glycobiology*, 26(9), 926–939. https://doi.org/10.1093/glycob/cww023

Tandrup, T., Bertelsen, A. B., Johansen, K. S., & Lo Leggio, L. (2018). Recent insights into lytic polysaccharide monooxygenases (LPMOs). *Biochemical Society Transactions*, 46(6), 1431–1447. https://doi.org/10.1042/BST201780544912

Terrapon, N., Lombard, V., Drula, E., Coutinho, P. M., & Henrissat, B. (2017). The CAZy Database: Principles and usage guidelines. In K. Aoki-Kinoshita (Ed.), *A Practical Guide to Using Glycomics Databases* (pp. 125–140). Springer. https://doi.org/10.1007/978-4-431-56454-6_6

Vu, V. V., Beeson, W. T., Span, E. A., Farquhar, E. R., & Marletta, M. A. (2014). A family of starch-active polysaccharide monooxygenases. *Proceedings of the National Academy of Sciences*, 111(38), 13822–13827. https://doi.org/10.1073/pnas.1408090111

Walton, P. H., & Davies, G. J. (2016). On the catalytic mechanisms of lytic polysaccharide monooxygenases. *Current Opinion in Chemical Biology*, 31, 195–207. https://doi.org/10.1016/j.cbpa.2016.03.014

Wang, B., Walton, P. H., & Rovira, C. (2019). Molecular mechanisms of oxygen activation and hydrogen peroxide formation in lytic polysaccharide monooxygenases. *ACS Catalysis*, 9(67), 4958–4969. https://doi.org/10.1021/acscatal.9b00778564

Wardman, J. F., Bains, R. K., Rahfeld, P., & Withers, S. G. (2022). Carbohydrate-active enzymes (CAZymes) in the gut microbiome. *Nature Reviews Microbiology*, 20(9), 542–556. https://doi.org/10.1038/s41579-022-00712-1

Wu, X., Zhang, S., & Zhang, Q. (2020). The contribution of specific subsites to catalytic activities in active site architecture of a GH11 xylanase. *Applied Microbiology and Biotechnology*, 104, 8735–8745. https://doi.org/10.1007/s00253-020-10865-9

Yi, W., Shi, J., Wang, L., Wang, D., Wang, Y., Song, J., Xin, L., & Jiang, F. (2024). Maternal PFOS exposure in mice induces hepatic lipid accumulation and inflammation in adult female offspring: Involvement of microbiome-gut-liver axis and autophagy. *Journal of Hazardous Materials*, 470, 134177. https://doi.org/10.1016/j.jhazmat.2024.134177

Zhu, B., & Yin, H. (2015). Alginate lyase: Review of major sources and classification, properties, structure-function analysis and applications. *Bioengineered*, 6(3), 125–131. https://doi.org/10.1080/21655979.2015.1030543.

3 Nano-CAZymes Synergy
A Novel Prospect

Manisha Chirania, Anil Kumar Verma, and Bhaskar Sharma

ABBREVIATIONS

AuNPs	Gold nanoparticles
CAZymes	Carbohydrate-active enzymes
CEs	Carbohydrate esterases
CNTs	Carbon nanotubes
EPR	Enhanced permeability and retention
Fe_3O_4	Magnetic iron oxide
FTIR	Fourier-transform infrared
GHs	Glycoside hydrolases
GO	Graphene oxide
GTs	Glycosyltransferases
HA	Hydroxyapatite
HRP	Horseradish peroxidase
IONPs	Iron oxide nanoparticles
K_m	Michaelis constant (indicator of enzyme affinity)
LCST	Lower critical solution temperature
LDH	Layered double hydroxides
LGA	Poly(lactic-co-glycolic acid)
LSPR	Localized surface plasmon resonance
MMP	Matrix metalloproteinases
MRSA	Methicillin-resistant *Staphylococcus aureus*
MWCNTs	Multi-walled carbon nanotubes
NADH	Nicotinamide adenine dinucleotide hydrogen
NCMPs	Nanocomposites
NCs	Nanoclusters
NPs	Nanoparticles
PAH	Polycyclic aromatic hydrocarbons
PEG	Polyethylene glycol
PLGA	Poly(lactic-co-glycolic acid)
PLs	Polysaccharide lyases
PNIPAM	Poly(N-isopropyl acrylamide)
PTX	Paclitaxel drug

DOI: 10.1201/9781003561682-3

RGD	Arginine-glycine-aspartic acid peptide
ROS	Reactive oxygen species
SEM	Scanning electron microscopy
SPIONs	Superparamagnetic iron oxide nanoparticles
TEM	Transmission electron microscopy
TiO_2	Titanium dioxide
TME	Tumour microenvironment
V_{max}	Maximum reaction velocity (indicator of enzyme efficiency)
ZnO	Zinc oxide

3.1 INTRODUCTION

Carbohydrate-active enzymes (CAZymes) comprise a broad group of enzymes involved in biological processes such as metabolism, synthesis, and degradation of carbohydrates. Owing to their essential role, CAZymes have garnered increasing interest from researchers across biological disciplines. These enzymes are pivotal in applications spanning health sciences to biotechnology (Chettri and Verma, 2023).

Classification of families of CAZymes and annotation of their constituent modules using primary sequences is now a common approach to studying, analysing, and understanding the ability of an individual organism or microbial ecosystem to produce or degrade complex carbohydrates (Drula et al., 2022).

In industry, CAZymes serve as key components in research and development, especially in the degradation of lignocellulosic biomass for the production of biofuels and other value-added products in biorefineries. They are widely used in the food and bakery industry for applications including fruit and vegetable preservation, fermentation, brewing, food supplement production, and bioremediation. They also play a crucial role in glycobiology (Chettri and Verma, 2023).

CAZymes also facilitate the degradation of complex polysaccharides, with glycoside hydrolases (GHs) being the significant enzymes that undertake most of the degradation processes associated with glycohydrolases and transglycosidases. In addition, other classes of CAZymes have more specialized functions, such as glycosyltransferases (GTs) involved in glycosidic bond formation, carbohydrate esterases (CEs) that remove ester bonds, and polysaccharide lyases (PLs) that break the glycosidic bonds in polysaccharides containing uronic acids in the β-position (Paudel et al., 2023). However, factors such as substrate specificity, limited activity, and instability pose challenges for the use of CAZymes in product manufacturing. Attempts to find new enzymes to overcome these limitations have mostly failed due to their time-consuming, costly, and labour-intensive nature. The need to keep industrial costs low, along with these limitations, has influenced the development of various enzyme engineering methods to improve the efficiency of CAZymes (Chettri et al., 2020). Combined strategies are, therefore, essential to meet the growing demand for cost-efficient bioenergy from abundant lignocellulosic biomass and to overcome environmental challenges.

Nanotechnology, therefore, represents a promising strategy that is rapidly evolving and attracting worldwide attention due to its wide range of applications in science, engineering, and technology. Hybrid nanomaterials, nanoclusters (NCs), and nanocomposites (NCMPs) uniquely combine physical, chemical, and biochemical properties, enabling applications ranging from drug delivery and degradation of pollutants to imaging and bioremediation. Through the use of nanotechnology – especially NCs and NCMPs – it is possible to develop environmentally friendly and efficient solutions that accelerate the degradation of organic pollutants and also aid in bioremediation (Chinni et al., 2023).

Nanomaterials used as catalysts facilitate the efficient conversion of lignocellulose into sugars and valuable chemicals by improving cell penetration. In addition, nanoparticles derived from microorganisms can shield fermenting strains, improving the stability and efficiency of biofuel production. These nano-based approaches not only increase the production of bioenergy but also contribute to environmental protection by promoting sustainable practices in biofuel production (Thanigaivel et al., 2022). The encapsulation of enzymes in nanoparticles can also improve the results of cancer treatment. This will facilitate targeted delivery in cancer research (Pourmadadi et al., 2024).

Therefore, different categories of nanoparticles are used to increase the efficiency of enzymes and to target them in industry, agriculture, and therapy.

3.2 ENHANCEMENT OF CAZYMES WITH NANOMATERIALS

In synthetic biology, the efficient use of enzymes is essential, especially in intricate, multi-step reactions. This need has led to much research into techniques that allow enzymes to be attached to larger scaffolds to increase stability – although this approach often means a compromise with catalytic efficiency. Recently, however, studies have shown that binding enzymes to nanoparticles can significantly improve their catalytic activity. This improvement is related to the unique microenvironment around the nanoparticles, which supports enzyme function in a way that larger scaffolds do not. Although the exact mechanisms behind this enhanced performance are still under investigation, it is evident that the nanoscale structure and interface of these materials are critical in supporting enzymatic activity. Consequently, the enhancement of CAZymes with nanomaterials has garnered significant attention, given the potential for improved stability, catalytic efficiency, reusability, and substrate specificity (Hooe et al., 2024) (Table 3.1).

3.2.1 Increased Enzyme Stabilization and Catalytic Activity

In both medicine and industry, stabilization of enzymes on carrier materials is crucial to improve their resistance under various conditions without compromising their activity. Environmental factors such as temperature, pH, and ionic strength have a direct effect on the function of enzymes. Therefore, numerous studies have focused on the stabilization of enzymes through techniques such as immobilization on solid supports, crosslinking, and entrapment.

Recently, nanoparticles have emerged as an effective method to optimize enzyme stability and activity, providing robust performance for various catalytic applications

TABLE 3.1
Summary of Enzyme-Nanomaterial Combinations and Their Catalytic Performance and Recyclability

Enzyme	Nanomaterial/Support	Immobilization Method	Application/ Reaction Type	Recyclability (%)	Notes on Stability/ Performance	Reference
Laccase	Multi-walled Carbon Nanotubes (MWCNTs)	Physical Adsorption	Catalysis under high-temperature conditions	100% after 9 cycles	Retained full activity after repeated uses	Ashkan et al. (2021)
Lipase	Graphene Oxide	Physical Adsorption	Hydrolysis at elevated temperatures	85% at 60°C, 65% at 70°C	High thermal stability observed	Ashkan et al. (2021)
Alpha-Amylase	Gold Nanorods	Adsorption	Expanded pH and temperature range for activity	Significant retention	High thermal and pH stability	Patil et al. (2024)
Cellulase	Styrene/Maleic Anhydride Copolymer Nanoparticles	Covalent Bonding	Hydrolysis of Carboxymethyl Cellulose (CMC)	80% after 10 cycles	Strong binding for high recyclability	Anboo et al. (2022)
Lipase	Poly(Glycidyl Methacrylate-co-Methyl Acrylate)/ Feather Polypeptide Nanofibrous Membrane	Covalent Bonding	Olive Oil Hydrolysis	62% after 7 cycles	Durable, but moderate recyclability	Anboo et al. (2022)
Lipase	Polyurethane/ Amidoxime/ Polyacrylonitrile/β-Cyclodextrin Nanofiber	Covalent Bonding	Hydrolysis of Olive Oil	45% after 5 cycles	Maintained reasonable activity in repeated use	Anboo et al. (2022)
Lipase	Chitosan-Mesoporous Silica SBA-15 Hybrid	Covalent & Crosslinking	Hydrolysis of Triacetin	85% after 10 cycles	Strong activity retention	Patil et al. (2024)

Lipase	Zinc-Doped Magnetic Nanoparticles	Adsorption	Hydrolysis of Fish Oil	50% after 20 cycles	Enhanced magnetic recovery potential	Patil et al. (2024)
α-Chymotrypsin	Magnetic Chitin-Nanofiber Composite	Crosslinking	General Catalytic Applications	84.9% after 20 days	Excellent stability for extended usage	Patil et al. (2024)
β-Galactosidase	Chitosan/Polyvinyl Alcohol Blend Nanofibers	Covalent Bonding	Extended storage applications	42% after 28 days	Moderate recyclability for long-term applications	Anboo et al. (2022)
Lipase	Titanate nanosheets (TNSs)	Electrostatic interaction	Hydrolysis of p-nitrophenyl acetate (pNPA)	Activity improved more than eight times compared to free lipase; heat tolerance enhanced	Not specified	Yamada et al. (2018)
Xylanase, Glucoamylase	Fe_3O_4-graphene oxide (GO) hybrids	Physical and covalent attachments	Enzyme immobilization for analytical and industrial processes	Increased surface area for enzyme loading and enhanced catalytic activity	Not specified	Soozanipour et al. (2018)
Protease	Activated alginate/dextrose (Alg/dex) beads	Covalent immobilization	Enzyme stabilization and activity improvement	Increased optimum temperature by 10 °C; lower activation energy; increased thermal stability	Retained 92.7% after 8 cycles; 52.4% after 12 cycles	Abdella et al. (2023)
Glucoamylase	Chitosan-based nanoparticles (including magnetic nanoparticles and graphene oxide)	Dripping granulation, one-step method, ionic crosslinking	Enzyme immobilization with high loading capacity	Increased loading capacity, optimized reaction conditions	Enhanced reusability and storage stability	Wang et al. (2019)

(Sharifi et al., 2020). With the development of nanotechnology, nanomaterials, as mentioned earlier, have become a famous sight in numerous disciplines such as medicine, pharmacy, and chemistry. These materials are highly suited for enzyme immobilization, as their surface area directly influences the amount of enzyme that can be effectively bound. This property allows nanomaterials to enhance enzyme activity and stability across various applications.

A range of nanostructured materials including nanoparticles, wires, tubes, fibres, plates, and rods, have already been successfully utilized for enzyme immobilization. Carbon allotropes, such as graphene and CNTs, along with materials like silica, calcium carbonate, titanium, alumina, magnetite, zirconia, and gold, show particular promise as carrier substrates for this purpose. For example, a study by Ashkan et al. (2021) reported the use of multi-walled carbon nanotubes (MWCNTs) to immobilize laccase, resulted in enhanced activity at elevated temperatures and complete retention of total activity after nine consecutive applications. This highlights the potential of CNTs to enhance enzyme performance. Similarly, graphene oxide has been used for lipase immobilization, retaining 85% of its original activity at 60°C and 65% at 70°C.

Magnetic nanoparticles, especially magnetite (Fe_3O_4), have attracted attention due to their mechanical properties and biocompatibility, contributing to increased enzyme stability and easier binding. Gold nanoparticles (AuNPs) have also proven effective; for instance, gold nanorods have been used to immobilize alpha-amylase, resulting in an expanded temperature and pH range for activity and retaining a substantial amount of the original activity after storage. Overall, the immobilization of enzymes increases stability, but factors such as the carrier material and environmental conditions can significantly affect enzyme activity and optimal performance (Ashkan et al., 2021).

Nevertheless, considerable efforts have been made to improve the catalytic properties of enzyme carrier systems by incorporating nanocatalytic species, also known as 'nanoenzymes' or nanomaterial-based artificial enzymes. Over the past decade, the increased activity and recovery of these nanomaterials with built-in enzyme-like properties under unfavourable reaction conditions, as well as their high stability and low cost, have attracted researchers. The development of nanoenzymes and integration with enzymes has many advantages. First and foremost, nanoenzymes not only serve as movement carriers for the enzymes but also provide counter and turbulence stability for enzymes under thermal, pH, and solvent changes. In addition, they enable the selective recovery and recycling of enzymes, saving both money and time.

In some cases, the increased activity of nanomaterials together with enzymes can trigger catalytic cascade reactions, expanding the use of immobilized enzymes, for example, in glucose biosensors. In addition, there are possibilities to reduce the overall loading of biocatalysts by hybridizing enzymes with nanoenzymes, which makes the process much cheaper. As a result, substantial resources have been allocated to the synthesis of inexpensive, multifunctional enzyme-nanozyme hybrids (Patil et al., 2024).

Various enzymes have been successfully immobilized on different nanomaterials to improve their catalytic performance and recyclability. For example, cellulase is immobilized on styrene/maleic anhydride copolymer nanoparticles by covalent bonding for the catalytic hydrolysis of carboxymethyl cellulose and achieves an impressive 80% recyclability after ten cycles. Laccase utilizes a polyurethane/amidoxime/polyacrylonitrile/β-cyclodextrin nanofiber membrane, also by covalent bonding, as a carrier for catalytic activity, with a recyclability of 47% after ten cycles. Lipase was immobilized in different configurations, including on a terpolymer poly(glycidyl methacrylate-co-methylacrylate)-g-polyethylene oxide nanofiber membrane via covalent bonding for the hydrolysis of olive oil, resulting in a recyclability of 45% after five cycles. Another lipase variant is incorporated into a poly(glycidyl methacrylate-co-methylacrylate)/feather polypeptide nanofibrous membrane and achieves a recyclability of 62% after seven cycles, also by covalent binding. In addition, the lipase is used with copper phosphate/CNTs for resolution reactions between 1-phenylethanol and vinyl acetate, achieving an excellent recyclability of 97% after eight cycles. The enzyme is also used with chitosan-mesoporous silica/SBA-15 hybrid nanomaterials via covalent and crosslinking methods for the hydrolysis of triacetin, achieving 85% recyclability after ten cycles.

Another configuration with SBA-15 nanomaterials of chitosan and mesoporous silica utilizes adsorption and crosslinking and achieves a recyclability of 82% after the same number of cycles. Zinc-doped magnetic nanoparticles are used for lipase with adsorption for the hydrolysis of fish oil and show a recyclability of over 50% after twenty cycles. α-Chymotrypsin is immobilized by crosslinking on a magnetic chitin-nanofiber composite, which serves as a support matrix for catalytic activity and achieves 84.9% recyclability after twenty days. Finally, β-galactosidase is immobilized by covalently binding to nanofibres made of a chitosan/polyvinyl alcohol mixture, showing a recyclability of 42% after twenty-eight days. These diverse combinations of enzymes and nanomaterials demonstrate the potential for improved catalytic efficiency and reusability in biotechnological applications (Anboo et al., 2022). Therefore, nanomaterials have proven to enhance enzyme stability, activity, and reusability in various applications. Effective carriers such as CNTs, graphene oxide, and magnetic nanoparticles provide ideal platforms for enzyme immobilization due to their high surface areas and supportive properties. As discussed, recent advancements in nanoenzymes introduce multifunctional benefits, enabling catalytic cascades and selective recovery while extending enzyme recyclability and reducing costs. These enzyme-nanozyme hybrids hold promise for more efficient and durable biocatalysis, illustrating the transformative impact of nanotechnology in industrial and biomedical enzyme applications.

3.3 TARGETED DELIVERY AND CONTROLLED RELEASE SYSTEMS FOR CARBOHYDRATE-ACTIVE ENZYMES

CAZymes represent a specialized class of enzymes responsible for the degradation, synthesis, and modification of carbohydrates (Zehcel et al., 2000; Jongkees & Withers, 2014). This group includes GHs, polysaccharide lyases, glycosyltransferases, and CEs, all of which are pivotal in various industrial applications such as biofuel production, food modification, and biomedical treatments. CAZymes play an essential role in the degradation of complex polysaccharides. Their applications range from the production of biofuels to therapeutic interventions for diseases related to carbohydrate metabolism.

Despite their broad potential, CAZymes often face challenges related to stability and loss of activity, especially under harsh environmental conditions. These enzymes, like many others, are susceptible to degradation in industrial and biological systems due to factors such as elevated temperatures, proteolytic breakdown, and pH fluctuations.

The use of nanotechnology for the targeted delivery and controlled release of CAZymes has proven to be a promising approach to alleviate these limitations. Nanoparticles can serve as carriers and ensure that CAZymes are accurately delivered to their target substrates or sites of action while maintaining activity in non-target environments. Controlled release strategies further improve the efficiency of CAZymes by ensuring sustained enzyme activity over extended periods, maximizing efficiency, and minimizing waste. This is particularly essential in continuous industrial processes and therapeutic applications where prolonged enzyme activity is essential for optimal results.

3.3.1 Nanoparticles for the Targeted and Controlled Delivery of CAZymes

The design of nanoparticles for enzyme delivery aims to improve enzyme stability, protect enzymes during transportation, and enable controlled release mechanisms. We discuss different categories of nanoparticles suitable for enzyme delivery systems and the associated strategies for targeted delivery and releasing CAZymes.

3.3.1.1 Polysaccharide-Based Nanoparticles

Polysaccharide-based nanoparticles have attracted attention due to their biodegradability, biocompatibility, and ability to form hydrogels that encapsulate enzymes (Meng et al., 2024). These nanoparticles improve the stability and activity of CAZymes by protecting them from degradation until a stimulus, such as a change in pH, ionic strength, or enzymatic cleavage, triggers their release. Their ability to modulate release in response to environmental changes makes them highly suitable for controlled release in industrial and therapeutic applications.

Chitosan nanoparticles: Chitosan, a natural cationic polysaccharide, forms nanoparticles by ionic gelation and is widely used for enzyme immobilization due to its bioadhesive properties and protective effects (Sabourian et al., 2020). The bioadhesive properties of chitosan enable effective interaction with biological tissue and other surfaces, which is helpful for enzyme immobilization. This adhesion not only stabilizes the enzyme but also improves its interaction with the substrate. The cationic nature of chitosan may allow it to form electrostatic bonds with the negatively charged surfaces of CAZymes, such as cellulases. This strong interaction can effectively immobilize the enzyme and prevent its denaturation, improving the operational stability of the enzyme in industrial environments such as biofuel production. This method can be carried out under mild conditions so that the activity of sensitive enzymes is maintained during encapsulation. In addition, chitosan nanoparticles can be tailored for pH-sensitive release, allowing CAZyme to be targeted in environments such as acidic biofuel processes or the gastrointestinal tract in therapeutic applications.

Alginate nanoparticles: Alginate, a polysaccharide derived from seaweed, is known for its biocompatibility and natural degradability. Alginate nanoparticles are produced by crosslinking with calcium ions, creating a gel matrix into which CAZymes can be encapsulated. The encapsulated enzymes can remain stable and protected until they are triggered by specific environmental changes, such as the presence of chelating agents or changes in ionic strength (Karim et al., 2022; Weng et al., 2023). In contrast to chitosan, which reacts quickly to pH changes, alginate nanoparticles are degraded more slowly. This property provides a slow, steady release of enzymes over time, which is particularly useful in industrial processes such as biofuel production, where long-term enzymatic activity is required. In biofuel production, alginate nanoparticles can be used to immobilize enzymes such as cellulases and hemicellulases to increase their efficiency in degrading plant material. Sustained release by alginate can improve the catalytic performance of these enzymes, leading to a higher process yield.

3.3.1.2 Polymeric Nanoparticles

Synthetic polymeric nanoparticles offer precise control over enzyme loading and release kinetics (Pourmadadi et al., 2024). They can be engineered to release CAZymes through polymer degradation in response to stimuli such as temperature, pH, or specific biomolecules. This approach enables sustained and controlled enzyme activity, which is critical for long-term applications in both industry and therapy. These nanoparticles are often engineered to be biodegradable and responsive to stimuli.

PLGA nanoparticles: Poly (lactic-co-glycolic acid) nanoparticles (PLGA) are one of the most commonly used polymeric materials for the encapsulation and delivery of enzymes. They are degraded by hydrolysis to lactic acid and glycolic acid, which are naturally metabolized by the body, making them safe for both industrial and therapeutic applications. The rate of degradation can be fine-tuned by adjusting the ratio of lactic acid to glycolic acid in the polymer, allowing precise control of the kinetics of enzyme release (Gustafson et al., 2022). PLGA-encapsulated beta-glucanases can be used in industrial applications such as biofuel production, where prolonged enzyme activity is required to degrade complex carbohydrates over an extended period efficiently. In biofuel production and other industrial bioprocesses, PLGA nanoparticles offer an effective way to achieve long-term, controlled enzyme activity. By encapsulating enzymes such as cellulases or hemicellulases, these nanoparticles can help to improve the efficiency of biomass conversion by maintaining the stability of the enzymes and prolonging their catalytic activity.

PNIPAM nanoparticles: Poly (N-isopropyl acrylamide) nanoparticles (PNIPAM) exhibit unique temperature-sensitive properties, making them ideal for industrial processes where enzyme release must be triggered by heat. PNIPAM particles remain stable at temperatures below their lower critical solution temperature (LCST). However, as soon as the temperature rises above this threshold, the particles collapse and release their encapsulated enzymes (Xu et al., 2020). This temperature-triggered release mechanism is beneficial in processes associated with temperature fluctuations, such as biomass processing. For example, amylases encapsulated in PNIPAM nanoparticles can be released when the temperature exceeds the LCST, allowing the controlled degradation of starch or other polysaccharides at specific points in the process. The ability to synchronize enzyme activity with temperature changes can optimize the overall efficiency of the reaction. In the production of biofuels or the bioprocessing of plant materials, encapsulation with PNIPAM can enable the targeted release of CAZymes when higher temperatures are introduced to ensure that the enzymes are active.

3.3.1.3 Metal and Metal Oxide Nanoparticles

Metal-based nanoparticles are very effective for enzyme immobilization due to their large surface area and their ability to improve enzyme stability, activity and recyclability. These nanoparticles are stable under a wide range of conditions and offer efficient recovery options, making them ideal for cost-sensitive industrial applications. Immobilization of enzymes on metal-based nanoparticles is usually achieved by surface adsorption, covalent bonding or encapsulation in porous structures. This approach ensures that the enzymes remain stable in response to environmental changes such as temperature and pH fluctuations, which is often the case in industrial processes (Liu and Dong, 2020).

Gold nanoparticles (AuNPs): Gold nanoparticles are often used as stable platforms for enzyme immobilization. They support both covalent and non-covalent binding of enzymes, which can increase the overall activity of the immobilized enzymes. AuNPs provide a stable and conductive surface that ensures that the enzymes retain their structure and functionality over an extended period, even under harsh industrial conditions. One of the unique advantages of AuNPs is their ability to increase enzyme activity through the localized surface plasmon resonance (LSPR) effect. This effect increases the local temperature and electron density near the surface of the nanoparticles, which in turn increases the reaction rate. For example, AuNPs have been used to immobilize enzymes such as cellulases, which improve their catalytic efficiency in biomass degradation processes due to this plasmonic enhancement (Al-Dulimi et al., 2020). The combination of AuNPs with magnetic cores leads to core-shell structures that not only provide the catalytic advantages of AuNPs but also allow magnetic separation and recovery of the immobilized enzyme. These structures have been effectively used in biomass processing, where cellulases immobilized on magnetic AuNPs exhibit improved catalytic efficiency and can be quickly recovered for reuse, reducing operational costs (Poorakbar et al., 2020).

Iron oxide nanoparticles (IONPs): Iron oxide nanoparticles (IONPs) are often used to immobilize enzymes due to their magnetic properties, which enable the recovery of enzyme-nanoparticle complexes after the reaction is complete. This property is particularly advantageous in large-scale industrial processes where the recovery and reusability of enzymes are critical to reducing costs. Enzymes such as xylanases and cellulases have been successfully immobilized on IONPs and show improved stability and reusability (Muley et al., 2020; Yadav et al., 2024; Kharazmi et al., 2020). The magnetic properties of IONPs also offer a unique advantage for targeted delivery of enzyme-nanoparticle complexes to specific sites. This ability can be used to achieve localized enzyme activity, allowing precision in catalytic processes where enzymes need to be concentrated at specific reaction sites. IONPs provide excellent protection for enzymes against denaturation under harsh conditions. This stability is particularly essential in therapeutic applications and biomass conversion, where high temperatures, fluctuating pH values and the presence of solvents can affect enzyme activity. The immobilization of CAZymes on IONPs allows these enzymes to retain their activity over multiple reaction cycles, improving the efficiency of industrial processes.

Titanium dioxide (TiO_2) and zinc oxide (ZnO) nanoparticles: Titanium dioxide and zinc oxide are other metal oxides that offer high surface area and chemical stability for enzyme immobilization. These materials are biocompatible and provide a robust platform for maintaining enzyme activity. The large surface area allows more enzyme molecules to be immobilized, which increases the overall catalytic efficiency (Liu and Dong, 2020).

3.3.1.4 Layered Double Hydroxides (LDHs)

Layered double hydroxides (LDHs) are metal hydroxides with a layered structure that makes them promising candidates for enzyme immobilization (Liu and Dong, 2020). LDHs can either intercalate enzymes between their layers or bind enzymes to their surface. This structure not only protects the enzymes from environmental influences but also enables controlled release, which improves their long-term stability and efficacy in various catalytic applications. LDHs are particularly useful in processes that require sustained enzyme activity and controlled release, such as industrial catalysis or environmental remediation.

3.3.2 Mechanisms for the Targeted Delivery of CAZymes via Nanoparticles

Recent advances in nanoparticle design have led to significant improvements in the protection, stability, targeted delivery, controlled release, and enzymatic activity of CAZymes in various applications, including industrial, medical, and environmental. The mechanisms by which nanoparticles facilitate the release of CAZymes can be broadly divided into two categories: active targeting and passive targeting (see Figure 3.1; Tiwari et al., 2023).

3.3.2.1 Active Targeting

Active targeting uses nanoparticles that are functionalized with specific ligands designed to bind to receptors on target cells or tissues. This mechanism enables more precise and efficient delivery of therapeutic agents, making it particularly valuable for medical applications. In the context of enzyme replacement therapies, active targeting is crucial to deliver therapeutic enzymes such as GHs to the sites where they are needed. PLGA nanoparticles, for example, can be conjugated with targeting ligands to ensure that the CAZymes are directed to diseased cells. This localized delivery not only increases the therapeutic effect but also minimizes systemic side effects, improving patient outcomes.

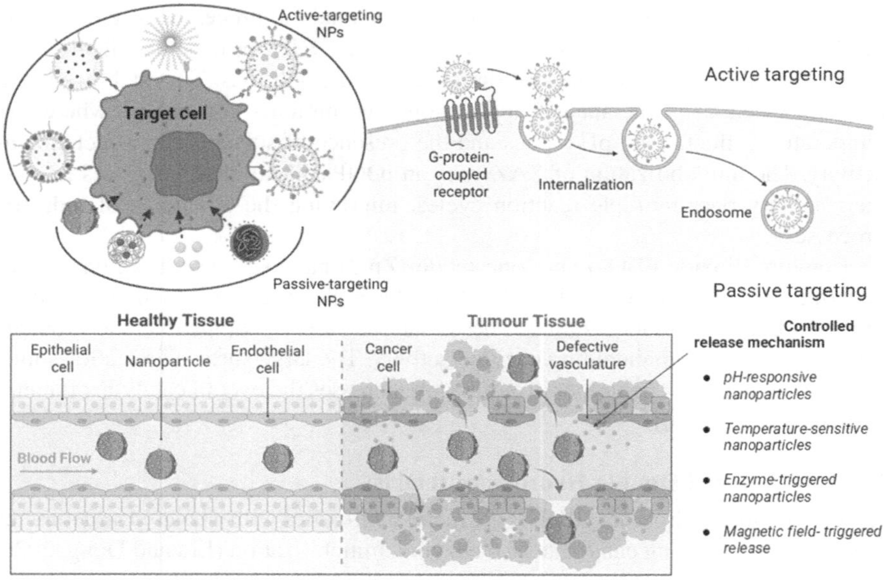

FIGURE 3.1 Active and passive targeting mechanisms of nanoparticles for enhanced drug delivery and controlled release within cells, illustrating strategies for precise intracellular targeting and improved therapeutic efficacy.

In addition to medical applications, active targeting can also be used in agriculture. HA and LDH can effectively bind to plant surfaces, enabling the targeted delivery of CAZymes to specific plant tissues. This targeting approach is particularly advantageous for biomass degradation, which is critical for biofuel production, as it increases the efficiency of enzyme action directly at the site of application (Sharma et al., 2022; Mitter et al., 2017).

3.3.2.2 Passive Targeting

Passive targeting relies on the inherent properties of nanoparticles, such as their size, surface charge and interaction with physiological environments, to facilitate the delivery of encapsulated CAZymes. An important mechanism underlying passive targeting is the enhanced permeability and retention (EPR) effect that occurs in tissues with leaky blood vessels, such as tumour microenvironments. In these scenarios, biocompatible nanoparticles, e.g. from chitosan or PLGA, can accumulate at the target site while minimizing the immune response. This is particularly useful for encapsulated CAZymes, which can reach higher concentrations in the areas where they are most needed. For example, studies have shown that chitosan nanoparticles encapsulating cellulases exhibit improved enzyme stability and protection against harsh environmental conditions (Bai et al., 2023). This improvement significantly increases the degradation processes of biomass and makes them more efficient for biofuel production. Smaller nanoparticles, including metal nanoparticles, carbon dots and CNTs, are promising candidates for introducing CAZymes directly into plant or animal cells. Their small size facilitates penetration into the tissue and improves the targeted delivery of enzymes (Gaur et al., 2021).

Nanoparticles not only improve the stability of the enzyme during circulation but also enable sustained release at the desired target site. For example, thermoreactive polymeric nanoparticles can release CAZymes in response to temperature changes, which is beneficial in both food processing and pharmaceutical applications. Encapsulated CAZymes tend to exhibit higher enzymatic activity than their free counterparts over more extended periods. This increased activity is particularly pronounced when the enzymes are encapsulated with protective materials such as polyethylene glycol (PEG). The PEG shielding helps to prevent recognition by the immune system and degradation, allowing the enzymes to maintain their functionality over an extended period (Tiwari et al., 2023).

3.3.3 Mechanisms for Controlled Release of CAZymes from Nanoparticles

The integration of advanced nanoparticle systems for CAZymes is crucial to improve their application efficiency in industrial, therapeutic and environmental fields. The ability to control the release of enzymes in response to specific environmental stimuli significantly optimizes the functionality of enzymes. We explore fundamental, sophisticated, controlled release mechanisms utilized by nanoparticles.

3.3.3.1 pH-Responsive Nanoparticles

pH-responsive nanoparticles are designed to undergo structural changes or degradation at specific pH values. Fluctuations in pH occur in many industrial and therapeutic environments, making pH-sensitive nanoparticles particularly suitable for enzyme delivery. These nanoparticles can be engineered by incorporating polymers with pH-sensitive functions that swell, dissolve or degrade at a given pH, triggering the release of the encapsulated CAZymes (Ding et al., 2022). For example, chitosan-based nanoparticles are known to degrade under slightly acidic conditions, rendering them ideal for applications in acidic biofuel processes or for drug delivery in the gastrointestinal tract, where the release of CAZymes can be triggered by changes in the pH in the environment. Poly (lactic-co-glycolic acid) nanoparticles can be modified to respond to pH changes. By adjusting the ratio of lactic acid to glycolic acid and the surface chemistry, these nanoparticles can release enzymes in response to different pH values. This makes them suitable for controlled release in targeted therapies or industrial processes that require specific conditions. pH-dependent systems ensure that enzymes are released in environments that are conducive to their activity, maximizing their efficiency and reducing waste. This specificity can lead to better catalytic outcomes, particularly in industrial applications such as biomass degradation and the delivery of therapies.

3.3.3.2 Temperature-Sensitive Nanoparticles

Temperature-sensitive nanoparticles provide a controlled release mechanism that adapts to industrial processes that are often subject to thermal fluctuations. This adaptation ensures that the enzymes are activated when they are needed, increasing the efficiency of the entire process. Thermoactive nanoparticles, such as those made of poly (N-isopropyl acrylamide) (PNIPAM), exhibit unique phase transition behaviour in response to temperature changes. Below a certain threshold (LCST), these nanoparticles remain soluble; however, when heated above the LCST, they collapse and release the encapsulated enzymes (Chang et al., 2023). This mechanism is particularly useful for CAZymes in industrial processes where elevated temperatures can activate enzyme release. Previous research has shown that cellulases encapsulated in PNIPAM nanoparticles are effectively released when the temperature is increased, which is particularly useful in lignocellulosic biomass degradation processes. Increasing the temperature increases the enzyme activity and thus optimizes the degradation of complex carbohydrates (Zhou et al., 2019).

3.3.3.3 Enzyme-Triggered Nanoparticles

Enzyme-triggered release systems are designed to respond to the presence of specific enzymes in the environment. These nanoparticles are coated or constructed with substrates that can be cleaved by the target enzymes (e.g. proteases). Upon enzymatic action, the outer matrix is degraded, leading to the release of the encapsulated CAZymes (Chapman and Stenzel, 2019). Nanoparticles containing peptide linkers that are susceptible to proteolytic cleavage can release CAZymes in environments where specific proteases are present, such as tumour tissues or the gastrointestinal tract. This mechanism enables precise release and maximizes the therapeutic potential of CAZymes in enzyme replacement therapies and targeted applications in agriculture. This approach not only ensures that the enzymes are released where they are needed but also enables a synergistic effect where the action of one enzyme can facilitate the release of another enzyme, optimizing substrate degradation and improving the efficiency of the entire process.

3.3.3.4 Magnetic Field-Triggered Release

Magnetic nanoparticles, such as iron oxide and superparamagnetic iron oxide nanoparticles (SPIONs), can be manipulated by external magnetic fields. The application of a magnetic field allows spatial control over the delivery and release of encapsulated enzymes, which can be precisely directed to specific sites (Vangijzegem et al., 2023). In industrial environments where specific enzyme action is required at specific sites, magnetic nanoparticles can be directed to the desired location using an external magnet. Once positioned, the magnetic field can be used to modulate enzyme activity to increase the efficacy of the enzyme-nanoparticle complex. The ability to control the release of enzymes spatially and temporally is particularly advantageous in processes that require high precision, such as targeted cancer therapies and localized agricultural applications. This control facilitates efficient enzymatic reactions, reduces the potential for side effects, and improves the overall outcome of the process.

The development of sophisticated mechanisms for the controlled release of CAZymes through nanoparticle systems can offer significant advantages in optimizing enzyme performance in various applications. By utilizing pH sensitivity, temperature sensitivity, enzyme-triggered release, and magnetic field modulation, researchers can tailor the release of CAZymes to specific operational requirements. This adaptability not only increases enzyme stability and activity but also improves the efficiency of enzymatic processes, paving the way for more effective biotechnological innovations in industry, medicine, and the environment.

3.3.4 Challenges and Limitations

Although nanoparticles hold great promise for the delivery and controlled release of CAZymes, several challenges must be overcome to realize their full potential.

3.3.4.1 Optimizing the Formulation of Nanoparticles

A key challenge is to ensure that the enzyme activity remains intact during the encapsulation process. Optimization of the nanoparticle formulation is critical to maximize the loading capacity of the enzymes and achieve desirable release profiles while maintaining the stability and functionality of the CAZymes.

3.3.4.2 Cost and Scalability

Scaling up nanoparticle-based delivery systems for industrial use remains a significant hurdle. The current cost of synthesizing nanoparticles is often prohibitive, and the development of cost-effective, scalable methods will be critical to the widespread adoption of these systems in industrial applications.

3.3.4.3 Biocompatibility and Environmental Impact

For food and medical applications, the biocompatibility of nanoparticle materials must be rigorously tested to ensure safety. In addition, the environmental impact of the use of nanoparticles in large-scale industrial processes needs to be carefully assessed, particularly concerning biodegradability and potential toxicity.

3.4 APPLICATIONS IN BIOENERGY, BIOMEDICINE, AND POLLUTANT DEGRADATION

The application of nanotechnology has transformed the improvement of enzymes, which has enormous implications for biofuel production and health sciences. In biofuel production, the incorporation of lignocellulosic materials into bioreactors is frequently challenged by the reliance on aggressive pretreatment methods. While these methods break down the rigid structures of lignocellulosic materials, they often fail to sufficiently degrade them, thereby impeding the hydrolysis of biomass into fermentable sugars. To address these problems, new processes using enzyme cascades have been developed, including modern technologies such as naked agglomeration, where nano-biomaterials are used to immobilize enzymes. This leads to an improvement in enzyme stability, activity, and targeting efficiency. The innovations in this area are essential for improving the processing of lignocellulosic biomass and thus increasing the yield of biofuels.

Similarly, nanomedicine has advanced by using nanocarriers to enable targeted therapy, as nanoparticles are smaller than cellular components. Due to their small size, large surface area, and biocompatibility, nanoparticles are also employed in other therapeutic applications, including tissue engineering and biosensing. Their physical properties allow for size adjustments, making them versatile for modifications across fields such as medicine and therapeutic development.

Enzyme improvement, in addition to nanotechnology, not only provides opportunities to improve biofuel production but also promotes innovation in biomedical treatment and environmental remediation. Therefore, it increases the enzyme efficiency to generate fuel, better drug delivery system and proper degradation of toxic compounds (see Figure 3.2; Hussain et al., 2024; Chettri et al., 2022).

3.4.1 Enhanced Biofuel Production

The increasing energy demand has fuelled the need for sustainable alternatives to fossil fuels, which has led to greater interest in the production of biofuels such as bioethanol and biobutanol. Although biobutanol has several advantages, its extraction and downstream processing is an expensive challenge. However, advances in nanotechnology offer promising solutions for the economical production of biobutanol. Stable nanobiocatalyst systems made from nanoparticles improve catalysis and modify chemical processes with improved texture and distribution of products. Nanoparticles also facilitate rapid downstream processing through adsorption, gas stripping, and liquid-liquid extraction processes that overcome the economic barrier of separation and purification. Nanomaterials are also integrated into nanocomposite membranes, enhancing enzyme separation and immobilization, which allows for enzyme reuse and improves their thermophysical properties. This advancement enables faster fuel production rates and reduces overall processing costs (Kapil et al., 2023). A significant application of nanotechnology in the biofuel industry is the immobilization of enzymes (biocatalysts) in the production of biodiesel and cellulosic ethanol through lipase catalysis. These techniques could help reduce the production costs of biofuels on a commercial scale.

Enzymatic hydrolysis is an effective method for modifying lignocellulosic, cellulosic, and starchy material, primarily using a cellulase mixture of endoglucanases, cellobiohydrolases, and β-glucosidase to degrade lignocellulosic material. However, the high cost of cellulases poses a challenge in bioethanol production from lignocellulosic biomass, making the immobilization of these enzymes essential for recovery, recycling, and cost reduction.

Nanotechnology significantly enhances enzymatic hydrolysis by enabling the co-immobilization of multiple enzymes on magnetic nanoparticles. This approach increases reaction yield and enzyme stability, reduces mass transfer resistance, and improves biocompatibility. For example, a study by Verma et al. (2020) reported that binding a thermostable β-glucosidase from *Aspergillus niger* to nanoscale magnetic

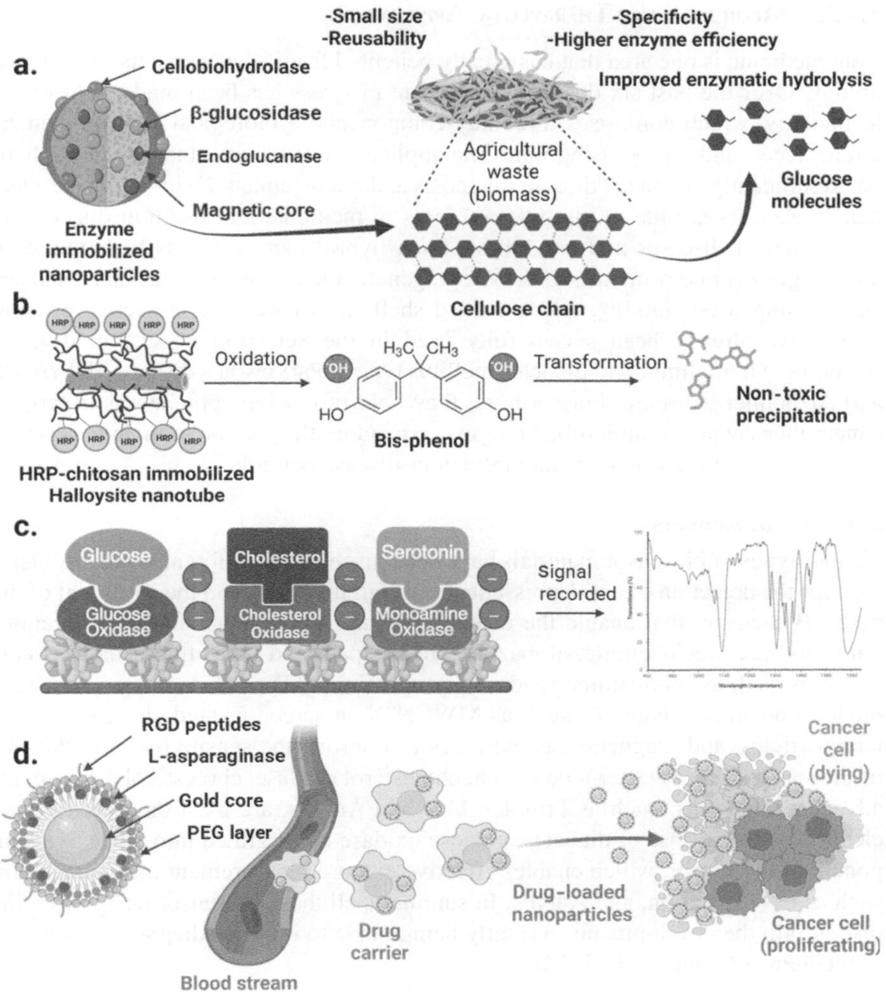

FIGURE 3.2 Roles of nanomaterial-linked enzymes in various applications: (a) enhancing biomass degradation for biofuel production, (b) facilitating pollutant degradation for environmental remediation, (c) enabling high-sensitivity biosensing for diagnostic purposes, and (d) advancing biomedicine, particularly in targeted cancer treatment.

nanoparticles using glutaraldehyde activation achieved 93% immobilization efficiency and enhanced cellobiose hydrolysis at 70°C. In addition, the use of Fe_3O_4 nanoparticles activated with carbodiimide resulted in a thermostable enzyme with a maximum activity of 62.7 U/mg at 50ºC, enabling six recycling cycles. These advances highlight the crucial role of nanotechnology in improving the efficiency and sustainability of enzymatic hydrolysis for biofuel production (Verma, 2020).

3.4.2 Medicinal and Therapeutic Applications

Nanomedicine is one area that has greatly benefited from advancements in nanotechnology. Over the past six decades, significant progress has been made in biosensor technology, which comprises three key components: a biological sensing element, a transducer, and a processing unit. The application of nanotechnology in medicine has significantly enhanced disease diagnosis and management. Hybrid enzyme-based nanobiosensors are the most advanced tools in medicine that use transducer elements to detect diseases and conditions. Such hybrid nanomaterials have numerous advantages over colorimetric sensors, e.g., genetic biosensors, such as increased sensitivity, improved stability, and extended shelf life of the enzyme-active proteins. They have already been successfully used in the detection of several diseases, including but not limited to diabetes mellitus type 2, Parkinson's disease, HIV/AIDS, and Alzheimer's disease. Furthermore, they hold promising applications in targeted cancer therapy and antimicrobial therapy, expanding their scope to include precision treatment options in oncology and infectious disease control.

3.4.2.1 Biosensors

Various types of biosensor materials have been invented for different health problems, e.g., for the detection of antidepressants, heart failure drugs, and the treatment of diabetes. Biosensors that enable the detection of antidepressants such as imipramine, amitriptyline, moclobemide, afobazole, phenazepam and tianeptine usually contain enzymes such as monoamine oxidase and tyrosinase. These biosensors are usually made of advanced materials such as MWCNTs on screen-printed electrodes, silver nanoparticles, and magnetic nanoparticles to improve the sensitivity and efficiency of the sensors. In cholesterol detection, cholesterol oxidase, cholesterol esterase, and advanced materials, such as Prussian blue and AuNPs, are used on screen-printed electrodes. In the case of diabetes, glucose oxidase is integrated into the active component of the sensor, which enables effective glucose measurement through a matrix such as CNTs, Nafion, and AuNPs. In summary, all these biosensor designs provide insight into the developments currently being made to improve diagnosis in the field of medicine (Anboo et al., 2022).

3.4.2.2 Antimicrobial Therapy

Infectious diseases caused by bacteria, fungi, parasites, viruses, and all other microorganisms have an impact on human health and other life forms, which is driving the search for successful treatments. Antimicrobials are widely used in various sectors such as healthcare, industry, households, and the marine environment. However, problems such as antimicrobial resistance continue to arise. A growing field addressing these challenges is nanotechnology, where drug delivery is improved through the use of nanoparticles that increase the bioavailability, solubility, and stability of active ingredients.

Chitosan and chitosan oligomers, for example, exhibit excellent properties such as non-toxicity, biodegradability, and biocompatibility, but also show extensive antimicrobial activity. Since the incorporation of chitosan in the production of nanoparticles, studies have shown antibacterial and antifungal properties through mechanisms such as membrane disruption and biofilm inhibition, with some processes showing the effects of pH and degree of deacetylation. In viral infections, nanoparticles facilitate the uptake of antigens, thereby enhancing the immune response. Chitosan-based nanomaterials have been successfully used to treat individual cases of infections across various fields (Oniha et al., 2024).

Another recent advance highlights the importance of nanoenzymes in improving antibacterial therapies and their application in wound healing. In particular, the immobilization of glucose oxidase in aerogels from nanoenzymes significantly improves anti-inflammatory treatment by combating methicillin-resistant *Staphylococcus aureus* (MRSA) infections. This innovative approach enables the removal of biofilms and the degradation of glucose at the wound site, which promotes healing in diabetic mice. In addition, the developed nanoenzymes can catalyse the formation of reactive oxygen species (ROS) on site and thus effectively fight bacterial infections. Iron-based nanoenzymes, with their extraordinary enzyme-like properties, enable multifunctional catalytic cascade reactions targeting disease biomarkers and open new avenues for diagnostic and therapeutic innovation (Shen et al., 2025). The integration of nanomaterials and enzymes presents a powerful approach in combating infections and enhancing therapeutic efficacy. This synergy not only boosts the stability and bioavailability of active compounds but also enables targeted treatments, such as improved wound healing and the disruption of antimicrobial-resistant biofilms. By harnessing the unique properties of enzymes and nanomaterials, new possibilities emerge for more effective and sustainable solutions to infectious diseases.

3.4.2.3 Cancer Treatment

Conventional cancer therapies are limited by resistance mechanisms. Radiotherapy encounters challenges such as DNA repair responses, cell cycle dysregulation, and the resilience of cancer stem cells, while chemotherapy is impeded by various drug resistance pathways. These factors often lead to treatment failure, recurrence, and poor patient outcomes, underscoring the need for better targeted approaches that specifically address the tumour microenvironment (TME) (Liu et al., 2021).

The TME, often referred to as the soil of cancer, comprises diverse and unique cellular and non-cellular elements that promote tumour growth and survival. The hypoxic, acidic, and nutrient-poor conditions, as well as the irregular vascular structures and dense stroma, induce specific triggers such as pH, temperature, and enzyme levels that nanobiotechnological therapies can exploit for targeted drug release. In addition, targeting overexpressed receptors on cancer cells can increase precision, reduce side effects, and improve therapeutic outcomes (Niculescu et al., 2022).

Nanomaterials can also be used to increase the bioavailability of enzymes. For example, L-asparaginase, an amidohydrolase enzyme, is known to be an effective anticancer drug, but its limited bioavailability restricts its more comprehensive pharmaceutical application. To improve its bioactivity, researchers loaded L-asparaginase together with the peptide Arg-Gly-Asp (RGD) to gold nanoparticles (GNPs), which target the enzyme more effectively in cancer cells, increasing its anticancer efficacy. This therapeutic conjugate, consisting of GNPs-PEG-L-asparaginase-RGD, was developed specifically for breast cancer cells, and showed promising therapeutic effects. The SH-PEG molecules were directly conjugated to GNPs via Au-S bonds without the need for amidation reactions or bifunctional crosslinkers. The conjugate showed significantly increased enzyme activity, with the RGD peptide effectively directing the enzyme to the cancer cells to exert its anticancer activity (Darvishi et al., 2022).

Enzymes, fundamental tools in nanobiotechnology, provide precise biological recognition and catalytic functions, making them ideal for targeted cancer therapies. Abnormal enzyme expression in cancer allows for the development of nanoparticles with enzyme-responsive binding, enabling targeted drug delivery. Matrix metalloproteinases (MMPs), such as MMP-2 and MMP-9, are frequently overexpressed in tumours and have been utilized to develop responsive nanoplatforms. For example, a recent study by Tian et al. (2022) reported on two paclitaxel (PTX) prodrugs linked to an octapeptide (AcGPLGIAGQ) that can be specifically degraded by MMP-2 in the tumour microenvironment, thereby effectively releasing PTX to inhibit tumour growth and proliferation (Tian et al., 2022).

Overall, the integration of nanobiotechnology with enzyme-based therapies offers a promising approach to overcoming the limitations of conventional cancer treatments. By exploiting the unique features of the tumour microenvironment (TME), such as pH, temperature, and specific enzyme expression, targeted drug delivery systems can be developed to enhance therapeutic precision and reduce side effects. Enzyme-responsive nanoparticles, like those targeting overexpressed MMPs or conjugated with specific peptides, enable more effective and selective treatment of cancer. This innovative approach has the potential to improve cancer therapy outcomes, addressing challenges like resistance mechanisms and offering a more personalized, efficient treatment strategy.

3.4.3 Environmental Remediation

A critical aspect of environmental remediation is the use of various processes such as adsorption/absorption, chemical reactions, photocatalysis, and filtration to remove pollutants from soil, water, or air. These processes are well suited to environmental remediation as nanotechnology-based materials have improved properties and performance. In general, materials with a high surface-to-volume ratio have a higher reactivity, which makes them attractive for use in remediation (Guerra et al., 2018).

3.4.3.1 Nanotechnology-Enhanced CAZymes in Pollutant Degradation

Nanotechnology and enzymes play a vital role in the degradation of environmental contaminants by increasing the effectiveness of environmental remediation measures. The unique properties of nanomaterials, such as their large surface area to volume ratio, significantly increase their reactivity and ability to interact with pollutants. Laccase, for example, an enzyme known for its ability to oxidize a variety of pollutants, can be effectively delivered and stabilized using nanotechnology. This combination, laccase, can break down pollutants in soil, water, and air more efficiently. Integrating laccase with nanomaterials enhances pollutant degradation by making the process more specific, targeted, and efficient, demonstrating the potential of this synergy in tackling environmental pollution challenges (Wong et al., 2019).

Similarly, a study by Deng et al. (2022) demonstrated that laccases can oxidize various polyphenols and aromatic compounds, offering a viable solution for the remediation of highly toxic polycyclic aromatic hydrocarbon (PAH) pollution. However, their effectiveness is often limited by low oxidation rates and recovery problems. To address this issue, the study explored immobilization of laccases on magnetic Fe_3O_4 particles coated with chitosan (Fe_3O_4@SiO_2-chitosan) to improve stability and reusability. The optimal immobilization conditions achieved a fixation capacity of 158 mg g^{-1} and retained 79.1% laccase activity. The immobilized laccase degraded 81.9% of anthracene and 69.2% of benzene pyrene within 48 hours, with simple magnetic recovery allowing reuse and activity remaining above 50% across three catalytic runs. These results highlight the potential of nanotechnology and immobilized laccase systems for effective PAH remediation (Deng et al., 2022).

The industrial revolution has significantly increased organic pollutants, highlighting the urgent need for effective wastewater treatment solutions. Various sources contribute to these pollutants, including PAHs (Borji et al., 2020), dyes, and volatile organic compounds (Das et al., 2020). Immobilized enzymes offer a promising approach to convert hazardous compounds into less toxic forms, especially when immobilized on nanoparticles.

For example, chitosan-halloysite hybrid nanotubes were fabricated by assembling chitosan onto halloysite, a natural nanotubular aluminosilicate. The structure of the hybrid nanotubes was characterized using nitrogen adsorption-desorption, Fourier-transform infrared (FTIR) spectroscopy, scanning electron microscopy (SEM), and transmission electron microscopy (TEM). These analyses confirmed that the hybrid nanotubes formed three-dimensional (3D) nanocomposites with a hierarchically porous architecture. The resulting structure demonstrated a strong capacity for immobilizing horseradish peroxidase (HRP) through cross-linking with glutaraldehyde. The enzyme loading achieved was as high as 21.5 mg/g, significantly exceeding the 3.1 mg/g observed with raw halloysite. After 35 days of storage, the immobilized HRP retained 100% activity, whereas free HRP retained only 27% of its original activity. Hence, the immobilized HRP exhibited high phenol removal efficiency from wastewater, demonstrating the potential of this hybrid-nanotube system for enzyme-based environmental applications (Zhai et al., 2013). Similarly, in a study, Abdollahi et al. (2018) explore the use of modified magnetic iron oxide nanoparticles (tyrosinase-MNPs) as a nano-biocatalyst for treating phenol-containing wastewater. The effectiveness of the nano-biocatalyst was assessed by examining factors such as catalyst dosage, pH, temperature, initial phenol concentration, and reusability. Results showed that the tyrosinase-MNPs effectively degraded phenol by more than 70% at a high concentration of 2500 mg/L across a wide range of pH and temperature conditions. The catalyst was highly reusable, achieving 100% phenol removal after three cycles and 58% after seven cycles. Additionally, the immobilized tyrosinase degraded phenol in real water samples by up to 78% in 60 minutes, with at least seven reuse cycles. These findings demonstrate the promising potential of tyrosinase-MNPs for efficient phenol removal from wastewater, offering a faster and more effective alternative to traditional treatment methods. Magnetic iron oxide nanoparticles carrying tyrosinase achieved over 70% phenol degradation and maintained their high activity `after several reuse cycles (Abdollahi et al., 2018). Furthermore, nanomaterials, specifically

Fe_3O_4 magnetic nanoparticles, offer significant advantages in wastewater treatment by enhancing the efficiency and reusability of enzymes. In this study, β-lactamase, an enzyme responsible for degrading β-lactam antibiotics, was immobilized on amino-modified magnetic nanoparticles. This combination not only facilitated the rapid degradation of antibiotics like penicillin G but also improved the enzyme's stability across a wide range of pH and temperature conditions. The immobilized enzyme demonstrated enhanced thermostability and storage stability compared to the free enzyme. With just 50 mg of the nanomaterial, 100 mL of penicillin G (5-50 mg/L) was completely degraded within 5 minutes. After 35 reuse cycles, the system retained over 95% degradation efficiency. This demonstrates the potential of the enzyme-nanoparticle system as a practical, cost-effective, and sustainable solution for removing antibiotic residues from pharmaceutical wastewater (Gao et al., 2018). The integration of nanomaterials with enzymes therefore enhances biocatalyst efficiency and stability, enabling effective degradation of pollutants like phenol, PAHs, and antibiotics. These advancements offer sustainable and cost-effective solutions for wastewater treatment, highlighting the potential of enzyme-loaded nanoparticles for practical environmental applications (Karthik et al., 2021).

3.5 OTHER APPLICATIONS

Nanoimmbolized enzymes also play an important role in the food industry. Food analysis often involves the detection of nutrients, contaminants, origin, and signs of adulteration. Enzymatic biosensors, a type of electrochemical biosensor, are frequently used in this area. They detect substrates either by monitoring the products of enzymatic reactions or by identifying enzyme inhibitors. Key enzymes include oxidases, peroxidases, and dehydrogenases, while target compounds include hydrogen peroxide and NADH (nicotinamide adenine dinucleotide (reduced form)). For example, oxidase-based biosensors help identify biogenic amines in foods, and L-lactate oxidase biosensors help evaluate freshness indicators such as L-lactic acid in tomatoes and infant formula.

Recent advances include the use of nanomaterials in biosensors to improve the response of the transducer and stabilize the receptors, such as CNT-modified biosensors for organophosphate neurotoxins, which show improved sensitivity and stability (Naqash et al., 2019). A study by Antony et al. (2016) described the immobilize α-amylase, an enzyme utilized in the food and fermentation industry for breaking down starch into maltose, onto nano-zinc oxide (ZnO) through electrostatic adsorption. The synergy between nano-ZnO and α-amylase improved the thermal stability, storage time, and reusability of the enzyme and showed better performance than the free enzyme. Kinetic analysis revealed a higher K_m value (Michaelis constant – an indicator of the enzyme's affinity) and a lower V_{max} value (maximum reaction rate-an indicator of the enzyme's efficiency) for the immobilized enzyme, indicating altered affinity and efficiency. Overall, this enzyme-nanoparticle system shows the potential for durable and reusable industrial biocatalysts (Antony et al., 2016). In addition to the food sector, nanotechnology is also helping with research into marine ecosystems and their enzymes. The incorporation of nanotechnology into marine enzymes offers strong alternatives for various industries by combining biological resistance with the improvement of efficiency and stability. Marine enzymes are being nanoscopically modified, even more than they have already been adapted to extreme oceanic conditions, so that they can withstand even the most hostile industrial conditions, such as high temperatures and changing pH levels. There are integration benefits in medicine, biofuels, food processing, and other areas where robust and environmentally friendly solutions are required. In this way, the potential of marine enzymes is further extended and enables applications that are more environmentally friendly and sustainable than before in all sectors (Enamala et al., 2021).

Overall, the integration of nano-immobilized enzymes in areas such as food analysis and marine biotechnology shows significant progress in improving stability, sensitivity, and reusability. However, further progress is essential to fully explore the synergy of nano-CAZymes and potentially open up even broader applications in sustainable industrial processes and environmentally sound solutions.

3.6 CONCLUSION AND FUTURE PROSPECTS

The convergence of nanotechnology and CAZymes has proven to be a revolutionary advance in numerous sectors, significantly improving the performance characteristics of enzymes. This synergistic approach has revolutionized the applications of enzymes through increased stability, improved catalytic efficiency, and increased reusability. The success of this integration is particularly evident in several areas, from increasing biofuel production through advanced enzyme immobilization to developing targeted drug delivery systems for cancer treatment. In addition, this technology has successfully enabled the development of advanced biosensors for medical diagnostics, facilitated the efficient degradation of environmental pollutants, and advanced applications in the food industry.

Looking ahead, several critical areas must be addressed to completely harness the potential of nano-enhanced CAZymes. The development of more biocompatible and intelligently responsive nanocarriers remains a priority, as does the need for economical production methods that enable large-scale applications. Research efforts should focus on understanding and optimizing the synergistic effects between different nanomaterials and CAZymes, especially in the development of multi-enzyme systems on single nanoplatforms. Investigating the orientation and conformational changes of enzymes on nanoparticles could provide valuable insights into improving the efficiency of the overall system.

Future advances in this area must address the practical challenges of scalability and cost efficiency. These include the development of recyclable nanoenzyme systems and research into sustainable sources of nanomaterials. In addition, the expansion of applications into new therapeutic areas, innovative environmental remediation techniques, and novel industrial applications offer exciting opportunities. However, these developments must be balanced with careful consideration of safety and regulatory requirements, including long-term stability studies and environmental impact assessments of nanomaterials.

The promising development of nano-enhanced CAZyme points to a future where these technologies could play a crucial role in addressing global challenges in sustainable energy production, healthcare, and environmental protection. The success of this endeavour will require continued collaboration between the various disciplines, combining expertise in enzymology, materials science, and nanotechnology. As research progresses, the integration of nano-enhanced CAZymes is expected to lead to increasingly sophisticated and practical solutions to complex industrial and environmental challenges without losing focus on sustainability and safety.

ACKNOWLEDGEMENT

The authors thank the Department of Microbiology, Sikkim University, for providing the computational infrastructure and central library facilities for procuring references and plagiarism analysis (Original Plagiarism Detection Software).

AUTHOR CONTRIBUTION STATEMENT

Manisha Chirania: Writing, original draft preparation, reviewing and editing. Anil Kumar Verma: Conceptualization, Supervision, reviewing and editing. Bhaskar Shrama: Writing, original draft preparation, reviewing and editing.

CONFLICT OF INTEREST

The author declares that the research was conducted without any commercial or financial relationships that could be construed as potential conflicts of interest.

FUNDING

This research did not receive any specific grant from funding agencies in the public, commercial, or not-for-profit sectors.

REFERENCES

Abdella, M. A. A., Ahmed, S. A., & Hassan, M. E. (2023). Protease immobilization on a novel activated carrier alginate/dextrose bead: Improved stability and catalytic activity via covalent binding. *International Journal of Biological Macromolecules*, 230, 123139. https://doi.org/10.1016/j.ijbiomac.2023.123139

Abdollahi, K., Yazdani, F., Panahi, R., & Mokhtarani, B. (2018). Biotransformation of phenol in synthetic wastewater using the functionalized magnetic nano-biocatalyst particles carrying tyrosinase. *3 Biotech*, 8, 1–10.

Al-Dulimi, A. G., Al-Saffar, A. Z., Sulaiman, G. M., Khalil, K. A., Khashan, K. S., Al-Shmgani, H. S., & Ahmed, E. M. (2020). Immobilization of l-asparaginase on gold nanoparticles for novel drug delivery approach as anti-cancer agent against human breast carcinoma cells. *Journal of Materials Research and Technology*, 9(6), 15394–15411.

Anboo, S., Lau, S. Y., Kansedo, J., Yap, P. S., Hadibarata, T., Jeevanandam, J., & Kamaruddin, A. H. (2022). Recent advancements in enzyme-incorporated nanomaterials: Synthesis, mechanistic formation, and applications. *Biotechnology and Bioengineering*, 119(10), 2609–2638. https://doi.org/10.1002/bit.28185

Antony, N., Balachandran, S., & Mohanan, P. V. (2016). Immobilization of diastase α-amylase on nano zinc oxide. *Food Chemistry*, 211, 624–630.

Ashkan, Z., Hemmati, R., Homaei, A., Dinari, A., Jamlidoost, M., & Tashakor, A. (2021). Immobilization of enzymes on nanoinorganic support materials: An update. *International Journal of Biological Macromolecules*, 168, 708–721.

Bai, Y., Jing, Z., Ma, R., Wan, X., Liu, J., & Huang, W. (2023). A critical review of enzymes immobilized on chitosan composites: Characterization and applications. *Bioprocess and Biosystems Engineering*, 46(11), 1539–1567.

Borji, H., Ayoub, G. M., Al-Hindi, M., Malaeb, L., & Hamdan, H. Z. (2020). Nanotechnology to remove polychlorinated biphenyls and polycyclic aromatic hydrocarbons from water: A review. *Environmental Chemistry Letters*, 18, 729–746. https://doi.org/10.1007/s10311-020-00979-x

Chang, K., Yan, Y., Zhang, D., Xia, Y., Chen, X., Lei, L., & Shi, S. (2023). Synergistic bonding of poly (N-isopropylacrylamide)-based hybrid microgels and gold nanoparticles used for temperature-responsive controllable catalysis of p-Nitrophenol reduction. *Langmuir*, 39(6), 2408–2421.

Chapman, R., & Stenzel, M. H. (2019). All wrapped up: Stabilization of enzymes within single enzyme nanoparticles. *Journal of the American Chemical Society*, 141(7), 2754–2769.

Chettri, D., & Verma, A. K. (2023). Biological significance of carbohydrate active enzymes and searching their inhibitors for therapeutic applications. *Carbohydrate Research*, 529, 108853.

Chettri, D., & Verma, A. K. (2022). Carbohydrate-based nanoparticles in therapeutics and biomedicine. *Current Trends in Biomedical Engineering and Biosciences*, 21(1), 556052. https://doi.org/10.19080/CTBEB.2022.21.556052

Chettri, D., Verma, A. K., & Verma, A. K. (2020). Innovations in CAZyme gene diversity and its modification for biorefinery applications. *Biotechnology Reports*, 28, e00525.

Chinni, S. V., Sanniraj, A., Koduru, J. R., & Lebaka, V. R. (2023). Nanobioremediation—an emerging eco-friendly approach for a sustainable environment. In J. R. Koduru, R. R. Karri, & N. M. Mubarak (Eds.), *Hybrid Nanomaterials for Sustainable Applications* (pp. 199–219). Elsevier.

Darvishi, F., Jahanafrooz, Z., & Mokhtarzadeh, A. (2022). Microbial L-asparaginase as a promising enzyme for treatment of various cancers. *Applied Microbiology and Biotechnology*, 106(17), 5335–5347.

Das, A., Jaswal, V., & Yogalakshmi, K. N. (2020). Degradation of chlorpyrifos in soil using laccase immobilized iron oxide nanoparticles and their competent role in deterring the mobility of chlorpyrifos. *Chemosphere*, 246, 125676. https://doi.org/10.1016/j.chemosphere.2019.125676

Deng, J., Wang, H., Zhan, H., Wu, C., Huang, Y., Yang, B., Mosa, A., & Ling, W. (2022). Catalyzed degradation of polycyclic aromatic hydrocarbons by recoverable magnetic chitosan immobilized laccase from Trametes versicolor. *Chemosphere*, 301, 134753.

Ding, H., Tan, P., Fu, S., Tian, X., Zhang, H., Ma, X., Gu, Z., & Luo, K. (2022). Preparation and application of pH-responsive drug delivery systems. *Journal of Controlled Release*, 348, 206–238.

Drula, E., Garron, M. L., Dogan, S., Lombard, V., Henrissat, B., & Terrapon, N. (2022). The carbohydrate-active enzyme database: Functions and literature. *Nucleic Acids Research*, 50(D1), D571–D577.

Enamala, M. K., Chavali, M., Pamanji, S. R., Tangellapally, A., Dixit, R., Singh, M., & Kuppam, C. (2021). Greener synthesis of enzymes from marine microbes using nanomaterials. In S. Thomas, A. Rane, K. Kanny, & V. K. Thakur (Eds.), *Handbook of Greener Synthesis of Nanomaterials and Compounds* (pp. 109–136). Elsevier.

Gao, X. J., Fan, X. J., Chen, X. P., & Ge, Z. Q. (2018). Immobilized β-lactamase on Fe 3 O 4 magnetic nanoparticles for degradation of β-lactam antibiotics in wastewater. *International Journal of Environmental Science and Technology*, 15, 2203–2212.

Gaur, M., Misra, C., Yadav, A. B., Swaroop, S., Maolmhuaidh, F. Ó., Bechelany, M., & Barhoum, A. (2021). Biomedical applications of carbon nanomaterials: Fullerenes, quantum dots, nanotubes, nanofibers, and graphene. *Materials*, 14(20), 5978.

Guerra, F. D., Attia, M. F., Whitehead, D. C., & Alexis, F. (2018). Nanotechnology for environmental remediation: Materials and applications. *Molecules*, 23(7), 1760.

Gustafson, K.T., Mokhtari, N., Manalo, E.C., Montoya Mira, J., Gower, A., Yeh, Y.S., Vaidyanathan, M., Esener, S.C., & Fischer, J.M. (2022). Hybrid silica-coated PLGA nanoparticles for enhanced enzyme-based therapeutics. *Pharmaceutics*, 15(1), 143.

Hooe, S.L., Breger, J.C. and Medintz, I.L. (2024). Enhancing enzymatic activity with nanoparticle display–an updated compendium and engineering outlook. *Molecular Systems Design & Engineering*, 9(7), 679–704.

Hussain, A., Parveen, F., Saxena, A. and Ashfaque, M. (2024). A review of nanotechnology in enzyme cascade to address challenges in pre-treating biomass. *International Journal of Biological Macromolecules*, 270, p.132466.

Jongkees, S. A., & Withers, S. G. (2014). Unusual enzymatic glycoside cleavage mechanisms. *Accounts of Chemical Research*, 47(1), 226–235.

Kapil, S., Vinayak, A., Pal, N., & Sharma, V. (2023). Application of nanotechnology in biobutanol production. In P. Verma, D. Kumar, & R. C. Kuhad (Eds.), *Advances and Developments in Biobutanol Production* (pp. 363–379). Woodhead Publishing.

Karim, A., Rehman, A., Feng, J., Noreen, A., Assadpour, E., Kharazmi, M. S., Lianfu, Z., & Jafari, S. M. (2022). Alginate-based nanocarriers for the delivery and controlled-release of bioactive compounds. *Advances in Colloid and Interface Science*, 307, 102744.

Karthik, V., Senthil Kumar, P., Vo, D. V. N., Selvakumar, P., Gokulakrishnan, M., Keerthana, P., Audilakshmi, V., & Jeyanthi, J. (2021). Enzyme-loaded nanoparticles for the degradation of wastewater contaminants: A review. *Environmental Chemistry Letters*, 19, 2331–2350.

Kharazmi, S., Taheri-Kafrani, A., Soozanipour, A., Nasrollahzadeh, M., & Varma, R. S. (2020). Xylanase immobilization onto trichlorotriazine-functionalized polyethylene glycol grafted magnetic nanoparticles: A thermostable and robust nanobiocatalyst for fruit juice clarification. *International Journal of Biological Macromolecules*, 163, 402–413.

Liu, D. M., & Dong, C. (2020). Recent advances in nano-carrier immobilized enzymes and their applications. *Process Biochemistry*, 92, 464–475.

Liu, Y. P., Zheng, C. C., Huang, Y. N., He, M. L., Xu, W. W., & Li, B. (2021). Molecular mechanisms of chemo- and radiotherapy resistance and the potential implications for cancer treatment. *MedComm*, 2(3), 315–340. https://doi.org/10.1002/mco2.55

Meng, Y., Qiu, C., Li, X., McClements, D. J., Sang, S., Jiao, A., & Jin, Z. (2024). Polysaccharide-based nano-delivery systems for encapsulation, delivery, and pH-responsive release of bioactive ingredients. *Critical Reviews in Food Science and Nutrition*, 64(1), 187–201.

Mitter, N., Worrall, E. A., Robinson, K. E., Li, P., Jain, R. G., Taochy, C., Fletcher, S. J., Carroll, B. J., Lu, G. Q., & Xu, Z.P. (2017). Clay nanosheets for topical delivery of RNAi for sustained protection against plant viruses. *Nature Plants*, 3(2), 1–10.

Muley, A. B., Mulchandani, K. H., & Singhal, R. S. (2020). Immobilization of enzymes on iron oxide magnetic nanoparticles: Synthesis, characterization, kinetics and thermodynamics. In: Challa V. Kumar (Ed.), *Methods in Enzymology* (Vol. 630, pp. 39–79). Academic Press / Elsevier.

Naqash, F., Masoodi, F. A., & Rather, S. A. (2019). Food enzymes and nanotechnology. *Enzymes in Food Biotechnology*, 769–784. https://doi.org/10.1016/b978-0-12-813280-7.00044-x

Niculescu, A. G., & Grumezescu, A. M. (2022). Novel tumor-targeting nanoparticles for cancer treatment-a review. *International Journal of Molecular Sciences*, 23(9), 5253. https://doi.org/10.3390/ijms23095253

Oniha, M. I., Oyesola, O. L., Taiwo, O. S., Oyejide, S. O., Akindana, S. A., Ajanaku, C. O., & Isibor, P. O. (2024). Nanochitosan-based fish disease prevention and control. In O. L. Oyesola, C. O. Ajanaku, & P. O. Isibor (Eds.), *Nanochitosan-Based Enhancement of Fisheries and Aquaculture: Aligning with Sustainable Development Goal 14 – Life Below Water* (pp. 113–138). Springer Nature Switzerland.

Patil, P. D., Karvekar, A., Salokhe, S., Tiwari, M. S., & Nadar, S. S. (2024). When nanozymes meet enzyme: Unlocking the dual-activity potential of integrated biocomposites. *International Journal of Biological Macromolecules*, 271, p.132357.

Paudel, L., Pardhe, B. D., Han, S. R., Lee, J. H., & Oh, T. J. (2023). Identification and evaluation of CAZyme genes, along with functional characterization of a new GH46 chitosanase from Streptomyces sp. *KCCM12257. International Journal of Biological Macromolecules*, 253, 127457.

Poorakbar, E., Saboury, A. A., Laame Rad, B., & Khoshnevisan, K. (2020). Immobilization of cellulase onto core-shell magnetic gold nanoparticles functionalized by aspartic acid and determination of its activity. *The Protein Journal*, 39, 328–336.

Pourmadadi, M., Dehaghi, H.M., Ghaemi, A., Maleki, H., Yazdian, F., Rahdar, A. and Pandey, S. (2024). Polymeric nanoparticles as delivery vehicles for targeted delivery of chemotherapy drug fludarabine to treat hematological cancers. *Inorganic Chemistry Communications*, 167, p.112819.

Sabourian, P., Tavakolian, M., Yazdani, H., Frounchi, M., van de Ven, T. G., Maysinger, D., & Kakkar, A. (2020). Stimuli-responsive chitosan as an advantageous platform for efficient delivery of bioactive agents. *Journal of Controlled Release*, 317, 216–231.

Sharifi, M., Sohrabi, M. J., Hosseinali, S. H., Hasan, A., Kani, P. H., Talaei, A. J., Karim, A. Y., Nanakali, N. M. Q., Salihi, A., Aziz, F. M., & Yan, B. (2020). Enzyme immobilization onto the nanomaterials: Application in enzyme stability and prodrug-activated cancer therapy. *International Journal of Biological Macromolecules*, 143, 665–676.

Sharma, B., Shrivastava, M., Afonso, L. O., Soni, U., & Cahill, D. M. (2022). Metal doped nitrogenous hydroxyapatite nanohybrids slowly release nitrogen to crops and mitigate ammonia volatilization: An impact assessment. *NanoImpact*, 28, 100424.

Shen, B., Yan, Z., Wang, Y., Zhu, L., Zhao, Q., & Jiang, L. (2025). Nanozyme-chitosan-aerogel immobilized enzyme-driven biocatalytic cascade for therapeutic engineering of diabetic wounds. *Carbohydrate Polymers*, 347, 122690.

Soozanipour, A., & Taheri-Kafrani, A. (2018). Enzyme immobilization on functionalized graphene oxide nanosheets: efficient and robust biocatalysts. *Methods in Enzymology*, 609, 371–403. https://doi.org/10.1016/bs.mie.2018.06.010

Thanigaivel, S., Priya, A. K., Dutta, K., Rajendran, S., Sekar, K., Jalil, A. A., & Soto-Moscoso, M. (2022). Role of nanotechnology for the conversion of lignocellulosic biomass into bipotent energy: A biorefinery approach for waste to value-added products. *Fuel*, 322, 124236.

Tian, H., Zhang, T., Qin, S., Huang, Z., Zhou, L., Shi, J., Nice, E. C., Xie, N., Huang, C., & Shen, Z. (2022). Enhancing the therapeutic efficacy of nanoparticles for cancer treatment using versatile targeted strategies. *Journal of Hematology & Oncology*, 15(1), 132.

Tiwari, H., Rai, N., Singh, S., Gupta, P., Verma, A., Singh, A. K., Kajal, Salvi, P., Singh, S. K., & Gautam, V. (2023). Recent advances in nanomaterials-based targeted drug delivery for preclinical cancer diagnosis and therapeutics. *Bioengineering*, 10(7), 760.

Vangijzegem, T., Lecomte, V., Ternad, I., Van Leuven, L., Muller, R. N., Stanicki, D., & Laurent, S. (2023). Superparamagnetic iron oxide nanoparticles (SPION): From fundamentals to state-of-the-art innovative applications for cancer therapy. *Pharmaceutics*, 15(1), 236.

Verma, M. L. (Ed.). (2020). *Nanobiotechnology for Sustainable Bioenergy and Biofuel Production*. CRC Press, Taylor & Francis Group.

Wang, D., & Jiang, W. (2019). Preparation of chitosan-based nanoparticles for enzyme immobilization. *International Journal of Biological Macromolecules*, 126, 1125–1132. https://doi.org/10.1016/j.ijbiomac.2018.12.243

Weng, Y., Yang, G., Li, Y., Xu, L., Chen, X., Song, H. and Zhao, C.X. (2023). Alginate-based materials for enzyme encapsulation. *Advances in Colloid and Interface Science*, 318, p.102957.

Wong, J. K. H., Tan, H. K., Lau, S. Y., Yap, P. S., & Danquah, M. K. (2019). Potential and challenges of enzyme incorporated nanotechnology in dye wastewater treatment: A review. *Journal of Environmental Chemical Engineering*, 7(4), 103261.

Xu, X., Liu, Y., Fu, W., Yao, M., Ding, Z., Xuan, J., Li, D., Wang, S., Xia, Y., & Cao, M. (2020). Poly (N-isopropylacrylamide)-based thermoresponsive composite hydrogels for biomedical applications. *Polymers*, 12(3), 580.

Yadav, A., Kumar Yadav, D., Rani, P., Bhardwaj, N., Gupta, A., & Bishnoi, N. R. (2024). Functionalized iron oxide nanoparticles for covalent immobilization of cellic CTec2 cellulase: Enabling enzyme reusability in cellulosic biomass conversion. *Biofuels*, 15(3), 363–373.

Yamada, A., Kamada, K., Ueda, T., Hyodo, T., Shimizu, Y., & Soh, N. (2018). Enhanced catalytic activity and thermal stability of lipase bound to oxide nanosheets. *RSC Advances*, 8(36), 20347–20352. https://doi.org/10.1039/c8ra03558j

Zechel, D. L., & Withers, S. G. (2000). Glycosidase mechanisms: Anatomy of a finely tuned catalyst. *Accounts of Chemical Research*, 33(1), 11–18.

Zhai, R., Zhang, B., Wan, Y., Li, C., Wang, J., & Liu, J. (2013). Chitosan–halloysite hybrid-nanotubes: Horseradish peroxidase immobilization and applications in phenol removal. *Chemical Engineering Journal*, 214, 304–309.

Zhou, Z., Ju, X., Zhou, M., Xu, X., Fu, J., & Li, L. (2019). An enhanced ionic liquid-tolerant immobilized cellulase system via hydrogel microsphere for improving in situ saccharification of biomass. *Bioresource Technology*, 294, 122146.

4 Exploring the Various Novel Sources of Carbohydrate-Active Enzymes (CAZymes)

Shreya Biswas and Arun Goyal

ABBREVIATIONS

AA	Auxiliary activities
AMCase	Acidic mammalian chitinase
C	Carbon
CAT	CAZyme Analysis ToolKit
CAZymes	Carbohydrate-active enzymes
CAZypedia	Encyclopaedia of CAZymes
CBM	Carbohydrate binding module
CBRF	CAZymes-based ranking of fungi
CE	Carbohydrate esterase
CO_2	Carbon dioxide
CUPP	Conserved unique peptide patterns
dbCAN	Database for automated carbohydrate-active enzyme annotation
DOC	Dissolved organic carbon
ECM	Extracellular matrix
GH	Glycoside hydrolase
GlcNAc	N-acetylglucosamine
GT	Glycosyltransferase
HPA	Human pancreatic a-amylase
HMOs	Human milk oligosaccharides
KEGG	Kyoto Encyclopaedia of Genes and Genomes
O_2	Oxygen
PCW	Plant cell wall
PL	Polysaccharide lyase
SOC	Soil organic carbon
SOM	Soil organic matter

DOI: 10.1201/9781003561682-4

4.1 INTRODUCTION

In nature, carbohydrates are the most stable, abundant and significant polymers (Holesh et al., 2025). However, compared to other polymers like polypeptides or nucleic acids, the complexity of carbohydrates is significantly greater. Every cell has a coating of carbohydrates on its surface, and this intricacy prepares the carbohydrates for essential functions in cellular signalling, particularly when they are conjugated with lipids or proteins (Wardman & Withers, 2024). Nature has created and employed a group of enzymes called carbohydrate-active enzymes (CAZymes) linked with biological processes requiring sugar biotransformation, from synthesis to metabolism (Contesini et al., 2021). They are responsible for the carbohydrate structure's variety and complexity, allowing it to perform numerous functions in nature. The study of genes and proteins that deal with the 'Sugar code' is very important due to various biological processes. 'Sugar code' is the use of different sugar moieties as receptors, markers or signalling molecules in cell membranes for particular reactions and pathways similar to nucleotide bases (Chettri & Verma, 2023). The remodelling of blood and organ antigens, the creation of longer-lasting biopharmaceuticals, the modification and analysis of protein glycosylation patterns to produce more effective biopharmaceuticals and many more are examples of applications of CAZymes in biomedical science. Furthermore, CAZymes frequently offer helpful basic research purposes for figuring out the functions of certain glycoconjugate and carbohydrate structures in biological systems (commonly referred to as the glyco-code) (Wardman & Withers, 2024). Therefore, as organisms have frequently evolved enzymes over millennia to perform the precise processes, nature has supplied us with ready-made enzymes for particular requirements

CAZymes have found a variety of applications beyond what nature intended (Wardman & Withers, 2024). CAZymes have a wide range of industrial uses, including functioning as a commercial enzyme source for research and development studies and degrading lignocellulosic biomass for the manufacture of biofuels and other value-added products in biorefineries. They are also used in the food and baking sectors to preserve fruits and vegetables, ferment, brew and produce nutraceuticals (Gavande et al., 2023). CAZymes are also utilized in animal feed and have been discovered to be relevant in the paper and pulp, textile and detergent sectors, as well as new applications in waste treatment and bioremediation. CAZymes are also important for the maintenance of extreme climate conditions (Tashkandi & Baz, 2023). The functional versatility of CAZymes extends beyond biomass degradation to play a pivotal role in the global carbon cycle, particularly under extreme climate conditions such as drought. During drought events, soil microbial activity often intensifies the decomposition of soil organic matter (SOM), accelerating CO_2 release into the atmosphere and disrupting the carbon sink function of terrestrial ecosystems. As key mediators of SOM breakdown, CAZymes govern the conversion of plant-derived polysaccharides into simpler carbon compounds, directly influencing soil organic carbon (SOC) pool dynamics (Sime et al., 2024; Q. Zhang et al., 2023). A deeper understanding of CAZyme activity under stress conditions offers promising avenues for climate change mitigation strategies. For instance, targeted manipulation of CAZyme-producing microbial communities or the deployment of CAZyme inhibitors could minimize carbon loss from soil systems during drought. Additionally, the integration of CAZyme function into predictive carbon flux models may enhance our capacity to forecast and manage ecosystem carbon storage under future climate scenarios. Such insights position CAZymes not only as tools for industrial bioconversion but also as molecular targets for sustaining soil carbon reservoirs and reducing atmospheric CO_2 emissions in a changing climate (Sime et al., 2024).

The vast application of CAZymes in health sciences and biotechnology has expanded the market and increased the funding in research and expansion in the domain of CAZymes (Wardman & Withers, 2024). Substantial research into the structure and functional mechanism of CAZymes is ongoing, as is the exponential growth in the number of CAZyme family members. Proper categorization and research of the reaction mechanism of CAZymes, as well as information preservation, is critical for identifying novel CAZymes and their families and gaining insight into the continuously sequenced genomes of diverse organisms (Ameri et al., 2022). In some circumstances, categorization may aid in predicting the carbohydrate substrate and consequently the glycobiology of the organism. Nature-sourced enzymes, however, might not always possess the precise activity or specificity that is required (Chettri & Verma, 2023). Such enzymatic variety may be obtained by generating synthetic gene libraries or by using techniques like functional metagenomics and proteomics (Wardman & Withers, 2024).

4.2 CLASSIFICATION OF CAZYMES

In the beginning, distinct CAZymes, such as glycoside hydrolase or polysaccharide lyase (PL), were categorized into several clans according to the conserved residues involved in the catalysis as well as the similarity of their sequences and structures (Cantarel et al., 2009). However, from 1990 onwards, there was a pressing need to categorize CAZyme clans into more functional and sequence-specific families due to the sharp growth in the number of carbohydrates and CAZymes that were described. Thus, the Carbohydrate-Active enZyme database (CAZy; www.cazy.org/) was created in 1998. Its goal is to categorize and arrange the enzymes that act on different types of carbohydrates. Based on primary sequence homology, the CAZy database divides CAZymes from diverse sources into distinct categories. Hundreds of enzyme members make up each group, and these members are further divided into many families according to the similarities in the protein sequences and their three-dimensional folding structure (Lombard et al., 2014). As a result, maintaining the structure and function relationship of a certain CAZymes becomes simple. For the core CAZymes that directly interact with carbohydrates, four distinct types of CAZymes have been identified: Glycoside hydrolase (GH), glycosyltransferase (GT), PL and carbohydrate esterase (CE) as depicted in Table 4.1. Due to their auxiliary activity function in relation to carbohydrates, two more groups, carbohydrate binding module (CBM) and auxiliary activities (AA) are in line with CAZy (Gavande et al., 2023). The categorization (i) better represents the structural aspects of these enzymes than their sole substrate specificity, (ii) aids in the discovery of evolutionary links between these

TABLE 4.1
CAZyme Classification and Applications

CAZyme	Enzyme Commission Number	Families	Activity	Substrates	Examples	Application	References
GH	EC 3.2.1.-	GH1-GH189	Catalyses the hydrolysis and/or trans-glycosylation of glycosidic linkages present between the two monomeric units of the polysaccharides	Cellulose, xylan, mannan, arabinan, pectin, galactan etc.	Endoglucanase, xylanase, mannanse, arabinofuranosidase, pectinase etc.	Biofuel; food and feed; paper and pulp; bioremediation; biofilm	Gavande et al. (2023), Cantarel et al. (2009), Yadav et al. (2024)
GT	EC 2.4.1.-	GT1-GT138	Synthesizes the glycosidic bond between the two monomeric units from phospho-activated sugar donors	Uridine diphosphate-glucose (UDP-glucose), UDP-glucuronoside etc.	Glycogen synthase, cellulose synthase, glucuronosyl-transferase etc.	Drug development; biocatalysts and synthetic biology; food industry; glycomics research	Yadav et al. (2024), Lairson et al. (2008)
PL	EC 4.2.2.-	PL1- PL44	Breaks glycosidic bonds of polysaccharides containing uronic acid by a β-elimination	Pectin, polygalacturonic acid, rhamnogalac-turonans, alginate etc.	Pectate lyase, alginate lyase, rhamnogalac-turonan lyase etc.	Wastewater treatment; food industry; drug delivery	Chakraborty et al. (2017), Linhardt et al. (1987)

CE	EC 3.1.1.-	CE1- CE20	Removes the O- and N-bonded ester-bonded modifications of complex polysaccharides	Pectin methyl esters, pectin acetyl esters etc.	Acetyl xylan esterase, pectin methyl esterase etc.	Animal feed; biomedical application	Yadav et al. (2024), Levasseur et al. (2013), Armendáriz-Ruiz et al. (2018)
AA	EC 1.-.-.-	AA1-AA17	Catalytic proteins involved in plant cell disintegration by helping GH, PL, and CE enzymes to get access to the carbohydrates of plant cell wall	Lignin, plant cell wall	Laccase, Mn^{2+} peroxidase, copper-dependent lytic cellulose monooxygenase LPMO etc.	Delignification; biofuel production; agriculture	Armendáriz-Ruiz et al. (2018), Levasseur et al. (2013)
CBM	-	CBM1-CBM106	Non catalytic proteins associated with catalytic CAZymes functioning by targeting and fostering a constant interaction with the substrate	Polysaccharides	Cellulose-binding CBM, Chitin-binding CBM etc.	Molecular probes; fibre modifications	Levasseur et al. (2013), Lombard et al. (2014)

enzymes and (iii) gives a useful framework for understanding mechanistic attributes (Cantarel et al., 2009).

4.3 SOURCE OF CAZYMES

Green biomanufacturing has gained popularity in recent years as a result of the severe energy problem and environmental degradation (X. Zhang et al., 2022). Search for green and environment friendly methods of industrial production has progressively become a new issue for humans. One such approach has been turning degradable biomass into fermentable sugars and ethanol, thereby replacing/blending of fossil fuel petrol by bioethanol. However, the transformation of biomass necessitates a large number of industrial enzymes, particularly CAZymes (The CAZypedia Consortium, 2018; Gavande et al., 2023). It is thus vital to get industrial enzymes with improved characteristics and cost-effectiveness of the process. Microorganisms confer additional benefits in the bioeconomy and green biomanufacturing due to their rapid growth and low cost. Filamentous fungi, which are abundant on Earth, play an important part in the global carbon cycle due to their exceptional ability to degrade organic matter and convert plant biomass into cost-effective fermentable sugar (Sime et al., 2024). Furthermore, filamentous fungi can create a variety of secondary metabolites that might be employed as antibiotics, such as penicillin or other medicines used in tumour treatment. CAZymes also been studied from plant and animal sources. But their higher cost of production restricts their development and use in this field (Table 4.2) (Chettri et al., 2022).

The CAZyme family includes enzymes from a variety of taxonomical groupings and in many cases, distinct kingdoms (Fig. 4.1) (Barrett, Jensen, et al., 2020). Compared to enzymes derived from plants and animals, microbial enzymes are more affordable, simpler and produced more quickly. Due to their superior productivity and ability to produce a wide variety of GH enzymes, fungi are typically selected over other microorganisms. Fungal enzymes are widely employed in the food and feed sectors, as well as in the manufacturing of biofuel, paper and pulp, textiles, detergents and beverages (T. Zhang et al., 2020). Apart from the CAZy database, many more databases have been created to facilitate the systematic research of CAZymes. These databases include dbCAN, which is similar to CAZy (Zheng et al., 2023); Conserved Unique Peptide Patterns or CUPP, a tool for functional annotation of CAZyme utilizing conserved peptide patterns (Barrett, Hunt, et al., 2020), CAZypedia, an encyclopaedia of CAZymes (The CAZypedia Consortium, 2018) available at www.cazypedia.org/; and CAZyme Analysis ToolKit (CAT; Park et al., 2010).

4.3.1 Bacterial

Bacterial CAZymes play pivotal roles in natural ecosystems, industrial applications and biotechnology (Zheng et al., 2023). The expansive array of bacterial CAZymes has revolutionized the understanding of microbial ecology and metabolic versatility while also driving advances in biomass utilization, health sciences and green chemistry. Bacteria exhibit exceptional diversity in their CAZyme repertoire, adapting to their ecological niches by metabolizing a wide variety of carbohydrate substrates

TABLE 4.2
Comparison of Expression Hosts for Recombinant CAZymes

Organism	Growth Condition	Genetic Transformation	Post-translational Modification	Expression Efficiency	Cost
Bacterial	Fast, high efficiency, easy scale-up	Well-defined and simple	None	High yield	Low
Fungal	Fast, high efficiency, easy scale-up	Well-established	Yes	Moderate yield	Low
Archaeal	Fast, high efficiency, easy scale-up	Well-established	None	Moderate yield	Low
Plant	Safe and efficient	Complex and time-consuming	Tailor-made glycans	High expression	Moderate cost and high contamination
Animal	Slow and limited large-scale production	Very complicated	Complex post-translational modifications	Low expression	High cost and high contamination

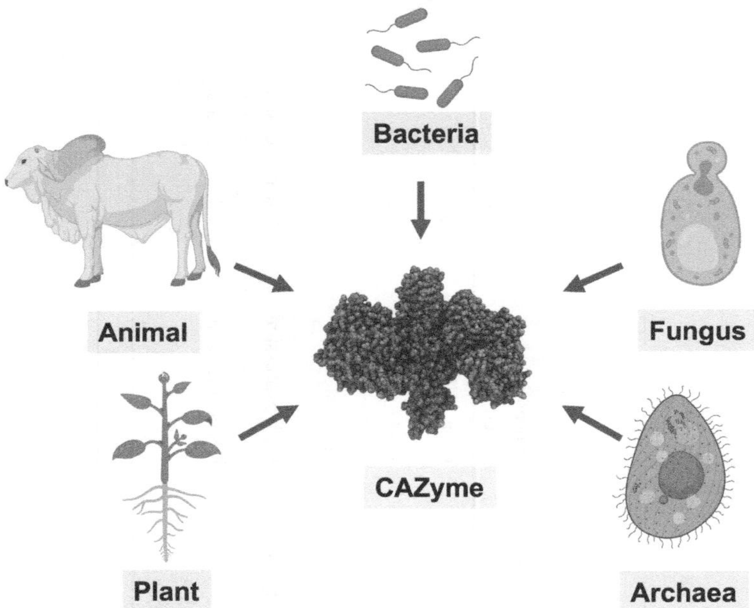

FIGURE 4.1 Various sources of carbohydrate-active enzymes (CAZymes).

(Lapébie et al., 2019). For example, soil-dwelling bacteria such as *Streptomyces* spp. and *Clostridium thermocellum* are renowned for their robust cellulases, enabling them to decompose plant-derived cellulose into fermentable sugars (Paudel et al., 2023; Thakur et al., 2021). Similarly, marine bacteria like *Vibrio* spp. and *Zobellia* spp. are rich in alginate lyases and agarases, reflecting their adaptation to seaweed-associated polysaccharides (Ausland et al., 2021; Bao et al., 2024). Gut bacteria, such as members of the 'Bacteroidetes' phylum, have evolved highly specialized CAZymes to degrade complex dietary fibres, aiding in host nutrition and health (Rodriguez & Martiny, 2020). The adaptability of bacterial CAZymes underscores their evolutionary success in diverse environments, offering a treasure trove of enzymes for industrial and therapeutic applications.

Among the CAZymes, GHs are the most extensively studied due to their essential roles in breaking glycosidic bonds (Gavande et al., 2023). These enzymes are pivotal in the bioconversion of plant biomass into simpler sugars, serving as a cornerstone for biofuel production. A model bacterial organism that has been researched is *Clostridium thermocellum* (now known as *Acetivibrio thermocellus*), which is highly specialized in breaking down plant cell wall (PCW) (Thakur et al., 2021). *Clostridium thermocellum* extracellularly secretes a multi-enzyme complex of molecular size of more than 2000 kDa, which is known as the cellulosome (Xu et al., 2016). Approximately, 78 glycosidase hydrolase gene sequences are found in *Clostridium thermocellum*, among which there are 37 glycosyltransferase sequences, 4 polysaccharide lyase sequences, 15 CE sequences and 97 carbohydrate-binding module sequences are among the several kinds of enzymes found in *Clostridium thermocellum* (Zhou et al., 2014). Cellulases, such as endo-glucanases (Cel8A, Cel5E, Cel5H, Cel9T, Cel124A etc.), cellobiolases (example: Cbh9A, Cel9K, Cel48S etc.), **β**-glucosidase (example: BglA) and exo-glucanases (example: Cel5O) have been identified from *Clostridium thermocellum*. Xylanases (example: *Ct*XynGH30, Xyn11A, Xyn5A, Xyn10Y etc.); mannanases (example: Man5A, Man26A, Man26B etc.) have also been characterised from the same bacterial source. Bifidobacterium bacterial strains have been identified as key players in carbohydrate degradation within the gut microbiome, demonstrating an impressive capacity for enzymatic activity that correlates with their adaptation to various hosts and environments (Lapébie et al., 2019). *Bifidobacterium adolescentis* and *Bifidobacterium longum* are known to release plant-based carbohydrates degrading CAZymes. The genome of *Bifidobacterium adolescentis* contains two **β**-xylosidase *viz.* XylB (GH120) and XylC (GH43) encoding genes (Kelly et al., 2021). Other species like *Bacteroide ovatus*, *Bacteroides xylanisolvens* and *Bacteroides intestinalis* hydrolyse xylan polysaccharides to give soluble xylo-oligosaccharides. These microorganisms also contain CAZymes that degrade pectin polysaccharides. *Bacteroides thetaiotaomicron* degrades extracellularly large pectin polymers to release mono- and oligo-saccharides which can be scavenged by certain bifidobacterial species, such as *Bifidobacterium longum*. It was reported that *Bifidobacterium breve* could cross-feed on the **β**-1,3-galacto-di/trisaccharides released from the hydrolysis of larchwood arabinogalactan by a GH2 family β-1,3-galactanase from *Bacteroides cellulosilyticus*. An **α**-arabinofuranosidase (GH51) produced by *Bifidobacterium longum* released arabinose from arabinoxylan, while arabinofuranosidaseA (GH43) from *Bifidobacterium adolescentis* removed arabinose residues from debranched mono-substituted arabinoxylan (Kelly et al., 2021). These gut bacterial systems have also shown to produce galactan-degrading CAZymes. A **β**-galactosidase (BgaA) and an exo-**β**-1,6-galactobiohydrolase (Bl1,6-Gal) from *Bifidobacterium breve,* exo-**β**-1,3 galactanase (Bl1,3Gal) and an endo-**β**-1,4-galactanase (galA) from *Bifidobacterium longum* have been also reported (Fernandez-Julia et al., 2022). These enzymes elicit hydrolytic activity towards galactan, thereby releasing galacto-oligosaccharides. Another bacterium releasing galactanase is *Bacillus* spp. All these bacterial systems initiate the digestion of a variety of complex polysaccharides by hydrolysing the glycosidic bonds to give different monosaccharides by utilising myriad of enzymes with complementary activities, mainly from diverse glycoside hydrolase (GH) families (Kelly et al., 2021). Another soil-dwelling bacteria, *Streptomyces* spp., is renowned for their robust cellulases enabling them to decompose plant-derived cellulose into fermentable sugars (Paudel et al., 2023).

PLs are critical for degrading algal polysaccharides such as alginate and ulvan, which are abundant in marine ecosystems. Bacteria like *Pseudoalteromonas* spp. and *Alteromonas* spp. harbour diverse PL families, enabling them to thrive on seaweed-derived substrates (Blanco-Cabra et al., 2020; Sawabe et al., 1997). The genomes of some bacterial species of *Bacteroides* spp. and '*Stenotrophomonas* spp. are also rich in PL CAZymes (Thakur et al., 2021). '*Clostridium thermocellum*' employs CEs to efficiently degrade lignocellulosic biomass, making it a model organism for consolidated bioprocessing in biofuel production (Levasseur et al., 2013). Esterases from CE2 family has been extensively studied from bacterium *Butyrivibrio proteoclasticus* and *Cellvibrio japonicus*. Bacterial glycosyl transferases (GTs) from *Campylobacter jejuni*, *Escherichia coli*, *Helicobacter pylori* and many more has also been studied for industrial purposes (Palcic, 2011). For instance, the exopolysaccharides synthesized by '*Leuconostoc*' spp. are widely used in the food industry as thickeners and stabilizers. Furthermore, the enzymatic mechanisms of GTs are being harnessed for the chemoenzymatic synthesis of complex carbohydrates, which are challenging to produce by traditional chemical methods (Lairson et al., 2008). Bacterial LPMOs, such as those from *Thermobifida fusca* and *Streptomyces* spp., have demonstrated exceptional activity against crystalline cellulose, paving the way for more sustainable bioconversion processes (Long et al., 2022). Even though the reported AAs mostly consist of fungal species, a few soil bacteria or aromatic-degrading bacteria from the digestive microbiota of animals have been reported for delignification, such as *Nocardia* spp., *Rhodococcus* spp., *Sphingobium* spp., *Streptomyces* spp. Other common bacterial sources of CAZymes include *Acetivibrio* spp., *Bacillus* spp., *Cellulomonas* spp., *Erwinia chrysanthemi*, *Ruminococcus* spp., *Thermonospora* spp., *Enterobacter* spp. and *Thermobifida fusca* (Menendez et al., 2015).

4.3.2 Fungi

Fungi are most effective in producing a cocktail of different enzymes (T. Zhang et al., 2020). Fungi release a range of enzymes that degrade organic waste found on the earth's surface; therefore, they serve an important part in sustaining our ecosystem's nutritional cycles. Filamentous fungi are more useful in the food, beverage and pharmaceutical sectors than other microbes since they can break down extrinsic plant material. They are also used to produce economically significant enzymes. Commercial-grade cellulases and a number of other highly catalytic enzymes are produced by the well-known and often used industrial fungi *Aspergillus niger*, *Trichoderma reesei* (*Hypocrea jecorina*) and *Neurospora crassa* (Zhao et al., 2014). The capacity of ascomycetes and basidiomycetous fungi to degrade plant biomass is firmly supported by recent fungal genomic investigations. The creation of public repositories such as JGI-MycoCosm, the 1000 fungal genome project and the Hungarian genome project have driven genome sequencing initiatives for various wood-decaying fungi (Kameshwar et al., 2019). This led to the creation of CAZymes-based ranking of fungi (CBRF) database, which is exclusive for CAZymes sourced from fungus. A microorganism's capacity to degrade plant biomass can be discovered by analysing its genetic composition in the initial phase. PCW-deconstructing enzymes have already been the subject of in-depth state-of-the-art reviews. The proteomic annotations of 103 fungi for various enzymes that break down PCWs have been compared (Zhao et al., 2014). Currently, there are 1057 fungal genome projects in the JGI-MycoCosm database, of which 443 have been published and made publicly available and 613 have not. The genome-wide distribution of CAZy annotations (CAZy groups and classes) of 443 published fungal genomes has been obtained on the CBRF website (T. Zhang et al., 2020). The findings showed that *Phanerochaete chrysosporium*, *Trametes versicolor*, *Laetiporus sulphureus* and *Hypoxylon* spp. have best ligninolytic capacity. Most significant lignin-hydrolysing CAZymes found in fungal sources are from AA1 and AA2 families. Similarly, *Phanerochaete carnosa, Neolentinus lepideus* and *Hypoxylon* spp. were shown to have the highest cellulolytic capacity. Similarly, CAZymes from families GH1, GH3, GH5, GH6, GH12, GH30, GH45, GH48, GH70, GH74, GH131, AA9 and AA16 are associated with cellulose-degradation in fungal sources. *Punctularia strigosozonata, Dacryopinax primogenitus* and *Hypoxylon* spp. were shown to have the best hemicellulolytic capacity. Hemicellulases, such as GH5, GH8, GH10, GH11, GH26, GH27, GH30, GH35, GH43, GH51, GH67, GH74, GH95, GH113, GH115, GH120, GH141 and CE1, CE2, CE3, CE4, CE5, CE6, CE15, CE16 and AA14 are predominantly found in fungal sources. Out of all the fungi, *Aspergillus sydowii*-C6d from camel rumen had the highest pectinolytic ability. *Aspergillus niger, Fibroporia radiculosa* and *Punctularia strigosozonata* also displayed superior pectin-degrading abilities indicating the presence of pectinolytic CAZymes (Tulsani et al., 2022). Fungal pectinolytic CAZymes are reported in GH28, GH53, GH78, GH88, GH105, GH106 and PL1, PL3, PL4, PL9, PL11, PL26 and CE8, CE12 families in CAZy database. *Aspergillus nidulans* and *Aspergillus niger* were reported to produce a broad spectrum of highly efficient cellulose and hemicellulose hydrolysing CAZymes (Gabriel et al., 2021). The fungi, *Aspergillus niger, Penicillium oxalicum, Aspergillus oryzae, Trichoderma reesei* and *Aspergillus nidulans* are the common hosts of CAZymes of filamentous fungi (T. Zhang et al., 2020). Other model fungi, like *Neurospora crassa*, can be used for both protein expression and mechanism study. Additionally, *Aspergillus niger* and *Aspergillus oryzae* are frequently used in the food sector, whereas *Trichoderma reesei* and *Penicillium oxalicum* are extensively used in cellulase expression and plant biomass breakdown. Furthermore, because the Gene Knockout Library of *Neurospora crassa* and *Aspergillus nidulans* is readily available they have been utilized for the investigation of various mechanisms in filamentous fungi and have greatly facilitated the future research. The intricate and time-consuming genetic alterations of filamentous fungi make them unsuitable for proteins that can be readily and efficiently generated by other hosts. The majority of filamentous fungus have the benefit of biomass breakdown because they are saprophytic. As a result, filamentous fungi are frequently employed to express CAZymes, which break down plant biomass in industrial settings and provide clean energy through green manufacturing (T. Zhang et al., 2020).

4.3.3 Archaea

An important area of study in microbiology and biogeochemistry is the Archaea, the least known branch of the tree of life (Qi et al., 2024). By facilitating distinct or uncommon metabolisms including methanogenesis, anaerobic methane oxidation, anaerobic alkanotrophy and chemolithotrophic ammonia oxidation, known Archaea play important roles in biogeochemical cycles. Although some Archaea groups may be easily grown in the lab, pure cultures account for less than 0.1% of Archaea species diversity. Over half of the archaeal taxonomic 'blind spots' have not yet been thoroughly characterized in the literature and are still awaiting discovery, despite the fact that the proliferation of archaeal genomes in recent decades has significantly broadened our perspective of the real diversity of Archaea. Archaeas are known as 'extremophiles' because they are prevalent in extreme physicochemical environment (Qi et al., 2024). More than 12000 archaeal species have been reported to contain CAZymes according to the database. These bacteria thrive in environments with severe pH and temperature and as a result, their enzymes, called extremozymes, have garnered a lot of interest in the biocatalysis and biotransformation sectors. In addition to being thermostable, extremozymes have exceptional resistance to high concentrations of organic solvents, salts, detergents and severe pH. Extremozymes are therefore perfect for industrial purposes with challenging chemical and physical circumstances, such as the pulp and paper industry or the conversion of starch and lignocellulose (Strazzulli et al., 2020). There are not many known archaeal isolates from the genera Metallosphaera, Acidianus and Sulfolobus. The potential metabolism of several archaeal lineages from different geothermal springs has been recently revealed by metagenomics. These include the chemolithotrophic Caldarchaeales (order), the symbiotic Aenigmatarchaeota (phylum) and Panguiarchaeales (order), the carbohydrate-metabolizing Bathyarchaeia (phylum), the ammonia-oxidizing Nitrosocaldaceae (family), the anaerobic methylotrophic class and several lineages that may be involved in the metabolism of methane and alkanes (Strazzulli et al., 2020). While lineages such as Micrarchaeia (class), Parvarchaeales (phylum) and Conexivisphaerales (family) were common in acidic springs, two Thermoproteia orders, Thermoproteales and Thermofilales, were more prevalent in alkaline springs. Carbohydrate metabolism has the greatest number of archeal sequences with functional annotations in the Kyoto Encyclopaedia of Genes and Genomes (KEGG) database. GTs were the most prevalent enzyme class among CAZymes that were discovered in the archaeal metagenomes. According to the CAZy database, the majority of GTs were members of the GT2 and GT4 families (Strazzulli et al., 2020). These two families are thought to be responsible for the majority of the microbial populations' carbohydrate-building activities, such as the synthesis of exopolysaccharides, glycoconjugates, cell envelope glycans, α- and β-glycans, etc. In Archaea, dolichyl-diphosphooligosaccharide protein glycotransferases or GT66, are involved in the glycosylation of proteins. Interestingly, GT66 is necessary in *Sulfolobus acidocaldarius*. Ceramide β-glucosyltransferases (GT21) and bifunctional trehalose-6-phosphate synthase/HAD hydrolase (GT20) are two other families that may be characterized from *Pyrobaculum arsenaticum*. The second most abundant CAZymes are GHs, followed by CE, CBM and redox AAs. Among GHs, the families, GH1, GH2, GH3, GH5, GH10, GH12, GH29, GH78, GH116, GH130 are hypothetically active on substrates containing equatorial β-D- and α-L-glycosidic linkages while the families GH4, GH13, GH15, GH31, GH36, GH38, GH57, GH65, GH77, GH99, GH109 and GH122 are specific for axial α-D-glycosidic bonds (Qi et al., 2024). Several enzymes involved in the hydrolysis of sugar appendages of hemicelluloses and in starch/glycogen mobilization and oligosaccharides of *N*-glycosylated proteins have been reported from *Ignisphaera aggregans*, *Desulfurococcaceae archaeon*, *Haloarcula taiwanensis* and many more. Archaea from the genera Pyrococcus, Pyrobaculum, Sulfolobus, Aeropyrum and Archaeoglobus are the source of a large number of hyperthermophilic esterases (Cabrera & Blamey, 2018). Additionally, halophilic archaea from the genera Halococcus and Haloarcula have been shown to contain these enzymes. The thermoacidophilic archaea *Sulfolobus solfataricus* and *Acidilobus saccharovorans* and the haloalkaliphilic archaeon *Haloarcula* spp. have also been shown to contain these enzymes. *Thermoproteus uzoniensis*, *Acidilobus saccharovorans*, *Metallosphaera cuprina* and *Ignisphaera aggregans* all produce endoglucanases. These enzymes have been identified in the hyperthermophilic archaeon *Pyrodictium abyssi* and the halophilic archaeon *Halorhabdus utahensis*. The halophilic archaeon *Halorhabdus utahensis* and the hyperthermophilic archaeon *Pyrodictium abyssi* have both been shown to have these enzymes. Furthermore, proteases from halophilic archaea from the genera Haloferax, Halobacterium, Natrinema and Natronomonas have also been identified. These enzymes function at high pH levels and are alkaline proteases; some of them are stable in organic solvents or at high temperatures (Cabrera & Blamey, 2018).

4.3.4 Plants

The most abundant renewable source of organic carbon on Earth is plant biomass. Its main component, the PCW, is a complex and extremely diverse macromolecular structure that envelops and shields the cell, making it essential to plant life. Several resistant polysaccharides, including cellulose (40–50%), hemicellulose (25–35%) and lignin (15–20%), make up PCW (Cardoso et al., 2022). One important stage in the carbon cycle is the microbial breakdown of plant structural carbohydrates to produce useable sugars. It has significant biotechnological significance, especially for the production of biofuels from lignocellulosic biomass, which is an essential goal for modern civilization. Therefore, knowing the biochemistry of PCW breakdown has rising industrial implications in addition to biological significance (Cardoso et al., 2022). Despite being present in all kingdoms of life, CAZymes are most prevalent in plants and plant-associated microbes because plants are the main source of carbohydrates (Huang et al., 2018). A wide variety of CAZymes mediate the breakdown of PCW carbs. Out of over 40 sequenced plant and algal genomes, only two (*Arabidopsis thaliana* and *Oryza sativa*) are covered by the CAZy database. This allowed for the creation of a database specifically for plant CAZymes (plantCAZyme) by including all sequenced bioenergy crops (poplar, switchgrass, sorghum) and evolutionarily significant organisms (moss, spike moss, algae) (Ekstrom et al., 2014). Two additional databases, pDAWG and Rice GT, only include a small number of CAZyme families and genomes. A very limited number of CAZyme families are also present in a few other databases, such as the Cell Wall Navigator database and the Cell Wall Genomics database. Multiple groups of CAZymes may be found in 23 dicots, 6 monocots, 2 mosses, 2 chlorophyte algae and 1 gymnosperm source (Cardoso et al., 2022). Among the plant sources the most characterized plant sources are *Eucalyptus grandis, Manihot esculenta, Phaseolus vulgaris, Sorghum bicolor, Cucumis sativus* and many more. CAZymes from 70 GH families, 58 GT families, 5 PL families and 12 CE families are distributed among all plant sources. Apart from that 7 AA families and 14 CBM families have also been identified from plant sources. Out of the 100 CAZyme families from rice (*Oryza sativa*) and 99 CAZyme families from Arabidopsis thaliana, 35 GH families have been reported to date (Ekstrom et al., 2014). Most of the cellulose-degrading plant-derived cellulases have been identified in GH1, GH2, GH3, GH9, GH16, GH55 and so on. Hemicellulose-degrading CAZymes are found in GH10, GH30, GH43, GH51 and so on.

4.3.5 Animals

Although microbial and plant-derived CAZymes have received a lot of attention, animal-derived CAZymes are just as important for their functions in physiology, metabolism and biotechnological applications (Wardman et al., 2022). CAZymes generated from animals are essential for functions including structural maintenance, cellular communication and digestion. The diversity, biological relevance, processes and uses of CAZymes derived from animals are examined in this essay, emphasizing their use in both industrial and natural systems. These enzymes are essential for the digestion of glycans and the metabolism of carbohydrates. The conversion of dietary carbohydrates into absorbable monosaccharides in mammals depends on GHs (Fernandez-Julia et al., 2022). The digestive enzymes, amylases, lactases and maltases are a few examples. Amylases make it possible for dietary carbohydrates to be broken down into simpler sugars that may be absorbed and used as fuel. Even though amylase can be found in both microorganism and animal sources, the first amylase gene to be studied from *Drosophila melanogaster* was **α**-amylase (Daïnou et al., 1987). Another source of amylase was found to be Japanese eel pancreas, which consists of 512 amino acids (Kurokawa et al., 2002). In 1998, human pancreatic a-amylase (HPA) was expressed in the methylotrophic yeast *Pichia pastoris* and the crystal structure has also been determined (Rydberg et al., 1999). Salivary and pancreatic amylases in humans and other mammals are essential for carbohydrate digestion. The small-intestine enzyme lactase-phlorizin hydrolase is in charge of hydrolysing the lactose carbohydrate (dimer of glucose and galactose) found in milk from mammals (Rings et al., 1994). The enzyme is essential for human and the majority of other animal feeding during the newborn period. Following this, specific activity of lactase drops to low adult levels. Large doses of lactose are no longer tolerated by the majority of adult humans and other mammals and consuming lactose can cause gastrointestinal problems. The enzyme, **β**-lactase has been studied from different mammals, such as human (Naim, 1991), rabbit (Keller et al., 1992), rat (Büller et al., 1987) and many more. Certain colonic microorganisms, such the lactic acid bacteria Lactobacillus and Bifidobacterium, have bacterial lactase, or β-galactosidase activity, which enables them to digest and use lactose in addition to their endogenous lactase activity (Forsgård, 2019). Lysozymes are CAZymes found in tears, saliva and other secretions of animals that hydrolyse the β-1,4-glycosidic bonds in peptidoglycan, a structural component of bacterial cell walls. By breaking down the cell walls of bacteria, lysozymes offer a first line of protection against bacterial infections. The highest concentration of this enzyme may be found in chicken egg whites, which contain around 0.3% lysozyme (Ferraboschi et al., 2021). Mammals' milk also contains lysozyme, either in the form of a free soluble protein or in the form of lysosomes and leucocytes. Since then, it has been well recognized that lysozyme is present in a variety of human secretions, including tears and saliva, as well as in the secretions of tissues and organs, including the placenta, sperm, leukocytes and blood. Lysozyme can also be found in insect's haemolymph, aquatic organisms, like fishes, clams, mussels, shrimp and sea cucumbers. A class of glycoside hydrolase enzymes called neurominidase or sialidase, breaks down the glycosidic bonds that hold neuraminic acids together. Numerous species include the vast family of neuraminidase enzymes. Mammalian neuraminidases, such as NEU1 and NEU2, regulate immune responses and lysosomal function (Warren & Spearing, 1960). For humans, the enzyme is present in the liver, submandibular-sublingual secretions and saliva. The oral cavity's polymorphonuclear leukocytes, epithelial cells or both may be the source of the salivary neuraminidases (Fukui et al., 1973). Another CAZyme found in animals like fish, amphibians and mammals are chitinases that contribute to the defence mechanisms against chitin-containing pathogens (Hamid et al., 2013). Acidic mammalian chitinase (AMCase) has been linked to allergy reactions and is involved in innate immunity in mammals. Chitinases degrade chitin, a polymer of N-acetylglucosamine (GlcNAc) found in the exoskeletons of insects and crustaceans. *Manduca sexta* (X. Zhu et al., 2001) and *Bombyx mori* (Koga et al., 1997) are two insects whose chitinases have been described. In ecdysis, when endo-chitinases randomly split the cuticle to chito-oligosaccharides, which are then hydrolysed to GlcNAc by exoenzymes, these enzymes serve crucial roles as degradative enzymes. For the creation of new cuticles, the GlcNAc monomer is recycled. Crustaceans such as prawns, krills and shrimp also contain chitinases (Bao et al., 2024). A crucial part of the extracellular matrix (ECM), hyaluronan, is broken down by hyaluronidases into smaller oligosaccharides. Mammal tissues express the enzymes hyaluronidase-1 (HYAL1) and hyaluronidase-2 (HYAL2). These enzymes aid in cell migration, wound healing and tissue remodelling. Body fluids (animal plasma, tissue fluids and semen), organs (kidney, liver, spleen and brain), animal venoms (snake venom, lizard poison, scorpion venom, spider venom, bee venom and ant poison) and some bacteria all contain hyaluronidase. Heparanase-1, another example of a CAZyme, is expressed in animals and controls inflammation, angiogenesis and the spread of cancer. Chronic inflammation and cancer are among the clinical diseases linked to dysregulation of heparanase activity. Human placenta and rat liver hepatocytes were the first to be reported to exhibit heparin sulphate breakdown by mammalian endo-glycosidic enzymes (Vlodavsky & Friedmann, 2001). Numerous normal and malignant cells and tissues have since been found to have heparanase activity, including platelets, cytotrophoblasts, mast cells, endothelial cells, macrophages, neutrophils, T and B lymphocytes, melanoma, lymphoma and carcinoma cells. Another family of enzymes are glycosyltransferases, which are essential for synthesizing complex glycans. In insects and mammals, GTs from family 47 (GT47) are involved in the production of complex glycoproteins (L. Zhang et al., 2023). Sialyl-transferases are a subclass of glycosyltransferases that catalyse the transfer of sialic acid residues to the terminal nonreducing sites of oligosaccharide chains of glycoproteins and glycolipids (Harduin-Lepers et al., 2005). ST3GAL and ST6GAL are sialyltransferases from human source, which are involved in the synthesis of sialylated glycoconjugates (Harduin-Lepers et al., 2001). Sialic acids play roles in immune evasion by pathogens, cancer metastasis and the regulation of inflammation. The variety of CAZymes seen in animals is indicative of their specific functions in preserving physiological functions, ranging from immunological protection and structural remodelling to digesting. Every enzyme plays a part in the complex web of carbohydrate metabolism and modification, ranging from lactases and amylases to glycosyltransferases and chitinases. In addition to highlighting the intricacy of animal biochemistry, these enzymes have a great deal of promise for both industrial and medicinal uses. Our knowledge of the involvement of CAZymes in health and illness will grow as a result of further study into their structure, function and regulation. This will open the door to creative medical and biotechnological approaches.

4.4 ENVIRONMENTAL SOURCES

4.4.1 Soil

Microbial remediation of the environment is a concept that has long been the subject of inquiry. Microbes are used in the biological process of microbial remediation to convert contaminants into less hazardous forms. In addition to effectively eliminating and lowering environmental pollutants, microbial remediation strategies may rebuild and repair ecosystems while preserving ecological stability (X. Wang et al., 2024). The soil ecosystem is incredibly diverse, including a wide range of microbiological habitats. The microbial variety found in soils is actually greater than that found in other habitats, such as water, air and other environmental niches. Notably, a variety of soil-dwelling microorganisms may be found in the forest soil habitat, which is one of the biggest and best settings in our ecosystem (Sime et al., 2024). These days, dependence on fossil fuels is becoming an issue, and the world's biggest environmental problem is the usage of fossil fuels to produce greenhouse gas emissions. Sustainable energy production and renewable biomass will require an alternative fuel production strategy. In order to mitigate the environmental issues of CO_2 emissions and hydrocarbon depletion, biofuel is one of the most promising renewable and sustainable energy sources (Sime et al., 2024). The degradation of lignocellulosic biopolymer plant materials is greatly aided by potential enzymes generated by microorganisms, particularly the microbial community that thrives in forest soil. Bioconversion is the process of converting biopolymers using microbial potential enzymes (X. Wang et al., 2024). The process of converting resistant plant material into fermentable sugar for the production of bioenergy is a complicated one that calls for the cooperation of a wide range of microbial enzyme consortia. Microbial enzymes have the greatest potential involvement in the synthesis of ethanol from forest biomass. By breaking down dead biomass, soil microorganisms play a vital role in maintaining the C balance and serve as a vital connection between soil C input and output (X. Zhang et al., 2022). This sequestered carbon's stability depends on the activities of microbial CAZymes, which break down organic matter, as well as the quantity of plant and microbial residues, which control the organic composition of the soil (Yu et al., 2024). The soil carbon profile may also be impacted by the selectivity of each enzyme towards various organic substrates and by changes in the CAZyme family's composition. The breakdown of polysaccharides and lignin is linked to the CAZymes glycoside hydrolases and AAs, respectively. While lysozymes and chitinases from GH families are associated with the breakdown of dead biomass from microbial communities, cellulases and hemicellulases from GH families have been identified as the primary enzymes breaking down plant biomass. Glycoside hydrolase families of GH2, GH10, GH30, GH36, GH39, GH43, GH51, GH67 and GH95 and esterase families of CE1, CE3, CE4, CE7, CE12 and CE15 were used for hemicellulose breakdown. Similarly, GH1, GH2, GH3, GH5, GH8, GH9, GH12 and GH116 were used for cellulose degradation. Through their respective CAZyme families demonstrated that the bacterial phyla Proteobacteria, Acidobacteria and Actinobacteria dominated the breakdown of dead plant and microbial biomass (X. Zhang et al., 2022). Environmental factors and the traits of the microbial community both directly and indirectly control the relative abundance of CAZyme subfamilies. The main microbial phyla, particularly Actinobacteria and Proteobacteria, are the primary source of soil CAZyme genes. Furthermore, the composition of CAZyme families in particular settings can be greatly influenced by the substrate choice of some special bacteria, such as Asticcacaulis and Cytophaga, which concentrate on decomposing complex organic molecules. Therefore, by altering the makeup of soil microbial communities, exogenous perturbations can also control the metabolism of carbohydrates (Yu et al., 2024). Common soil bacteria and fungi play important role in producing various CAZymes and decomposing organic matter. Agrobacterium, Bacillus, Pseudomonas, Corynebacterium, Alcaligenes, Micrococcus, Clostridium, Flavobacterium, Arthrobacter, Xanthomonas and Mycobacterium are among the common bacterial genera that have been identified from the soil. Some of the common fungal genera found in soil include *Rhizoctonia* spp., *Amanita* spp., *Thelephora* spp., *Verticillium* spp., *Tricholoma* spp., *Descomyces* spp., *Phytophthora* spp., *Torrendia* spp. and *Pythium* spp. (Sime et al., 2024). The primary CAZyme families involved for the breakdown of plant-driven biomass have been reported by many studies. AA1, AA2, AA3, AA4, AA5 and AA6 were employed for lignin degradation. Numerous CBMs, including CBM5, CBM6, CBM9, CBM32, CBM48, CBM50 and CBM66 were also found in soil microbials.

4.4.2 Hot Springs

One of the primary sources of thermophiles and functional macromolecules is hot springs. The hot spring microbiota and its genomic and macromolecule composition may be influenced by physicochemical characteristics, such as salinity, in addition to the crucial variables of temperature and pH (Liew et al., 2022). Organic carbon from the biomass of nearby dense forests affects the sediments of geothermal spring clusters, creating distinct ecosystems with a thermotolerant biomass-degrading microbial community and a range of putative CAZymes (Dixit et al., 2021). Due to the enzymatic adaptations to the harsh circumstances in these habitats, hot springs are an especially good setting for more study into lignocellulolytic enzymes. In geothermal systems, where thermostable enzymes have been selected to be active across a broad temperature and pH range, extreme conditions are prevalent, but they are rare in most natural systems. In the processing of lignocellulose, thermostability is preferred because it promotes favourable enzyme kinetics, enhanced substrate solubility and a decreased chance of contamination by undesirable organisms. Investigating environmental metagenomes of microbial communities from thermal settings over a broad range of pH and temperature offers a special chance to broaden the repertoire of known extremozymes with lignocellulolytic activities (Reichart et al., 2021). More than 1000 hot springs' metagenome data have been entered into the NCBI database and heated springs in Argentina, India, Japan, the United States, Russia, New Zealand, China and other nations have been the subject of in-depth microbiota studies. Even though these samples were collected from the same hot spring, the microbial diversity in the sediment, soil and water may differ. Additionally, the scientific community is interested in biofilms at hot locations (Liew et al., 2022). The hot springs were categorized based on temperature and pH level. The temperature categories were mesophilic (<45.0°C), thermophilic (45.0–79.9°C) and hyperthermophilic (>80.0°C) and the pH categories were acidophilic (<6.0) and alkaliphilic (>8.0). Streptomyces was the most prevalent genus in the mesophilic group, closely followed by Roseiflexus and Pseudomonas. The thermophilic bacteria are commonly found in the genera Bacillus, Sulfurihydrogenibium, Thermocrinis, Clostridium and Thermus. The archeal genera, Hydrogenobaculum, Sulfolobus and Acidolobus were found in very acidic environment, whereas members of the crenarchaeotal genus Sulfolobus are prevalent in acidic hot springs (Reichart et al., 2021). Both bacterial and archaeal species are extreme thermophiles, while the great majority of hyperthermophiles belong to the Archaea domain. Only archaea can be detected in the maximum temperature range (103–113°C). Thermotogales and Aquificales are the only class among the archaea, which make up the majority of hyperthermophiles. *Thermotoga maritima, Pyrococcus furiosus* and *Sulfolobus solfataricus* are biotechnologically significant due to their extremely heat-stable enzymes and ability to function under harsh circumstances (Reichart et al., 2021). The ability to carry out biotechnological activities at higher temperatures is frequently favourable, which explains the application interest in thermophilic CAZymes. For example, greater temperatures cause a drop in viscosity and an increase in diffusion coefficient, which in turn leads to increased reaction rates in chemical processes involving organic solvents (Park et al., 2017). The CAZymes of Acidobacteria phylum has been characterized to contain cellulose, hemicellulose and oligosaccharide-degrading CAZymes. These include CAZymes from various families such as GH1, GH2, GH3, GH5, GH8,GH9, GH10, GH12, GH16, GH26, GH30, GH43, GH48, GH51, GH116 and many more. *Chloroflexi* spp. from hot spring sources contains many oligosaccharide-degrading enzymes, whereas *Dictyoglomus* spp. isolates have been reported to yield multiple hemicellulases, especially xylanases (Dixit et al., 2021). Firmicutes (phylum) are a primary aim for biomass transformation because they enable consolidated bioprocessing of lignocellulose in a single organism without the requirement for pretreatment, and this phylum contains some of the most thermophilic cellulolytic organisms discovered thus far. The anaerobic class Clostridia can maintain their cellulose-binding ability throughout breakdown by using a multi-enzyme complex known as a cellulosome. *Caldicellulosiruptor thermophilum* may then ferment the sugars that have been broken down from cellulose, resulting in the production of lactate, ethanol and acetate. CAZymes from different GT families (GT1, GT4, GT28, GT35, GT51 etc.) were also reported from the given sources. Among the CEs, CE1, CE4 and CE8 have been accounted in these microbial sources (Dixit et al., 2021).

4.4.3 Marine

The oceans are the planet's largest ecosystem, home to over 80% of all plant and animal species, and they make up 70% of the planet's surface (Bäumgen et al., 2021). In particular, because the seas contain the largest carbon pool in the carbon cycle, marine systems have a remarkable influence on the barometrical CO_2 concentration. However, the increased amount of climatic CO_2 causes the waters to retain it more effectively, which lowers the pH value (Bäumgen et al., 2021). Thus, the $(CO_3)^{-2}$ concentration within the surface water is diminished, making the sea acidification an exasperating impact for the sea-carbonate chemistry, which is of awesome significance for marine calcifying living beings like molluscs, shellfish, echinoderms, corals, expansive calcareous green growth, foraminifera and a few phytoplankton. Also, the increasing amount of algal population includes a tremendous potential for biotechnological applications of value-added items and biorefinery processes. A broad spectrum of CAZymes are also associated with the decomposition of macroalgal and marine polysaccharides (Bäumgen et al., 2021). Benthic macroalgae and pelagic phytoplankton use photosynthesis to produce significant amounts of organic matter on continental shelves, which are productive marine systems. A significant portion of organic matter generated from algae is made up of polysaccharides, which play crucial roles in microbial metabolism and nutrient cycles. As a result, a variety of bacteria have evolved to specialize in biological interactions such as colonizing algal surfaces and breaking down algal polysaccharides. Flavobacteria, including the Flavobacteriaceae and Cryomorphaceae groups, play a significant role in the breakdown of marine polysaccharides. Marine flavobacteria have specialized genetic machinery that allows them to break down a variety of polysaccharides, much like the human gut does (Wolter et al., 2021). For example, *Zobellia galactanivorans* breaks down laminarin, alginate, agar and carrageenan through intricate biochemical and regulatory processes. Similar activities have been reported in the flavobacterial genera *Gramella forsetii* and *Formosa agariphila*. All of the organic stuff from the water column ends up on the ocean floor, which is also thought to be the planet's main carbon sink. Thus, the storage of enormous quantities of carbon is regulated by microbes that inhabit marine sediments. The quantity of microorganisms in soil and seawater is almost comparable to the number of these bacteria that reside on and under the sea bottom (López-Sánchez et al., 2024). The most abundant CAZyme found in marine sources is of GH13, which helps in the degradation of amylopectin. Marine habitat is rich in various carbohydrate polymeric structures, such as agar, alginate, carragenan, cellulose, laminarin, mannan, pectin, porphyrin, ulvan, xylan etc. (Bäumgen et al., 2021). GH family of 16 and 82 hydrolyses carrageenan; GH127 and GH129 enzymes are α-galactosidases, whereas GH2 enzymes are β-galactosidases. GH86 and GH16 have been identified to degrade porphyrin polysaccharide. Endo-acting laminarinases have been grouped into GH16, GH17, GH55, GH64 and GH81 families, while the GH3 family contains exo-acting laminarinases. Several ulvan lyases from the families GH105, PL24, PL25, PL28 and PL40 have been identified from different marine microbial sources. Currently, β-agarases are members of the GH16, GH50, GH86 and GH118 families, whereas α-agarases are members of the GH96 and GH117 families. The porphyranases that were discovered were found in GH16 and GH86. Alginate lyases have been characterized from families PL7, PL14, PL15 and PL17 (Hehemann et al., 2014).

4.5 GUT

It has long believed that gut microorganisms are essential to human health and that dysbiosis of the gut microbiota is linked to a number of illnesses, such as autism, diabetes, obesity and inflammatory bowel disease (Ye et al., 2019). Examining the carbon and energy resources is one method to comprehend what maintains a stable gut microbiota, as they are necessary for their existence and procreation. The composition of gut microorganisms is significantly influenced by human indigestible polysaccharides, which are important energy sources for the gut microbiota. Human milk oligosaccharides (HMOs), the third most prevalent solid component in human milk, are indigestible to newborns throughout the early stages of life. Utilization of these carbon sources is made possible by the digestion of these polysaccharides by a variety of CAZymes that are produced by gut microorganisms. Numerous distinct CAZymes are expressed by the gut microbiota due to the wide variety of human indigestible polysaccharides contained in our diets (Ducarmon et al., 2024). Since the microbiota plays a role in the metabolic and nutritional processes that directly affect the host's health, there have been a number of noteworthy developments in recent years about the gut microbiome of diverse living forms (Chettri et al., 2022). Synergistic interactions are formed between microorganisms in microbiomes and their host. It also shapes their composition and affects their hosts' physiology in addition to the environment in which they live. One such instance is the gut microbiota of distinct creatures, such as ruminants, termites and other herbivores, that rely on different plant components and polysaccharides for nourishment. Because the host does not have the enzymes needed for their metabolism, the microorganisms are essential for the utilisation of the polysaccharides that remain trapped in this carbon source (Shen et al., 2020). Plant polysaccharides make up the gut of ruminants, where microbial fermentation occupies 70% of the stomach's volume. This is where the active microbial degradation of dietary plant polysaccharides occurs.

4.5.1 Ruminants

The rumen chamber in the digestive system of ruminants has developed into an effective fermentation unit with a remarkable capacity to use plant biomass (Shen et al., 2020). Specialized microflora in the rumen have a complicated mechanism for the hydrolysis of plant polymers into their soluble oligomers, which the host would otherwise be unable to digest. Bacteria make up over 95% of the entire microbial community and assist far more to ruminal fermentation than any other microbial domain, including bacteria, archaea, fungi, protozoa and bacteriophages. Cellulases and hemicellulases, which are accountable for the breakdown of resistant plant polysaccharides, are among the CAZymes produced by these organisms. Rumen can therefore be a great niche for investigating CAZymes for application in cost-effective and competent conversion of biofuels and value-added products (Bohra et al., 2019). Numerous bacterial species that aid in the breakdown of cellulose and lignin are found in the rumen of adult ruminants. *Fibrobacter succinogenes*, *Prevotella ruminicola, Ruminococcus flavefaciens* and *Ruminococcus albus* are a few frequent bacteria, with *F. succinogenes* being the most prevalent and important in the breakdown of plant cellulose. Bacteria in the animal's digestive tract are linked to a higher percentage of enzymes that target oligosaccharides. These enzymes include β-glucuronidase, β-galactanase, β-mannosidase, β-xylanase, β-xylosidase, α-L-arabinofuranosidase, β-galactosidase, β-glucosaminidase and α-arabinanase (Chettri et al., 2022). The most dominant bacteria found in the rumen of the goat is *Butyrivibrio* spp. It was discovered that rumen bacteria, *Butyrivibrio* spp. encodes a wide range of CAZymes. Together, 4,421 GHs, 3,605 GTs, 1,706 CBMs, 1,283 CEs and 110 PLs were identified. These enzymes were predicted to play a role in the depolymerization and transportation of the insoluble plant molecules (Palevich et al., 2019). The goat rumen consists of many microbes, including several species, such as those from the genera Butyrivibrio, Ruminococcus and Clostridium, that are capable of effectively breaking down lignocellulose. Numerous lignocellulolytic enzymes, including cellulases, hemicellulases and ligninases, are produced by these microbes and break down complex lignocellulosic material into simpler sugars that the bacteria may use as an energy source. Cellulases contain CAZymes from families of GH5, GH6, GH7, GH9, GH44, GH45, GH48, GH88 and GH95, whereas hemicellulases are reported mostly from families of GH8, GH10, GH11, GH12, GH26, GH28 and GH53. The majority of the GH family of CAZymes and belonged to the genus Enterococcus, especially *Enterococcus_B* and *Enterococcus_C* (Rabapane & Matambo, 2024). CBM67, CE1 (trehalose 6-O-mycolyltransferase), GH92, GH89 (N-acetylglucosaminidases), GH78, GH43_8, GH33 (sialidases), GH38, GH32, GH29 (fucosidases), GH2 (galactosidases), GH18, GH16, GH29 and GH28, intramolecular β-galactosidases (GH98), sulfatases (GH20, GH2 and GH42) and core glycoside hydrolases (GH101, GH129, GH84, GH85 and GH89) are among the diverse CAZy families identified (K. Zhang et al., 2024). The two main bacterial cellulases are GH5 and GH9 and GH44. GH5, GH6, GH9, GH45 and GH48 are major cellulases that come from fungus. Fungal cellulosomes are known to include large amounts of the proteins CAZymes GH48 and GH618. GH5, GH10, GH11, GH26 and GH43 are examples of hemicellulases that are present in the gut (Peng et al., 2021). Results of ruminal fluid microbiota study based on 16S rRNA gene sequence show that Bacteroidetes is the predominant phylum in the rumen of both cows and buffalo (Chettri et al., 2022). Prevotella, Ruminococcus and Butyrivibrio were the three other most abundant genera. A complete set of 142 CAZyme families was discovered in the rumen metagenome of cattle, whereas 141 CAZyme families were found in buffaloes (Malik et al., 2024). The most prevalent families in the metagenome were GH43, GH3, GH13, GT2 and CE1, and the abundances of these genes were similar in buffaloes and cattle. GH43, GH13 and GH3 made up more than one-third of the overall abundance in the GH class. The main contributors of CAZymes were *Prevotella* spp., *Bacteroides* spp., *Alistipes* spp., *Clostridium* spp. and *Ruminococcus* spp., which were present in all five CAZyme groups. Rumen microbiota is also rich in oligo degrading enzymes from GH1, GH2, GH3, GH13, GH18, GH20, GH27, GH29, GH31, GH32, GH35, GH38, GH42, GH43, GH52, GH92, GH94, GH97, GH130 and debranching enzymes from GH23, GH33, GH51, GH54, GH62, GH67, GH77, GH78, GH84, GH103 and GH127 (L. Wang et al., 2019).

4.5.2 Lower Vertebrates

The gastrointestinal tract (GI) of vertebrates is a unique habitat made up of ecologically conserved interactions and an evolutionary process between the resident microflora and the host (Cabral et al., 2022). But just as the GI tract's anatomy varies from species to species, so does the gut microbiota. Proteobacteria, Firmicutes and Bacteroidetes genera make up around 90% of the gut microbiota of fish. Actinobacteria, Lactobacillus, Staphylococcus and Bacillus are a few prevalent bacterial genera. In fish, there is little to no CAZyme hydrolysis of the β-glycosidic bonds of non-starch polysaccharides (derived from plant diets) (Wu et al., 2025). Nonetheless, cellulase and xylanase activity were demonstrated in vitro by a number of carbohydrate-active Bacillus species that were isolated from the hindgut of herbivorous fish (Facimoto et al., 2023). The existence of genes encoding particular extracellular CAZymes that aid fish in digesting the non-starch polysaccharides of plant diets provided evidence in support of this finding. A high population of microbial communities, up to 1011 CFU/g in the hindgut, is seen in the gastrointestinal tract of birds. Similar to other vertebrates, Firmicutes, Bacteroidetes, Actinobacteria and Proteobacteria dominate the avian gastrointestinal tract microbiota. Only around 17% of vertebrate species are known to exist in reptiles, according to research. The phyla Proteobacteria, Bacteroidetes, Firmicutes and Fusobacteria are the most prevalent in the gut of reptiles. The gut microbiome of amphibians is little understood; however, typical bacterial species include *Pseudomonas* spp., *Staphylococcus* spp., *Enterobacter* spp., *Aeromonas* spp. etc. The primary constituents of the gut microbiome of amphibians are Firmicutes, Proteobacteria and Bacteroidetes, same as in humans and mice. Since amphibians and reptiles are mostly carnivorous, there is little evidence to support the occurrence of CAZyme-producing bacteria in these species (Ghazanfar et al., 2025).

4.5.3 Higher Vertebrates

The mammal gut is home to over a thousand different types of microorganisms (Wardman et al., 2022). The gut microbiome is made up of these species and their around 4.2 million genes. Our cells and the microbiome work closely together to support a number of metabolic processes, many of which the human genome is unable to carry out. These include regulating the immune system, absorbing nutrients, defending against infections and – above all – efficiently harvesting energy from the food. The breakdown of the carbohydrate components in our diet is one of the most important processes in the effective harvesting of energy. The main source of energy and one of the essential 'building blocks' of life are carbohydrates. The majority of the bacteria found in the human gut are species of the following genera: Escherichia, Enterobacter, Clostridium, Klebsiella, Lactobacillus, Bacteroides, Butyrivibrio, Bifidobacterium and Prevotella (Chettri et al., 2022). Other microorganisms and yeasts may also be present. The majority of CAZymes are formed by bacteria such as *Butyrivibrio fbrisolvens*, *Enterobacter cloacea*, *Escherichia coli*, *Klebsiella pneumoniae*, *Clostridium hathewayi*, *Prevotella copri*, *Prevotella salivae* and *Bacteroides* spp. Dietary fibres, especially pectin, have become essential elements that provide an intriguing and complex benefit to human health. In the large intestine, pectin is fermented by beneficial bacteria using CAZymes, despite the fact that humans are incapable of digesting it (Yüksel et al., 2024). In response to dietary pectin, gut bacteria produce CAZymes as an adaptive mechanism. Glycosidic linkages in both carbohydrate and non-carbohydrate compounds are broken down by CAZymes. Most of the pectin-degrading CAZymes found in gut habitat are from various families, such as esterases (CE1, CE4, CE6, CE8, CE12 and CE20); hydrolases (GH1, GH2, GH3, GH5, GH10, GH28, GH30, GH35, GH42, GH43, GH51, GH53, GH78, GH93, GH105, GH106, GH127, GH137, GH138, GH139, GH140, GH141, GH142, GH143, GH147 and GH165) and lyases (PL1, PL9, PL10 and PL11) (Yüksel et al., 2024). Additionally, CAZymes involved in both complex (such as cellulose, glycan, starch and glycogen) and simple (such as fructose, galactose and glucose) carbohydrate breakdown in the gut using CAZymes from GH113, GH13, GH25, GH26, GH36, GH53, GH72, GH73, GH77 and GH94 families (Bhattacharya et al., 2015). Overall, the Bacteroides genera demonstrated a wide range of carbohydrate metabolizing abilities. The genus, Capnocytophaga from Bacteroidota phylum, produces GH6 glucanase/xylanase and GH164 mannanase participating in plant polysaccharide hydrolysis. Another genus, Odoribacter possesses GH159 galactofuranosidase GH159 associated with the levan/inulin binding module, CBM66 and galactose binding module, CBM62 involved in the hydrolysis of plant and animal glycan, respectively. Similarly, the genus Parabacteroides produces α-mannanase from GH92 family and glucosidase/xylosidase from GH116 family acting on plant polysaccharides and a chondroitin lyase of family PL35 acting on animal glycans (chondroitin). The starch binding module, CBM25 as well as xylan-binding modules, CBM13 and CBM9, have been reported from this microbiota in the gut system (Bhattacharya et al., 2015). Another genus, Paraprevotella of phylum Bacteroidota, produces enzymes (GH43, GH51, GH115, GH141, GH146, PL11 and PL26), hydrolysing plant polysaccharides (Yüksel et al., 2024). The enzymes that act on both plant and animal glycans are from GH30 and GH98, whereas GH99 is specific for only animal glycans and chitin with chitin-binding domain, CBM12 and levan/inulin-binding module, CBM66. In addition, the Firmicutes phylum displays a high level of phylogenetic and CAZyme diversity. Cellulosilyticum (genus) has the highest number of CAZymes, including the families (GH9, GH10, GH12, GH26, GH74, PL9 and GH5) for plant glycan, GH113 for fungal polysaccharide degradation and CBM25 for starch binding (Fernandez-Julia et al., 2022). Clostridium (genus), belonging to the same phylum, also produces GH14 that interacts with plant and animal glycans, whereas Leuconostoc (genus) GH81 interacts with plant polysaccharides. Eubacterium (genus), on the other hand, possesses GH70, which is vital in dextran consumption and can modulate biofilm growth. The genus, Enterococcus, contains glucosaminidases from GH73 family for utilising peptidoglycans and chitinases from GH18 family along with its carbohydrate-binding module, CBM12 for chitin and GH126 amylase for starch and glycogen. The genus Roseburia possesses CAZymes that are required to hydrolyse chitobiose/cellobiose (GH94) and pectin (GH53), as well as rhamnose (CBM67) and starch/glycogen/amylose (CBM42). CAZyme for rhamnogalacturonan (GH140), chondroitin (GH84), fungal polysaccharides (GH55) and other β-glucosides (GH1) is abundant in Coprobacillus (genus). In addition to possessing mannan-degrading CAZyme (GH125), Lactobacillus (genus) may manufacture fructan from sucrose (GH68) and glycosaminoglycans (GT14). Ruminococcus (genus), a member of the Firmicutes phylum, contains GH11, GH44 and GH48 enzymes for hydrolysis of plant and animal polysaccharides and GH1 associated with CBM4 for chitin/cellulose/xylan (Onyango et al., 2021).

4.6 CONCLUSION

Various sources of CAZymes are pivotal in advancing our understanding of microbial ecology and biotechnology. These enzymes play critical roles in carbohydrate degradation, influencing nutrient cycling in ecosystems and offering vast potential in industrial applications. Their diverse mechanisms and specificity make them invaluable for various bioprocessing tasks, such as biomass conversion and waste treatment. As the demand for sustainable solutions increases, the exploration of CAZymes becomes more pressing, necessitating innovative approaches to enzyme discovery and characterization. The availability of high-quality genomic data from diverse bacterial sources, such as those deposited in the European Nucleotide Archive, facilitates deeper understanding of CAZyme functionality and relationships within microbial communities. Future research on metagenomic techniques to uncover novel CAZymes from underexplored communities, particularly in extreme environments, have become of great importance. Additionally, integrating systems biology and synthetic biology would enhance the ability to optimize these enzymes for industrial purposes. A multidisciplinary approach will not only expand our understanding of CAZymes but also provide sustainable alternatives to traditional chemical processes, ultimately leading to a more environmentally friendly future.

ACKNOWLEDGEMENTS

The authors thank the Department of Biosciences and Biotechnology,IIT Guwahati, for providing the computational infrastructure and central library facilities forprocuring references and plagiarism analysis (Turnitin Plagiarism Detection Software).

AUTHOR CONTRIBUTION STATEMENT

Shreya Biswas: Writing, original draft preparation,reviewing, and editing. Arun Goyal: Reviewing and editing.

FUNDING

This research did not receive any specific grant from funding agencies in the public, commercial, or not-for-profit sectors.

CONFLICT OF INTEREST STATEMENT

The authors state that the research was conducted without any commercial or financial relationships that could be construed as potential conflicts of interest.

REFERENCES

Ameri, R., García, J. L., Derenfed, A. B., Pradel, N., Neifar, S., Mhiri, S., Mezghanni, M., Jaouadi, N. Z., Barriuso, J., & Bejar, S. (2022). Genome sequence and Carbohydrate Active Enzymes (CAZymes) repertoire of the thermophilic *Caldicoprobacter algeriensis* TH7C1T. *Microbial Cell Factories*, *21*(1), 91. https://doi.org/10.1186/s12934-022-01818-0

Armendáriz-Ruiz, M., Rodríguez-González, J. A., Camacho-Ruíz, R. M., & Mateos-Díaz, J. C. (2018). Carbohydrate esterases: An overview. In G. Sandoval (Ed.), *Lipases and Phospholipases: Methods and Protocols* (pp. 39–68). Springer. https://doi.org/10.1007/978-1-4939-8672-9_2

Ausland, C., Zheng, J., Yi, H., Yang, B., Li, T., Feng, X., Zheng, B., & Yin, Y. (2021). dbCAN-PUL: A database of experimentally characterized CAZyme gene clusters and their substrates. *Nucleic Acids Research*, *49*(D1), D523–D528. https://doi.org/10.1093/nar/gkaa742

Bao, K., Yang, M., Sun, Q., Zhang, K., & Huang, H. (2024). Genome analysis of a potential novel vibrio species secreting pH- and thermo-stable alginate lyase and its application in producing alginate oligosaccharides. *Marine Drugs*, *22*(9), 414. https://doi.org/10.3390/md22090414

Barrett, K., Hunt, C. J., Lange, L., & Meyer, A. S. (2020). Conserved unique peptide patterns (CUPP) online platform: Peptide-based functional annotation of carbohydrate active enzymes. *Nucleic Acids Research*, *48*(W1), W110–W115. https://doi.org/10.1093/nar/gkaa375

Barrett, K., Jensen, K., Meyer, A. S., Frisvad, J. C., & Lange, L. (2020). Fungal secretome profile categorization of CAZymes by function and family corresponds to fungal phylogeny and taxonomy: Example Aspergillus and penicillium. *Scientific Reports*, *10*(1), 5158. https://doi.org/10.1038/s41598-020-61907-1

Bäumgen, M., Dutschei, T., & Bornscheuer, U. T. (2021). Marine polysaccharides: Occurrence, enzymatic degradation and utilization. *ChemBioChem*, *22*(13), 2247–2256. https://doi.org/10.1002/cbic.202100078

Bhattacharya, T., Ghosh, T. S., & Mande, S. S. (2015). Global profiling of Carbohydrate active enzymes in human gut microbiome. *PLOS ONE*, *10*(11), e0142038. https://doi.org/10.1371/journal.pone.0142038

Blanco-Cabra, N., Paetzold, B., Ferrar, T., Mazzolini, R., Torrents, E., Serrano, L., & LLuch-Senar, M. (2020). Characterization of different alginate lyases for dissolving *Pseudomonas aeruginosa* biofilms. *Scientific Reports*, *10*(1), 9390. https://doi.org/10.1038/s41598-020-66293-2

Bohra, V., Dafale, N. A., & Purohit, H. J. (2019). Understanding the alteration in rumen microbiome and CAZymes profile with diet and host through comparative metagenomic approach. *Archives of Microbiology*, *201*(10), 1385–1397. https://doi.org/10.1007/s00203-019-01706-z

Büller, H. A., Montgomery, R. K., Sasak, W. V., & Grand, R. J. (1987). Biosynthesis, glycosylation, and intracellular transport of intestinal lactase-phlorizin hydrolase in rat. *Journal of Biological Chemistry*, *262*(35), 17206–17211. https://doi.org/10.1016/S0021-9258(18)45511-1

Cabral, L., Persinoti, G. F., Paixão, D. A. A., Martins, M. P., Morais, M. A. B., Chinaglia, M., Domingues, M. N., Sforca, M. L., Pirolla, R. A. S., Generoso, W. C., Santos, C. A., Maciel, L. F., Terrapon, N., Lombard, V., Henrissat, B., & Murakami, M. T. (2022). Gut microbiome of the largest living rodent harbors unprecedented enzymatic systems to degrade plant polysaccharides. *Nature Communications*, *13*(1), 629. https://doi.org/10.1038/s41467-022-28310-y

Cabrera, M. Á., & Blamey, J. M. (2018). Biotechnological applications of archaeal enzymes from extreme environments. *Biological Research*, *51*(1), 37. https://doi.org/10.1186/s40659-018-0186-3

Cantarel, B. L., Coutinho, P. M., Rancurel, C., Bernard, T., Lombard, V., & Henrissat, B. (2009). The Carbohydrate-Active EnZymes database (CAZy): An expert resource for glycogenomics. *Nucleic Acids Research*, *37*(suppl_1), D233–D238. https://doi.org/10.1093/nar/gkn663

Cardoso, V., Brás, J. L. A., Costa, I. F., Ferreira, L. M. A., Gama, L. T., Vincentelli, R., Henrissat, B., & Fontes, C. M. G. A. (2022). Generation of a library of carbohydrate-active enzymes for plant biomass deconstruction. *International Journal of Molecular Sciences*, *23*(7), Article 7. https://doi.org/10.3390/ijms23074024

Chakraborty, S., Rani, A., Dhillon, A., & Goyal, A. (2017). Polysaccharide lyases. In A. Pandey, S. Negi, & C. R. Soccol (Eds.), *Current Developments in Biotechnology and Bioengineering* (pp. 527–539). Elsevier. https://doi.org/10.1016/B978-0-444-63662-1.00023-3

Chettri, D., Nad, S., Konar, U., & Verma, A. K. (2022). CAZyme from gut microbiome for efficient lignocellulose degradation and biofuel production. *Frontiers in Chemical Engineering*, *4*. https://doi.org/10.3389/fceng.2022.1054242

Chettri, D., & Verma, A. K. (2023). Biological significance of carbohydrate active enzymes and searching their inhibitors for therapeutic applications. *Carbohydrate Research*, *529*, 108853. https://doi.org/10.1016/j.carres.2023.108853

Contesini, F. J., Frandsen, R. J. N., & Damasio, A. (2021). Editorial: CAZymes in biorefinery: From genes to application. *Frontiers in Bioengineering and Biotechnology*, *9*. https://doi.org/10.3389/fbioe.2021.622817

Daïnou, O., Cariou, M. L., David, J. R., & Hickey, D. (1987). Amylase gene duplication: An ancestral trait in the *Drosophila melanogaster* species subgroup. *Heredity*, *59*(2), 245–251. https://doi.org/10.1038/hdy.1987.119

Dixit, S., Gaur, M., Subudhi, E., Sahoo, R. K., Dey, S., Mahapatra, L. D., Mandal, S. D., Senthil Kumar, N., & Anirudh, H. (2021). Bacterial diversity and CAZyme potential revealed in pandanus rich thermal spring cluster of India: A non-cultivable 16S rRNA sequencing approach. *Frontiers in Microbiology*, *12*. https://doi.org/10.3389/fmicb.2021.760573

Ducarmon, Q. R., Karcher, N., Tytgat, H. L. P., Delannoy-Bruno, O., Pekel, S., Springer, F., Schudoma, C., & Zeller, G. (2024). Large-scale computational analyses of gut microbial CAZyme repertoires enabled by Cayman (p. 2024.01.08.574624). *bioRxiv*. https://doi.org/10.1101/2024.01.08.574624

Ekstrom, A., Taujale, R., McGinn, N., & Yin, Y. (2014). PlantCAZyme: A database for plant carbohydrate-active enzymes. *Database*, *2014*, bau079. https://doi.org/10.1093/database/bau079

Facimoto, C. T., Clements, K. D., White, W. L., & Handley, K. M. (2023). Bacteroidia and Clostridia genomes collectively encode for a progressive cascade of marine polysaccharide degradation along the hindgut of the herbivorous fish *Kyphosus sydneyanus* (p. 2023.12.23.570891). *bioRxiv*. https://doi.org/10.1101/2023.12.23.570891

Fernandez-Julia, P., Commane, D. M., van Sinderen, D., & Munoz-Munoz, J. (2022). Cross-feeding interactions between human gut commensals belonging to the Bacteroides and Bifidobacterium genera when grown on dietary glycans. *Microbiome Research Reports*, *1*(2), 12. https://doi.org/10.20517/mrr.2021.05

Ferraboschi, P., Ciceri, S., & Grisenti, P. (2021). Applications of lysozyme, an innate immune defense factor, as an alternative antibiotic. *Antibiotics*, *10*(12),1534. https://doi.org/10.3390/antibiotics10121534

Forsgård, R. A. (2019). Lactose digestion in humans: Intestinal lactase appears to be constitutive whereas the colonic microbiome is adaptable. *The American Journal of Clinical Nutrition*, *110*(2), 273–279. https://doi.org/10.1093/ajcn/nqz104

Fukui, Y., Fukui, K., & Moriyama, T. (1973). Source of neuraminidase in human whole saliva. *Infection and Immunity*, *8*(3), 329–334. https://doi.org/10.1128/iai.8.3.329-334.1973

Gabriel, R., Mueller, R., Floerl, L., Hopson, C., Harth, S., Schuerg, T., Fleissner, A., & Singer, S. W. (2021). CAZymes from the thermophilic fungus *Thermoascus aurantiacus* are induced by C5 and C6 sugars. *Biotechnology for Biofuels*, *14*(1), 169. https://doi.org/10.1186/s13068-021-02018-5

Gavande, P. V., Goyal, A., & Fontes, C. M. G. A. (2023). Chapter 1 – Carbohydrates and Carbohydrate-Active enZymes (CAZyme): An overview. In A. Goyal & K. Sharma (Eds.), *Glycoside Hydrolases* (pp. 1–23). Academic Press. https://doi.org/10.1016/B978-0-323-91805-3.00012-5

Ghazanfar, M., Shakir, H. A., Khan, M., Ali, S., Hidayatullah, I. M., Franco, M., Alessa, A. H., Dar, M. A., & Irfan, M. (2025). Chapter 10 – Exploiting the gut bacteria for the production of biofuels from lignocellulose. In D. Zhu, M. A. Dar, & M. Shahnawaz (Eds.), *Biofuels and Sustainability* (pp. 157–172). Elsevier Science Ltd. https://doi.org/10.1016/B978-0-443-21433-2.00022-0

Hamid, R., Khan, M. A., Ahmad, M., Ahmad, M. M., Abdin, M. Z., Musarrat, J., & Javed, S. (2013). Chitinases: An update. *Journal of Pharmacy & Bioallied Sciences*, *5*(1), 21–29. https://doi.org/10.4103/0975-7406.106559

Harduin-Lepers, A., Mollicone, R., Delannoy, P., & Oriol, R. (2005). The animal sialyltransferases and sialyltransferase-related genes: A phylogenetic approach. *Glycobiology*, *15*(8), 805–817. https://doi.org/10.1093/glycob/cwi063

Harduin-Lepers, A., Vallejo-Ruiz, V., Krzewinski-Recchi, M.-A., Samyn-Petit, B., Julien, S., & Delannoy, P. (2001). The human sialyltransferase family. *Biochimie*, *83*(8), 727–737. https://doi.org/10.1016/S0300-9084(01)01301-3

Hehemann, J.-H., Boraston, A. B., & Czjzek, M. (2014). A sweet new wave: Structures and mechanisms of enzymes that digest polysaccharides from marine algae. *Current Opinion in Structural Biology*, *28*, 77–86. https://doi.org/10.1016/j.sbi.2014.07.009

Holesh, J. E., Aslam, S., & Martin, A. (2025). Physiology, carbohydrates. In *StatPearls*. StatPearls Publishing. www.ncbi.nlm.nih.gov/books/NBK459280/

Huang, L., Zhang, H., Wu, P., Entwistle, S., Li, X., Yohe, T., Yi, H., Yang, Z., & Yin, Y. (2018). dbCAN-seq: A database of carbohydrate-active enzyme (CAZyme) sequence and annotation. *Nucleic Acids Research*, 46(Database issue), D516–D521. https://doi.org/10.1093/nar/gkx894

Kameshwar, A. K. S., Ramos, L. P., & Qin, W. (2019). CAZymes-based ranking of fungi (CBRF): An interactive web database for identifying fungi with extrinsic plant biomass degrading abilities. *Bioresources and Bioprocessing*, *6*(1), 51. https://doi.org/10.1186/s40643-019-0286-0

Keller, P., Zwicker, E., Mantei, N., & Semenza, G. (1992). The levels of lactase and of sucrase-isomaltase along the rabbit small intestine are regulated both at the mRNA level and post-translationally. *FEBS Letters*, *313*(3), 265–269. https://doi.org/10.1016/0014-5793(92)81206-2

Kelly, S. M., Munoz-Munoz, J., & van Sinderen, D. (2021). Plant glycan metabolism by bifidobacteria. *Frontiers in Microbiology*, *12*. https://doi.org/10.3389/fmicb.2021.609418

Koga, D., Sasaki, Y., Uchiumi, Y., Hirai, N., Arakane, Y., & Nagamatsu, Y. (1997). Purification and characterization of *Bombyx mori* chitinases. *Insect Biochemistry and Molecular Biology*, *27*(8), 757–767. https://doi.org/10.1016/S0965-1748(97)00058-1

Kurokawa, T., Suzuki, T., Ohta, H., Kagawa, H., Tanaka, H., & Unuma, T. (2002). Expression of pancreatic enzyme genes during the early larval stage of Japanese eel *Anguilla japonica*. *Fisheries Science*, *68*(4), 736–744. https://doi.org/10.1046/j.1444-2906.2002.00487.x

Lairson, L. L., Henrissat, B., Davies, G. J., & Withers, S. G. (2008). Glycosyltransferases: Structures, functions, and mechanisms. *Annual Review of Biochemistry*, *77*, 521–555. https://doi.org/10.1146/annurev.biochem.76.061005.092322

Lapébie, P., Lombard, V., Drula, E., Terrapon, N., & Henrissat, B. (2019). Bacteroidetes use thousands of enzyme combinations to break down glycans. *Nature Communications*, *10*(1), 2043. https://doi.org/10.1038/s41467-019-10068-5

Levasseur, A., Drula, E., Lombard, V., Coutinho, P. M., & Henrissat, B. (2013). Expansion of the enzymatic repertoire of the CAZy database to integrate auxiliary redox enzymes. *Biotechnology for Biofuels*, *6*(1), 41. https://doi.org/10.1186/1754-6834-6-41

Liew, K. J., Liang, C. H., Lau, Y. T., Yaakop, A. S., Chan, K.-G., Shahar, S., Shamsir, M. S., & Goh, K. M. (2022). Thermophiles and carbohydrate-active enzymes (CAZymes) in biofilm microbial consortia that decompose lignocellulosic plant litters at high temperatures. *Scientific Reports*, *12*(1), 2850. https://doi.org/10.1038/s41598-022-06943-9

Linhardt, R. J., Galliher, P. M., & Cooney, C. L. (1987). Polysaccharide lyases. *Applied Biochemistry and Biotechnology*, *12*(2), 135–176. https://doi.org/10.1007/BF02798420

Lombard, V., Golaconda Ramulu, H., Drula, E., Coutinho, P. M., & Henrissat, B. (2014). The carbohydrate-active enzymes database (CAZy) in 2013. *Nucleic Acids Research*, *42*(D1), D490–D495. https://doi.org/10.1093/nar/gkt1178

Long, L., Hu, Y., Sun, F., Gao, W., Hao, Z., & Yin, H. (2022). Advances in lytic polysaccharide monooxygenases with the cellulose-degrading auxiliary activity family 9 to facilitate cellulose degradation for biorefinery. *International Journal of Biological Macromolecules*, *219*, 68–83. https://doi.org/10.1016/j.ijbiomac.2022.07.240

López-Sánchez, R., Rebollar, E. A., Gutiérrez-Ríos, R. M., Garciarrubio, A., Juarez, K., & Segovia, L. (2024). Metagenomic analysis of carbohydrate-active enzymes and their contribution to marine sediment biodiversity. *World Journal of Microbiology and Biotechnology*, *40*(3), 95. https://doi.org/10.1007/s11274-024-03884-5

Malik, P. K., Trivedi, S., Kolte, A. P., Mohapatra, A., Biswas, S., Bhattar, A. V. K., Bhatta, R., & Rahman, H. (2024). Comparative rumen metagenome and cazyme profiles in cattle and buffaloes: Implications for methane yield and rumen fermentation on a common diet. *Microorganisms*, *12*(1), 47. https://doi.org/10.3390/microorganisms12010047

Menendez, E., Garcia-Fraile, P., Rivas, R., Menendez, E., Garcia-Fraile, P., & Rivas, R. (2015). Biotechnological applications of bacterial cellulases. *AIMS Bioengineering*, *2*(3), 163–182. https://doi.org/10.3934/bioeng.2015.3.163

Naim, H. Y. (1991). Expression of a full-length cDNA coding for human intestinal lactase-phlorizin hydrolase reveals an uncleaved, enzymatically active, and transport-competent protein. *Journal of Biological Chemistry, 266*(19), 12313–12320. https://doi.org/10.1016/S0021-9258(18)98898-8

Onyango, S. O., Juma, J., De Paepe, K., & Van de Wiele, T. (2021). Oral and gut microbial carbohydrate-active enzymes landscape in health and disease. *Frontiers in Microbiology, 12*. https://doi.org/10.3389/fmicb.2021.653448

Palcic, M. M. (2011). Glycosyltransferases as biocatalysts. *Current Opinion in Chemical Biology, 15*(2), 226–233. https://doi.org/10.1016/j.cbpa.2010.11.022

Palevich, N., Kelly, W. J., Leahy, S. C., Denman, S., Altermann, E., Rakonjac, J., & Attwood, G. T. (2019). Comparative genomics of rumen *Butyrivibrio* spp. uncovers a continuum of polysaccharide-degrading capabilities. *Applied and Environmental Microbiology, 86*(1), e01993–19. https://doi.org/10.1128/AEM.01993-19

Park, B. H., Karpinets, T. V., Syed, M. H., Leuze, M. R., & Uberbacher, E. C. (2010). CAZymes Analysis Toolkit (CAT): Web service for searching and analyzing carbohydrate-active enzymes in a newly sequenced organism using CAZy database. *Glycobiology, 20*(12), 1574–1584. https://doi.org/10.1093/glycob/cwq106

Park, S. , Lee, B., & Park, K. (2017). Extremophilic carbohydrate active enzymes (CAZymes). *Journal of Nutritional Health & Food Engineering, 7*(1), 230–237. https://doi.org/10.15406/jnhfe.2017.07.00230

Paudel, L., Pardhe, B. D., Han, S.-R., Lee, J. H., & Oh, T.-J. (2023). Identification and evaluation of CAZyme genes, along with functional characterization of a new GH46 chitosanase from *Streptomyces* sp. *KCCM12257. International Journal of Biological Macromolecules, 253*, 127457. https://doi.org/10.1016/j.ijbiomac.2023.127457

Peng, X., Wilken, S. E., Lankiewicz, T. S., Gilmore, S. P., Brown, J. L., Henske, J. K., Swift, C. L., Salamov, A., Barry, K., Grigoriev, I. V., Theodorou, M. K., Valentine, D. L., & O'Malley, M. A. (2021). Genomic and functional analyses of fungal and bacterial consortia that enable lignocellulose breakdown in goat gut microbiomes. *Nature Microbiology, 6*(4), 499–511. https://doi.org/10.1038/s41564-020-00861-0

Qi, Y.-L., Chen, Y.-T., Xie, Y.-G., Li, Y.-X., Rao, Y.-Z., Li, M.-M., Xie, Q.-J., Cao, X.-R., Chen, L., Qu, Y.-N., Yuan, Z.-X., Xiao, Z.-C., Lu, L., Jiao, J.-Y., Shu, W.-S., Li, W.-J., Hedlund, B. P., & Hua, Z.-S. (2024). Analysis of nearly 3000 archaeal genomes from terrestrial geothermal springs sheds light on interconnected biogeochemical processes. *Nature Communications, 15*(1), 4066. https://doi.org/10.1038/s41467-024-48498-5

Rabapane, K. J., & Matambo, T. S. (2024). Profiling the dynamic adaptations of CAZyme-Producing microorganisms in the gastrointestinal tract of South African goats. *Heliyon, 10*(17). https://doi.org/10.1016/j.heliyon.2024.e37508

Reichart, N. J., Bowers, R. M., Woyke, T., & Hatzenpichler, R. (2021). High potential for biomass-degrading enzymes revealed by hot spring metagenomics. *Frontiers in Microbiology, 12*. https://doi.org/10.3389/fmicb.2021.668238

Rings, E. H. H. M., van Beers, E. H., Krasinski, S. D., Verhave, M., Montgomery, R. K., Grand, R. J., Dekker, J., & Büller, H. A. (1994). Lactase; Origin, gene expression, localization, and function. *Nutrition Research, 14*(5), 775–797. https://doi.org/10.1016/S0271-5317(05)80212-X

Rodriguez, C. I., & Martiny, J. B. H. (2020). Evolutionary relationships among bifidobacteria and their hosts and environments. *BMC Genomics, 21*(1), 26. https://doi.org/10.1186/s12864-019-6435-1

Rydberg, E. H., Sidhu, G., Vo, H. C., Hewitt, J., Ct, H. C. F., Wang, Y., Numao, S., Macgillivray, R. T. A., Overall, C. M., Brayer, G. D., & Withers, S. G. (1999). Cloning, mutagenesis, and structural analysis of human pancreatic α-amylase expressed in *pichia pastoris*. *Protein Science, 8*(3), 635–643. https://doi.org/10.1110/ps.8.3.635

Sawabe, T., Ohtsuka, M., & Ezura, Y. (1997). Novel alginate lyases from marine bacterium *Alteromonas* sp. Strain H-4. *Carbohydrate Research, 304*(1), 69–76. https://doi.org/10.1016/S0008-6215(97)00194-8

Shen, J., Zheng, L., Chen, X., Han, X., Cao, Y., & Yao, J. (2020). Metagenomic analyses of microbial and carbohydrate-active enzymes in the rumen of dairy goats fed different rumen degradable starch. *Frontiers in Microbiology, 11*. https://doi.org/10.3389/fmicb.2020.01003

Sime, A. M., Kifle, B. A., Woldesemayat, A. A., & Gemeda, M. T. (2024). Microbial carbohydrate active enzyme (CAZyme) genes and diversity from Menagesha Suba natural forest soils of Ethiopia as revealed by shotgun metagenomic sequencing. *BMC Microbiology, 24*(1), 285. https://doi.org/10.1186/s12866-024-03436-9

Strazzulli, A., Cobucci-Ponzano, B., Iacono, R., Giglio, R., Maurelli, L., Curci, N., Schiano-di-Cola, C., Santangelo, A., Contursi, P., Lombard, V., Henrissat, B., Lauro, F. M., Fontes, C. M. G. A., & Moracci, M. (2020). Discovery of hyperstable carbohydrate-active enzymes through metagenomics of extreme environments. *The FEBS Journal, 287*(6), 1116–1137. https://doi.org/10.1111/febs.15080

Tashkandi, M., & Baz, L. (2023). Function of CAZymes encoded by highly abundant genes in rhizosphere microbiome of *Moringa oleifera*. *Saudi Journal of Biological Sciences, 30*(3), 103578. https://doi.org/10.1016/j.sjbs.2023.103578

Thakur, A., Sharma, K., Mutreja, R., & Goyal, A. (2021). Thermostable Enzymes from Clostridium thermocellum. In H. Thatoi, S. Mohapatra, & S. K. Das (Eds.), *Bioprospecting of Enzymes in Industry, Healthcare and Sustainable Environment* (pp. 251–267). Springer. https://doi.org/10.1007/978-981-33-4195-1_12

The CAZypedia Consortium. (2018). Ten years of CAZypedia: A living encyclopedia of carbohydrate-active enzymes. *Glycobiology, 28*(1), 3–8. https://doi.org/10.1093/glycob/cwx089

Tulsani, N. J., Jakhesara, S. J., Hinsu, A. T., Jyotsana, B., Dafale, N. A., Patil, N. V., Purohit, H. J., & Joshi, C. G. (2022). Genome analysis and CAZy repertoire of a novel fungus *Aspergillus sydowii* C6d with lignocellulolytic ability isolated from camel rumen. *Electronic Journal of Biotechnology, 59*, 36–45. https://doi.org/10.1016/j.ejbt.2022.06.004

Vlodavsky, I., & Friedmann, Y. (2001). Molecular properties and involvement of heparanase in cancer metastasis and angiogenesis. *The Journal of Clinical Investigation, 108*(3), 341–347. https://doi.org/10.1172/JCI13662

Wang, L., Zhang, G., Xu, H., Xin, H., & Zhang, Y. (2019). Metagenomic analyses of microbial and Carbohydrate-Active Enzymes in the rumen of holstein cows fed different forage-to-concentrate ratios. *Frontiers in Microbiology, 10*. https://doi.org/10.3389/fmicb.2019.00649

Wang, X., Chi, Y., & Song, S. (2024). Important soil microbiota's effects on plants and soils: A comprehensive 30-year systematic literature review. *Frontiers in Microbiology, 15*. https://doi.org/10.3389/fmicb.2024.1347745

Wardman, J. F., Bains, R. K., Rahfeld, P., & Withers, S. G. (2022). Carbohydrate-active enzymes (CAZymes) in the gut microbiome. *Nature Reviews Microbiology, 20*(9), 542–556. https://doi.org/10.1038/s41579-022-00712-1

Wardman, J. F., & Withers, S. G. (2024). Carbohydrate-active enzyme (CAZyme) discovery and engineering via (Ultra)high-throughput screening. *RSC Chemical Biology, 5*(7), 595–616. https://doi.org/10.1039/D4CB00024B

Warren, L., & Spearing, C. W. (1960). Mammalian sialidase (neuraminidase). *Biochemical and Biophysical Research Communications, 3*(5), 489–492. https://doi.org/10.1016/0006-291X(60)90161-3

Wolter, L. A., Mitulla, M., Kalem, J., Daniel, R., Simon, M., & Wietz, M. (2021). CAZymes in *Maribacter dokdonensis* 62–1 from the patagonian shelf: Genomics and physiology compared to related flavobacteria and a co-occurring *alteromonas* strain. *Frontiers in Microbiology*, *12*. https://doi.org/10.3389/fmicb.2021.628055

Wu, D. G., Harris, C. R., Kalis, K. M., Bowen, M., Biddle, J. F., & Farag, I. F. (2025). Comparative metagenomics of tropical reef fishes show conserved core gut functions across hosts and diets with diet-related functional gene enrichments. *Applied and Environmental Microbiology*, *91*(2), e02229–24. https://doi.org/10.1128/aem.02229-24

Xu, Q., Resch, M. G., Podkaminer, K., Yang, S., Baker, J. O., Donohoe, B. S., Wilson, C., Klingeman, D. M., Olson, D. G., Decker, S. R., Giannone, R. J., Hettich, R. L., Brown, S. D., Lynd, L. R., Bayer, E. A., Himmel, M. E., & Bomble, Y. J. (2016). Dramatic performance of Clostridium thermocellum explained by its wide range of cellulase modalities. *Science Advances*, *2*(2), e1501254. https://doi.org/10.1126/sciadv.1501254

Yadav, V., Biswas, S., & Goyal, A. (2024). Enzymes of industrial significance and their applications. In P. Verma (Ed.), *Industrial Microbiology and Biotechnology: An Insight into Current Trends* (pp. 277–307). Springer Nature. https://doi.org/10.1007/978-981-97-1912-9_11

Ye, L., Das, P., Li, P., Ji, B., & Nielsen, J. (2019). Carbohydrate active enzymes are affected by diet transition from milk to solid food in infant gut microbiota. *FEMS Microbiology Ecology*, *95*(11), fiz159. https://doi.org/10.1093/femsec/fiz159

Yu, Z., Zhang, W., He, H., Li, Y., Xie, Z., Sailike, Ah., Hao, H., Tian, X., Sun, L., Liang, Y., Fu, R., & Yang, P. (2024). The CAZyme family regulates the changes in soil organic carbon composition during vegetation restoration in the Mu Us desert. *Geoderma*, *452*, 117109. https://doi.org/10.1016/j.geoderma.2024.117109

Yüksel, E., Voragen, A. G. J., & Kort, R. (2024). The pectin metabolizing capacity of the human gut microbiota. *Critical Reviews in Food Science and Nutrition*, *65*(25), 4823–4845. https://doi.org/10.1080/10408398.2024.2400235

Zhang, K., He, C., Wang, L., Suo, L., Guo, M., Guo, J., Zhang, T., Xu, Y., Lei, Y., Liu, G., Qian, Q., Mao, Y., Kalds, P., Wu, Y., Cuoji, A., Yang, Y., Brugger, D., Gan, S., Wang, M., … Chen, Y. (2024). Compendium of 5810 genomes of sheep and goat gut microbiomes provides new insights into the glycan and mucin utilization. *Microbiome*, *12*(1), 104. https://doi.org/10.1186/s40168-024-01806-z

Zhang, L., Prabhakar, P. K., Bharadwaj, V. S., Bomble, Y. J., Peña, M. J., & Urbanowicz, B. R. (2023). Glycosyltransferase family 47 (GT47) proteins in plants and animals. *Essays in Biochemistry*, *67*(3), 639–652. https://doi.org/10.1042/EBC20220152

Zhang, Q., Xu, X., Duan, J., Koide, R. T., Xu, L., & Chu, J. (2023). Variation in microbial CAZyme families across degradation severity in a steppe grassland in northern China. *Frontiers in Environmental Science*, *11*. https://doi.org/10.3389/fenvs.2023.1080505

Zhang, T., Liu, H., Lv, B., & Li, C. (2020). Regulating Strategies for producing carbohydrate active enzymes by filamentous fungal cell factories. *Frontiers in Bioengineering and Biotechnology*, *8*. https://doi.org/10.3389/fbioe.2020.00691

Zhang, X., Huang, Z., Zhong, Z., Li, Q., Bian, F., & Yang, C. (2022). Metagenomic insights into the characteristics of soil microbial communities in the decomposing biomass of Moso bamboo forests under different management practices. *Frontiers in Microbiology*, *13*. https://doi.org/10.3389/fmicb.2022.1051721

Zhao, Z., Liu, H., Wang, C., & Xu, J.-R. (2014). Erratum to: Comparative analysis of fungal genomes reveals different plant cell wall degrading capacity in fungi. *BMC Genomics*, *15*(1), 6. https://doi.org/10.1186/1471-2164-15-6

Zheng, J., Hu, B., Zhang, X., Ge, Q., Yan, Y., Akresi, J., Piyush, V., Huang, L., & Yin, Y. (2023). dbCAN-seq update: CAZyme gene clusters and substrates in microbiomes. *Nucleic Acids Research*, *51*(D1), D557–D563. https://doi.org/10.1093/nar/gkac1068

Zhou, C., Ma, Q., Mao, X., Liu, B., Yin, Y., & Xu, Y. (2014). New insights into clostridia through comparative analyses of their 40 genomes. *BioEnergy Research*, *7*(4), 1481–1492. https://doi.org/10.1007/s12155-014-9486-9

Zhu, X., Zhang, H., Fukamizo, T., Muthukrishnan, S., & Kramer, K. J. (2001). Properties of *Manduca sexta* chitinase and its C-terminal deletions. *Insect Biochemistry and Molecular Biology*, *31*(12), 1221–1230. https://doi.org/10.1016/S0965-1748(01)00070-4

5 Integrating Carbohydrate-Active Enzymes in Biorefineries

Pioneering Sustainable Development through Circular Economy

Satwika Das, Anbunithi K., Rajendran Omana Rajesh, Shahin Mohammed S. K., and Ashish A. Prabhu

ABBREVIATIONS

AA	Auxiliary activities
CAGR	Compound annual growth rate
Cas9	CRISPR-associated protein 9
CAZymes	Carbohydrate-active enzymes
CBM	Carbohydrate-binding module
CE	Carbohydrate esterases
CO_2	Carbon dioxide
CRISPR	Clustered regularly interspaced short palindromic repeats
GH	Glycoside hydrolases
GHG	Greenhouse gases
GT	Glycosyl transferases
IEA	International Energy Agency
LCB	Lignocellulosic biomass
NDRC	National Development and Reform Commission
OSW	Onion skin waste
PCW	Plant cell wall
PHB	Polyhydroxybutyrate
PL	Polysaccharide lyases
SDG	Sustainable development goals
TSMNP–Zn	Terpyridine-immobilized silica-coated magnetic nanoparticles–zinc
UNO	United Nations Organization
USD	US dollar

DOI: 10.1201/9781003561682-5

5.1 INTRODUCTION

Global change towards sustainability has developed a huge interest in biorefineries via the conversion of biomass into value-added products. They have played a significant role in shifting towards a circular economy, a view based on the efficient utilization of resources and waste minimization. Biorefineries can be defined as integrated systems where biomass is converted into a range of bio-based products. The ultimate aim of biorefineries is to utilize renewable resources effectively and minimize the environmental drawbacks. Biorefineries can be categorized based on product focus, energy-driven or material-driven, and type of feedstock.

The circular economy goal is to eliminate waste by using resources continuously, in contrast to the more traditional linear economy, often described as "take-make-dispose." The products are designed to last, be repaired, and be recycled in the circular economy. Biorefineries align closely with principles like resource efficiency, waste valorization, and sustainable development. Recent studies highlighted the importance of biorefineries in the growth of the circular bioeconomy (Ubando, Felix, and Chen 2020). They serve as a strategic channel for switching biomass-based industries towards sustainability. Integrating biomass feedstocks, including lignocellulosic materials and algae biorefineries, can create numerous outputs with very low environmental effects (Ubando, Felix, and Chen 2020). Endorsement of biorefineries can remarkably contribute to achieving net-zero emissions by lowering the greenhouse gases corresponding to traditional fossil fuel emissions.

A lignocellulose feedstock biorefinery aims to transform lignocellulosic biomass (LCB), composed mainly of cellulose, hemicellulose, and lignin, into biofuels and biochemicals. This approach is necessary for achieving sustainable energy solutions since lignocellulosic materials are inedible, abundant, and do not raise concerns with the food vs. feed debate. The stages of product conversion commence with the lignocellulosic materials (e.g., straw, corn stover), which are then composed and refined to improve their digestibility. Further enzymatic hydrolysis is performed with the help of cellulases, hemicellulases, and other auxiliary enzymes; both cellulose and hemicellulose are broken down into fermentable sugars. These released sugars are utilized by other microbes to yield biofuels such as ethanol or other platform chemicals (Nath 2024). The final step involves the recovery of desired products with significantly low waste generation. The interest in employing lignocellulosic feedstocks lowers the dependence on fossil fuels and reduces the environmental impact. Lignocellulosic biorefineries can valorize agricultural waste streams and provide for circular economy principles by biotransformation of waste into essential resources. Along with biofuels, this biorefinery system permits the production of various other commercially important byproducts from biomass (James and Lee 1997).

Carbohydrate polymers are the most abundant photosynthesis-fixed carbon on the planet and are utilized by almost all existing life forms, whether autotrophic or heterotrophic, for the generation of energy and sustenance of life. The utilization of agricultural residues, forest and farmland wastes, and other biomass has gained popularity in the current scenario for the generation of a plethora of commercially important products like biofuels and other platform chemicals, which find application in a range of industries. Plant biomass primarily consists of cellulose, hemicellulose, and lignin, thereby often being referred to as LCB. These complex carbohydrates can be broken down into simpler components via a range of physical, chemical, and biological methods. The physico-chemical methods, like acidic and alkaline pretreatments, steam explosion, and thermal degradation, generate toxic by-products and other intermediates and enhance the cost of the process due to energy requirements. The biological methods prove to be a sustainable alternative, where microbes are utilized for the breakdown of these compounds. Fungi, bacteria, and yeast strains are known to produce a multitude of hydrolytic enzymes that aid in the degradation of carbohydrates and conjugates into simpler useful components and fermentable sugars. Either the microbes or the extracted enzymes from microbial sources are used for the hydrolysis of the biomass for the generation of easily bio-available components. The key players in the enzymatic degradation are a group of specialized enzymes, known as the carbohydrate-active enzymes (CAZymes). These enzymes catalyze the hydrolysis of polysaccharides into simple sugars, enabling their subsequent biotransformation into biofuels like bioethanol, biobutanol, biohydrogen, and biogas. Apart from carbohydrate degradation, CAZymes are also involved in the degradation of bacterial exopolysaccharides, biosynthesis of various biomolecules, and even in the glycosylation of proteins and lipids. The multiple roles of CAZymes lead to a multitude of applications in health, nutrition, biotechnology, and beyond (Garron and Henrissat 2019).

Plant biomass, rich in lignocellulosic materials, can be used as a cost-effective source for the production of these CAZymes. The effective breakdown of the complex polymers present in the plant-derived biomass not only improves sugar yields but also gives way to the overall sustainability of biorefinery systems by lowering waste (James and Lee 1997).

In brief, biorefineries represent an important innovation in the pursuit of sustainable development. In refining biomass to valuable products by fusing the principles of circular economy, they provide a way towards reducing the environmental impact and increasing resource efficiency. The incorporation of CAZymes adds further to increasing the activity of biorefineries through conversion effectiveness and product yield. Their integrated actions facilitate the effective conversion of complex biomass into fermentable sugars that act as precursors for sustainable biofuels and biochemicals. As research continues to improve these enzymatic processes and optimize biorefinery systems, the prospects for sustainable biomass application will expand substantially, and the role of biorefineries will become increasingly important in forming a sustainable future grounded in renewable resources.

5.2 CARBOHYDRATE-ACTIVE ENZYMES IN BIOMASS CONVERSION

5.2.1 Key and Emerging CAZymes

The CAZymes are functionally diverse and can be typically classified into families based on their sequence similarities and enzyme activities (Benini 2020). Based on their sequences, a total of 680000 sequences have been placed in 215 families, permanently available in the CAZy database. The CAZymes include glycoside hydrolases (GHs), glycosyl transferases (GTs), polysaccharide lyases (PLs), carbohydrate esterases (CEs), and auxiliary activities (AAs).

GHs (EC 3.2.1.-), also known as glycosidases and glycosyl hydrolases, are a group of enzymes that catalyze the glycosidic bond hydrolysis, leading to the formation of a sugar hemiacetal or hemiketal and the corresponding aglycone. GH family enzymes are either exo-acting, cleaving the substrate at the end of the chain, or endo-acting, cleaving in the middle. The exo-acting GHs most frequently cleave the substrates at the non-reducing ends. The GHs have been classified into more than 100 families. The GH families have conserved catalytic mechanisms and are often classified based on that. GHs catalyze either by retaining the anomeric configuration or by inverting it. In the case of the GH97 family, both inverting and retaining GHs are present. The inverting enzymes use glutamate as a general base, and the retaining enzymes use aspartate as a catalytic nucleophile. The GHs of the GH4, GH109, GH177, and GH179 families can hydrolyze both α- and β-glycosides, using an NAD-dependent hydrolysis mechanism (Bains et al. 2024).

GTs (EC 2.4.x.y) are involved in the biosynthesis of carbohydrates, glycosides, disaccharides, and polysaccharides. These enzymes catalyze the transfer of sugar moieties from activated donor molecules (glycosyl donors) to specific acceptor molecules (glycosyl acceptors), forming glycosidic bonds. Similar to the GHs, GT family enzymes also act via the inverting or the retaining mechanisms. Some GTs also catalyze the transfer of glycosyl groups to inorganic phosphate, water, and protein residues. Glycosyl transfer to tyrosine, serine, or threonine residues leads to the formation of O-linked glycoproteins, and transfer to asparagine residues leads to N-linked glycoproteins. In some cases, GTs might also transfer glycosyl to lipids, generating glycolipids (Yang et al. 2018). The GTs that use sugar-nucleotide donors are called Leloir enzymes, after Luis F. Leloir, the scientist who discovered sugar nucleotides. The GTs using non-nucleotide donors are known as non-Leloir glycosyltransferases. One key feature of GTs is that based on the type of GT expressed, the human ABO blood grouping is done.

PLs (EC 4.2.2.-) cleave specific activated glycosidic linkages in acidic polysaccharides. Unlike hydrolysis used by the other CAZymes, PL family enzymes use an eliminase mechanism, leading to unsaturated oligosaccharide products. The PLs are classified into 18 families. All PLs commonly identify uronic acid residues in the polysaccharides and, via β-elimination, produce unsaturated saccharides with C=C at the non-reducing terminal uronate residues. The uronate recognition and the method of catalysis suggest structural similarity between the members of the PL families. The PL11 family has been reported to show little to no structural homology to any of the other PLs. Some PLs have also been identified to cleave hexose-1,4-α-sequences as well. Generally, PLs come from a wide range of pathogenic and non-pathogenic microbial sources (Linhardt, Galliher, and Cooney 1987). The pectin lyase from plant pathogenic bacteria has been exclusively studied, and it is believed that degradation of plant cell walls (PCWs) is targeted using the PLs produced by the saprophytic or pathogenic organisms (Sutherland 1995).

CEs cleave and release alkyl or acyl groups attached by ester linkages to carbohydrates (Armendáriz-Ruiz et al. 2018). According to the CAZy database, the CEs were classified into 16 families; however, later, it was found that the CE10 family enzymes were active against non-carbohydrate substrates, thereby bringing the total number of CE families to 15. The CEs remove the acylated moieties of the polysaccharides, thereby facilitating the access of GHs, leading to successful saccharification of biomass. Several CE families, like CE1, CE4, CE7, CE11, and CE14, have also been found to contain enzyme targets for drug design and other biomedical applications (Nakamura, Nascimento, and Polikarpov 2017).

AAs broadly classify redox enzymes that act in conjunction with the CAZymes. Lignin is found together with the carbohydrates in plant biomass, and the lignin-degrading enzymes, as well as lytic polysaccharide monooxygenases (LPMOs), have been included in the AA family in order to accommodate a range of enzymatic mechanisms and substrates. Members of the previously classified GH61 family and CBM33 (carbohydrate-binding module (CBM)) family were identified to be LPMOs, which often act in concert with the lignin fragments in biomass degradation. The lignolytic enzymes do not act on the carbohydrates, but the close association of lignin and carbohydrates in the PCW brings about cooperation between the lignin-hydrolyzing enzymes and the classical CAZymes (Levasseur et al. 2013). AA families are constructed based on amino acid sequence similarities and structural similarities. These widespread catalytic modules enhance access of GH, PL, and CE family enzymes to the PCW carbohydrates by breaking down the lignin, and thereby facilitate complete saccharification of the plant biomass. In addition to the catalytic domains, the CBMs and dockerin molecules mediate the binding of the catalytic domains for efficient enzymatic action.

5.2.2 Mechanisms of Action in Plant Biomass Degradation

In the PCW, the cellulosic fraction is arranged in the form of an elementary fibril, insoluble in water and highly crystalline. It is encased by a hemicellulose monolayer and wrapped by a lignin–hemicellulose matrix. This entire structure is known as a cellulose microfibril and, due to physical interactions and covalent bonding, is very difficult to hydrolyze. Thus, the concerted action of many hydrolytic enzymes is required for the degradation of plant biomass. The microbial sources for CAZymes required for the degradation of plant biomass are listed in Table 5.1.

Plant biomass broadly consists of cellulose, hemicellulose, lignin, and some other compounds like pectin. Enzymes like cellulases, hemicellulases, pectinases, lignolytic enzymes, and LPMOs are the most adept in achieving complete saccharification.

The cellulase enzyme is a consortium of three different enzymes – endo-β-1,4-glucanases (endocellulases, EC 3.2.1.9.1), cellobiohydrolases (exocellulases, EC 3.2.1.91), and β-1,4-glucosidases (EC3.2.1.21), which act in synergy to hydrolyze the cellulose molecules. In bacteria, the cellulases are bound to the cell walls as extracellular aggregates, called cellulosomes, and are produced constitutively. Fungal cellulases are secreted as extracellular enzymes, and only in the presence of cellulose. Fungal cellulase production rates are higher compared to other organisms; however, bacterial cellulases are more stable due to their high natural diversity (Datta 2024).

The complex structure of the hemicellulosic fraction of plant matter requires the concerted action of many enzymes for its complete degradation. A large part of the

TABLE 5.1
Microbial Sources of CAZymes Used in the Degradation of Plant Biomass

Enzyme	Microorganism	Examples
Cellulases	Bacteria	*Cellulomonas* sp., *Cytophaga* sp., *Cellovibrio* sp., *Micromonospora* sp., *Acidothermus* sp., *Paenibacillus* sp, *Streptomyces* sp., *Pseudomonas* sp.
	Fungi	*Trichoderma reesei*, *Aspergillus* sp., *Penicillium* sp., *Humicola* sp., *Neurospora crassa.*
	Yeast	*Aureobasidium microstictum*, *Trichosporon pullulans*, *Rhodotorula glutinis*, *Candida* sp., *Trichosporon* sp.
Hemicellulases	Bacteria	*Bacillus halodurans*, *Bacillus subtilis*, *Bacillus stearothermophilus*, *Clostridium thermocellum*, *Roseithermus sacchariphilus*, *Stenotrophomonas maltophila*, *Thermoactinomyces vulgaris*, *Micrococcus* sp.
	Fungi	*Aspergillus niger*, *Aspergillus terreus*, *Fusarium* sp., *Phanerochaete chrysosporium*, *Trichoderma pleuroticola*, *Trichoderma viride*
	Yeast	*Aureobasidium pullulans*, *Cryptococcus albidus*, *Cryptococcus adeliae*, *Cryptococcus flavus*, *Pichia stipitis*, *Pichia capsulata*, *Pseudozyma* sp., *Candida* sp.
Pectin-degrading enzymes	Bacteria	*Enterobacter* spp., *Bacillus licheniformis*, *Bacillus stearothermophilus*, *Bacillus cereus*, *Bacillus* sp., *Erwinia* sp.
	Fungi	*Penicillium chrysogenum*, *Aspergillus oryzae*, *Aspergillus flavus*, *Moniliella* sp., *Trichoderma harzinum*, *Schizophyllum commune*, *Fusarium* sp.
	Yeast	*Wickerhamomyces anomalus*, *Saccharomyces cerevisiae*, *Kluyveromyces marxianus*, *Filobasidium capsuligenum*, *Pichia anomala*
Lignin-degrading enzymes	Bacteria	*Streptomyces viridosporus*, *S. paucinobilis*, *S. griseus*, *Rhodococcus jostii*
	Fungi	*Ustilago maydis*, *Panaeolus papilionaceus*, *Coprinopsis friesii*, *Phanerochaete chrysosporium*, *Ganoderma* sp.
	Yeast	*Candida guillermondii*, *C. tropicalis*, *Cryptococcus albidus*, *Cryptococcus* sp.

polymers are either insoluble or are closely packed within the insoluble cellulose matrix. The hemicellulases' mode of action is thus often enhanced by the action of cellulases. Complete degradation of the hemicellulosic fraction requires the synergistic action of many enzymes like xylanases (EC 3.2.1.8), mannanases (EC 3.2.1.78), glucuronidases (EC 3.2.1.31), xylosidases (EC 3.2.1.37), arabinofuranosidases (EC 3.2.1.55), and hemicellulolytic esterases (EC 3.1.1.7-) (Shallom and Shoham 2003).

Pectin is a polysaccharide, often found in plant biomass associated with the main lignocellulosic framework, between the plasma membrane and the cell wall. Pectinolytic enzymes, or pectinases, are most commonly obtained from fungi; polygalacturonases are the most relevant ones. The fungal pectinases have generally been reported to be acidic, while bacterial pectinolytic enzymes are alkaline, known as alkaline pectinases. Pectinases belong to different CAZy classes based on their mode of action; polygalacturonases (EC 3.2.1.15) are GHs, while pectate lyase belongs to the PL enzyme family.

The lignolytic group of enzymes consists mainly of laccase (EC 1.10.3.2), lignin peroxidase (EC 1.11.1.14), manganese peroxidase (EC 1.11.1.14), and versatile peroxidase (EC 1.11.1.16). The peroxidases break down lignin via H_2O_2-dependent oxidative depolymerization, and laccase degrades lignin by direct or indirect oxidation.

Endoglucanases, exoglucanases, and β-glucosidases are the three categories into which cellulolytic enzymes are classified based on enzymatic mechanisms and substrate specificity. A complex synergy among cellulase enzymes orchestrates the efficient bioconversion of cellulose into glucose, as demonstrated in two distinct stages. Endoglucanases, also known as carboxymethyl cellulases, act on β-1,4-glycosidic bonds randomly inside the chain, giving rise to long-chain cello-oligosaccharides. Exoglucanases, also called cellobiohydrolases, act on cellulose and cello-oligosaccharides by breaking the β-1,4-glycosidic bonds from the ends, releasing glucose and cellobiose units. The exo- and endoglucanases act to decrease the degree of cellulose polymerization during the liquefaction phase. β-Glucosidase, also known as cellobiase, then catalyses the conversion of cellobiose to glucose and also acts to further break down cellodextrins.

The endoglucanase, exoglucanase, and β-glucosidase act in synergy via acid catalysis mechanisms, which can be either inverting or retaining mechanisms. The spatial arrangement of catalytic sites is a significant factor in the selection of the mechanism. The hydrolysis in the inverting mechanism is direct and occurs through sequential deprotonation, without the formation of an enzyme-substrate complex during the reaction. In contrast, the retaining mechanism involves the carboxylic ends of amino acid residues acting as nucleophilic bases, which attack glycosidic bonds, break them, and form an enzyme-substrate complex with an oligomeric fragment (Kuhad et al. 2016).

Hemicellulose is a heterogeneous and highly complex polymer that requires multiple enzymes to be hydrolyzed completely. Xylan, the largest component of the hemicellulosic fraction, can be broken down by endoxylanases (endo-β-1,4-xylanase, or 1,4-β-*D*-xylan xylanohydrolase), which hydrolyzes the glycosidic linkages within the structure of xylan, releasing xylooligosaccharides, and β-xylosidases (1,4-β-*D*-xylan xylohydrolase), which act on the β-1,4-glycosidic bonds on the non-reducing ends and release xylose units. Mannanases (endo-1,4-β-mannanase/1,4-β-*D*-mannanohydrolase) and β-mannosidases (β-*D*-mannosidase/1,4-β-*D*-mannoside mannohydrolase) act in the same way on mannan-based hemicellulose fractions to release mannose units. Other debranching enzymes like α-*L*-arabinofuranosidase and α-*D*-glucuronidase hydrolyse α-1,2, α-1,3, or α-1,5 bonds to remove side groups and other substituents, and esterases like *p*-coumaryl esterase, phenolic acid esterases, and feruloyl esterase hydrolyze ester bonds in the hemicellulose structure for hydrolysis. The conjugated action of cellulases and hemicellulases is required for the proper digestion of plant matter. In industrial utilization of plant biomass, agricultural residues, and fruit and vegetable wastes, hemicellulases and cellulases are often used together for complete saccharification. The amount of reducing sugar released acts as an indicator for the degree of degradation of the cellulosic and hemicellulosic fraction by the enzymes (de Souza and Kawaguti 2021).

The PCW consists of pectin, which hinders the action of the cellulases and hemicellulases. Pectin is often hydrolyzed by the combined action of protopectinases, which hydrolyze glycosidic bonds to solubilize the insoluble pectin polymers; esterases, which remove methoxyl and acetyl esters, forming polygalacturonic acid; exo- and endo-hydrolases like polygalacturonase and polymethylgalacturonase, which split glycosidic linkages to generate D-galacturonic acid monomers; and oligomethyl-galacturonate and lyases, which catalyze trans-eliminative breakdown of pectate and pectinate polymers (Masuelli 2022).

Among all the components of plant cells, lignin is the most complex and the hardest to break down. Industries have been using thermochemical methods to remove lignin via pretreatments and steam explosion; however, these approaches are not environmentally friendly, cost-effective, or sustainable in the long run. Hence, the ligninolytic enzymes produced by microorganisms are being explored to establish an eco-friendly bioprocess. Microbial lignin-degrading enzymes can be classified into two major groups: the lignin-degrading auxiliary enzymes and the lignin-modifying enzymes. These auxiliary enzymes are not able to digest the lignin polymers independently and require the involvement of other enzymes like peroxidases, which use H_2O_2-based oxidative mechanisms. These enzymes include cellobiose dehydrogenase, aryl alcohol oxidases, glyoxal oxidase, glucose oxidase, and pyranose 2-oxidase. The lignin-modifying enzymes, also called the ligninolytic enzymes, are able to degrade lignin independently and include laccase and heme-containing peroxidases like lignin (LiP), manganese (MnP), versatile peroxidase (VP), and feruloyl esterase, using natural and synthetic mediators (Kumar and Chandra 2020).

Thus, for the complete degradation of plant biomass and utilization of the nutritive potential, the enzymes targeting all the components are required to function in synergy in order to establish a sustainable bioprocess and generate a circular bio-economy.

5.3 INTEGRATION OF CAZYMES IN BIOREFINERY PROCESSES

The utilization of CAZymes in biorefineries is an important achievement in terms of sustainability for converting biomass into value-added products. CAZymes consist of a wide range of enzymes and have broad applications among those enzymes that modify, degrade, and biosynthesize carbohydrates, specifically for enzymes involved in the depolymerization of complex polysaccharides to fermentable sugars. This enzyme activity is adequate for effectively processing lignocellulosic raw materials, which are widely available in plant biomass and support the production of biofuels and biochemicals using renewable resources (Contesini, Frandsen, and Damasio 2021).

Renewable energy sources have been outlined against bioconversion technologies and the combinatorial use of CAZymes in biomass refining. CAZymes make the degradation of cellulose and hemicellulosic structures easier using some of their action mechanisms like glycoside hydrolysis and polysaccharide lysis, along with various supplementary activities allowing lignocellulose degradation altogether.

The following monosaccharides could be directly utilized for microbial fermentation, producing biofuels, bioplastics, or other valuable products (Chettri et al. 2022). In recent times, a lot of developments in enzyme engineering and synthetic biology have facilitated the establishment of customized CAZyme cocktails optimized for a specific biomass that improves the overall efficiency. (Bornscheuer et al. 2012). These kinds of engineered enzymes exhibit enhanced activity and stability in industrial conditions, making them suitable for large-scale biorefinery applications. Additionally, the integration of CAZymes with pretreatment processes like steam explosion or acid hydrolysis has further increased the utilization of cellulose and hemicellulose fractions, in that way making it easier for more efficient enzymatic hydrolysis (Himmel et al. 2007).

These enzymes have a broader scope beyond these traditional applications; new ones like metagenomic analysis are finding more fresh candidates for novel CAZymes from different microbial groups, especially in the gut microbiome of herbivorous animals, where these enzymes are supposed to have increased stability and activity under industrial conditions, making them a suitable candidate for biorefineries. Biorefineries under such conditions will have a positive impact on the reduction of cost and improved efficiency of biomass conversion as they keep coming up with different and novel diversities in the genetic biochemical pool of CAZymes (Chettri et al. 2022).

Currently, CAZymes are at the top of the revolution in biorefinery, which is increasing the adaptation towards a sustainable and renewable bioeconomy; their ability to convert a large number of various sources of biomass into valuable products not only addresses the challenges of depending on fossil fuels but also contributes to waste valorization and environmental sustainability.

5.3.1 CAZymes in Waste Valorization

The role of CAZymes in waste valorization is emerging as many industries use them to include sustainability in converting waste into useful products. CAZymes can degrade the biomass, synthesize the biomass into other materials, or modify various materials from carbohydrates. Thus, they serve in breaking down polysaccharides derived from LCB, mainly residues from agriculture, forestry by-products, and neglected organic waste. CAZymes facilitate the efficient fermentation of these substrates into fermentable sugars and other vital compounds for making biofuels, bioplastics, and other valuable biochemicals through their enzymatic degradation process (Joshi et al. 2024). Alongside the recovery of sugars, lignocellulosic matrix degradation should boost the yield production of bioactive compounds like oligosaccharides for food and pharmaceutical applications. Acquired metagenomic tools advancements allow researchers to discover a whole range of potentially useful CAZymes, thus attesting to a considerable source of novel enzymes that can be utilized for industrial applications (Nizami et al. 2017). At the same time, CAZymes in biorefineries make possible the more efficient combination of specific enzymatic cocktails, which are specific and optimized for a particular feedstock as they relate to valorization processes (Chettri, Verma, and Verma 2020). However, this is a step further in research into CAZymes in this regard, but this use could likely co-invent contributions to a circular economy by changing waste into valuable resources without undermining the environmental impacts of waste disposal and fossil fuel dependency (Chettri, Verma, and Verma 2020). CAZymes play a vital role in waste valorization by enabling the efficient conversion of LCB into valuable products (see Table 5.2). Their ability to degrade complex carbohydrates is enhanced by ongoing research into novel enzyme discovery and engineering techniques. The use of CAZymes in waste valorization processes overcomes the difficulty of biomass conversion while also contributing to environmental sustainability. CAZymes contribute to the reduction of landfill use and greenhouse gas emissions connected with waste disposal by turning trash into profitable goods (Chandukishore et al. 2023). Additionally, the generation of biofuels and biochemicals from waste materials can reduce dependency on fossil fuels, further boosting global sustainability initiatives. As the demand for sustainable solutions grows, the application of CAZymes in biorefineries will be crucial for promoting a circular economy that maximizes resource recovery while minimizing environmental impact.

5.3.2 Case Studies of CAZyme-Driven Circular Economy Models

Biorefineries based on CAZymes are already established with great potential. One example is Beta Renewables, Italy, where they synthesize 75 million liters of ethanol per year from cellulose commercially. The company claims that this is the largest ethanol biorefinery in the world, working with a patented PROESATM process using CAZymes (Novozymes). The company used LCB like wheat straw, rice straw, and Arundo donax as a carbon source to produce ethanol. As a circular economy concept, the leftover lignin is used in an electric power plant to generate energy that is sold to the local grid (C. O. G. Silva, Vaz, and Filho 2018). Like this, different CAZyme-driven circular economy models are established in biorefineries and explained below.

5.3.2.1 Circular Model 1

Naranjo and co-workers (Naranjo, Cardona, and Higuita 2014) have used banana waste as feedstock for the production of polyhydroxybutyrate (PHB). The banana consists of 85% pulp and 15% peel in its total dry weight. The banana starch was hydrolyzed by gelatinization, liquefaction, and saccharification steps. For gelatinization, banana waste was pre-heated at 90 °C in water, followed by liquefaction at 90 °C for 3 h using α-amylase. Finally, the saccharification was performed using glucoamylase at 70 °C for 48 h to derive glucose. Likewise, the banana peel was pretreated at 170 °C at 5 atm to expose the cellulose and hemicellulose, followed by depolymerization of the hemicellulose to xylose by auto-hydrolysis. The saccharification of cellulose was performed by a cocktail of cellulases. The glucose obtained from whole banana waste was used as a carbon source for producing PHB and ethanol by *Burkholderia sacchari* IPT101 and *Pichia stipitis,* respectively. After the synthesis of PHB, cell

TABLE 5.2
CAZymes Used in Waste Valorization

Biomass	Enzymes	Organism	Product	Applications	Reference
Mixed Food Waste	Glucoamylase	*Aspergillus fumigatus, A. terreus*	Glucose	Used as a substrate for the sustainable production of several value-added products	Das et al. (2024), Chandukishore et al. (2023)
Sugarcane Bagasse	Xylanase	*Pseudozyma brasiliensis*	Xylooligo-saccharides	Production of animal feed, paper manufacturing, xylitol production	Haldar and Purkait (2020)
	Cellulases	*T. reesei, A. niger*	Bioethanol	Additive in petroleum	Sukumaran et al. (2009)
	Hemicellulases	*Candida guillermondii*	Xylitol	Used as an artificial sweetener, used in pharmaceuticals	S. S. Silva et al. (2007)
Rice Straw	Cellulases	*T. reesei, A. niger*	Bioethanol	Additives in the petroleum industry	Sukumaran et al. (2009)
Rice Husk	Cellulase	*Xanthomonas campestris*	Exopolysaccharides	Food and flavoring Tablagents	
Wood Chips	Hemicellulases	*Candida guillermondii*	Xylitol, Arabitol, Mannitol, Sorbitol	Artificial sweetener, used as a prebiotic also	Yamaguchi et al. (2014)
Wheat Bran	Amylase	*Penicillium decumbens*	Maltose	Intravenous sugar supplement	Sun et al. (2008)
	Cellulase	*Xanthomonas campestris*	Xanthan, Exopolysaccharides	Food and flavoring agents	Öner (2013)
Water Hyacinth	Cellulases	*A. niger*	Bioethanol	Additive in petroleum	Haldar and Purkait (2020)
	Hemicellulase	*C. tropicalis*	Bioethanol, Xylitol	Petroleum, sweetener in foods	Shankar et al. (2020)
Grape Stalks	Pectinases	*Aspergillus awamori*	Chitosan	Used in cosmetics and medicine	Botella et al. (2007)

lysis of IPT101 was carried out with sodium hypochlorite, and the biopolymer was extracted after centrifugation. The resulting PHB was spray-dried after evaporation, and 98% (w/w) purity of PHB was obtained. Similarly, approximately 7–10% (w/w) of bio-ethanol was obtained after fermentation, which was distilled in distillation columns and obtained 99.7% of pure ethanol by weight.

The life cycle analysis of this experiment showed that scenario 3 (simulation model) with wastewater recycling and recovery of the process at the mass integration level drastically reduced water consumption and CO_2 generation. The addition of one water recycling plant to the existing bioprocess will be the best solution to recycle and reuse the water from this biorefinery. The balanced organic matter could be used as an organic fertilizer for a better circular bioeconomy process. This integrated biorefinery concept was selected for techno-economic analysis, which also showed scenario 3 with the highest profit for glucose, ethanol, and PHB as 316, 238, and 31.5 kg/ton, banana, respectively. Overall, this process shows 30.6% less energetic integration of the total global energy requirements and 35% mass integration in water savings that decrease the production cost and the impact on the environment.

5.3.2.2 Circular Model 2

In another biorefinery concept, Choi et al. used onion skin waste (OSW) as a feedstock for the synthesis of quercetin (flavonoid) and sugar. The optimized enzyme cocktail of 0.72 mg of cellulase, 0.16 mg of pectinase, and 1.0 mg of xylanase per gram of dry OSW derived almost 98.5% of glucose. From this enzyme hydrolysate, the solid residue was removed and used for further extraction of quercetin. The quantification of quercetin in each OSW extract (enzyme hydrolysis residue and non-treated OSW) was analyzed using image analysis software. The data revealed that enzymatically hydrolyzed OSW showed 1.61-fold of quercetin percentage compared to non-treated OSW. The quercetin was purified using terpyridine-immobilized silica-coated magnetic nanoparticles–zinc (TSMNP–Zn) with 90% adsorption efficiency. This approach for the extraction of quercetin and other biosugars from OSW could be potentially applied for obtaining value-added products from agricultural waste in future biorefineries (Choi et al. 2015).

5.3.3 Engineering and Optimization of CAZymes

In this context, CAZymes refers to a group of diverse enzymes involved in the hydrolysis of complex carbohydrates that facilitate the breakdown of plant biomass into fermentable sugars. Engineering and optimization of CAZymes are very critical in advancing biorefinery applications, particularly for the conversion of complex LCB into biofuels and various valuable products. This optimization is crucial for the conversion of materials like agricultural residues, forestry waste, and municipal waste into more valuable products to contribute to a sustainable and circular economy (Islam and Roy 2018).

Recently, enzyme engineering has drastically enhanced the properties of CAZymes. Direct evolution and rational design (see Figure 5.1) are some of the basic techniques that are used to improve the activity of an enzyme. Directed evolution is

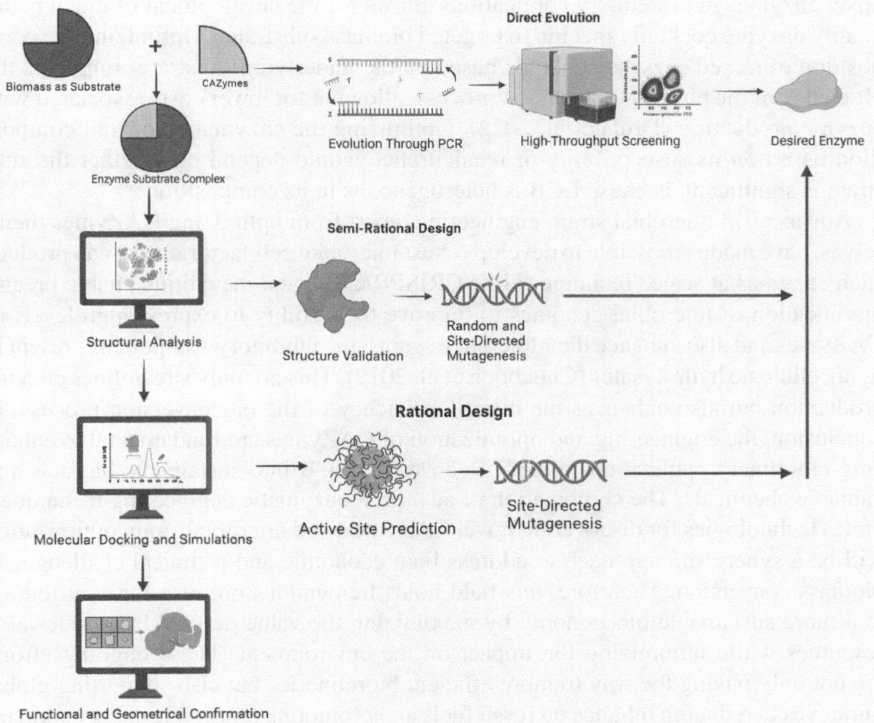

FIGURE 5.1 Common methods used in engineering and optimization of CAZymes through In-Silico and In-Vitro.

a form of natural selection that involves the generation of an enzyme variant library and screening it to identify improved activity and stability under specific conditions (Wardman and Withers 2024). This approach has resulted in the discovery of CAZymes with enhanced properties, like withstanding severe conditions that are usually encountered in industrial processes. Rational design is a method that uses structural information of enzymes to enhance their function by doing some mutations. It is possible to understand the three-dimensional structure of CAZymes with techniques like X-ray crystallography or cryo-electron microscopy, and also identification of key amino acids that influence the enzyme activity is also possible (Chettri, Verma, and Verma 2020).

Omics technologies, such as genomics, transcriptomics, proteomics, and metabolomics, have made it easier to identify novel CAZymes and characterize them. Such types of technologies give researchers insight into vast microbial diversities in diverse environments like soil or decaying plants, leading to the discovery of enzymes that are not yet characterized, which also hold unique properties (André et al. 2014). For example, metagenomics can detect CAZyme genes in environmental samples that can then be expressed in suitable host organisms for further study. The use of novel enzymes in biorefinery applications allows for the development of much better quality enzyme cocktails specific to targeted biomass substrates. Optimizing the composition of mixed enzyme cocktails based on the underlying substrates improves the efficiency of the biomass conversion process, allowing for lower costs associated with enzyme production (Drula et al. 2022). Optimizing the enzymatic cocktail composition based on its susceptibility or recalcitrance would depend on whether the substrate is significant, because LCB is heterogeneous in its composition.

Advances in microbial strain engineering, apart from optimizing CAZymes themselves, have made it possible to develop robust microbial cell factories that can produce such enzymes at scale. Techniques like CRISPR/Cas9 genome editing enable precise modification of microbial genomes to improve their ability to express high levels of CAZymes and also enhance their tolerance against the inhibitory compounds present in lignocellulosic hydrolysates (Chuaboon et al. 2019). This not only streamlines enzyme production but also enhances the overall efficiency of the bioconversion process. In conclusion, the engineering and optimization of CAZymes are fundamental to enhancing biorefinery applications aimed at converting LCB into sustainable biofuels and valuable chemicals. The combination of advanced enzymatic engineering techniques, omics technologies for discovering novel CAZymes, and microbial strain optimization will be a synergistic approach to address both economic and technical challenges in biomass conversion. Therefore, this field holds tremendous promise for contributing to a more sustainable bioeconomy by maximizing the value derived from renewable resources while minimizing the impact on the environment. These ongoing efforts are not only paving the way to more efficient biorefineries but also supporting global initiatives in reducing reliance on fossil fuels and promoting renewable energy sources.

5.4 CHALLENGES AND OPPORTUNITIES IN SCALING CAZYME-BASED BIOREFINERIES

5.4.1 Economic and Technical Barriers

In biobased biorefineries, microorganisms that synthesize CAZymes efficiently degrade PCW polymer from non-edible renewable LCB obtained from forest, agriculture, and agro-industry to synthesize value-added products for a circular economy. Even though the diversity and complexity of the lignin polymer and cellulose microfibrils make the biomass degradation by CAZymes recalcitrant in biorefineries (Srivastava et al. 2015). This makes the scale-up of the process energy-intensive and uneconomic. Although biorefineries offer green advantages and sustainability, they have very low commercial success at industrial production levels due to irregular LCB supply, complex or inefficient pretreatment and saccharification, and scale-up challenges that lead to high capital and operating investment (Usmani et al. 2021). On the other side, the collection, transportation, and storage of biomass at the biorefinery site will be a huge challenge that makes the sustainable processes more energy-intensive, with the emission of CO_2 and other greenhouse gases. Hence, it is better to construct biorefineries using CAZymes near LCB cultivation places for eco-friendly, economically viable processes. Moreover, C-C bond formation during the recondensation of lignin in traditional fractionation processes creates complexity in the depolymerization of LCB (Guadix-Montero and Sankar 2018) and lowers the enzyme activity. For better depolymerization and CAZyme activity, most of the existing research focused on the beta-O-4 bond in the lignin should be reverted to other bonds (Sánchez-González, Martín-Martínez, and Dobado 2017) to aid in improving CAZyme-based biorefineries.

To improve the existing CAZyme-based biorefineries, newly isolated microorganisms that grow on biomass naturally will be potent with high CAZyme activity. Among the organisms, fungi are the predominant ones with extensively diverse enzymes industrially scaled in biorefineries (van den Brink and de Vries 2011). Efforts like phylogenomics and comparative genomics, coupled with the characterization of CAZymes, aid in understanding the evolution of genomics in fungal genomes. To get additional knowledge for highly active enzymes, a detailed comparison of genes coding for CAZymes in fungi with different ecologies or substrates is essential for sustainable biorefineries. Also, the discovery of novel enzymes that degrade lignin polymer and cellulose microfibrils is mandatory to enhance the activity of the existing enzymatic cocktails at an industrial scale (Hage and Rosso 2021). Nevertheless, research on the characterization of the specificity of the enzymes is necessary to know the involvement of the loci in carbohydrate metabolism at a molecular level (Ameri et al. 2022). Metabolic and media engineering with system biology approaches to improving the activity of the existing CAZymes is crucial for better trade-off profit from biorefineries.

5.4.2 Policy and Market Impacts

The International Energy Agency (IEA) published the data that globally a total of 30.6 Gt of CO_2 was emitted in 2020, which is distant from the United Nations Organization's (UNO) net-zero goals by 2050. Hence, many countries have taken initiatives to meet UNO's Sustainable Development Goals. The European Union (EU) launched a European Bioeconomy Strategy for efficiently utilizing LCB and organic waste for industrial value-added products in a sustainable circular way. In addition, the EU has created the European Green Deal and supports the Paris Agreement on climate change to establish no net GHG emissions by 2050 in Europe (Usmani et al. 2021). Likewise, the US federal government revealed a National Bioeconomy Blueprint with five major strategies: investment of shares into bioeconomy innovation, academic support, scaling up technologies from lab to industry, reducing regulatory barriers to deploy environment-friendly technologies, and developing public-private partnerships (The White House 2012). These frameworks led to almost 2500 bioproducts (approximately 65.4 billion litres of biofuels), contributing to 2.5% of the US economy in 2017 with 4 million job opportunities (Guo and Song 2019). Also, the US has plans to enlarge the bioeconomy size to a billion tons by 2030 (Gong et al. 2024). In India, the Ministry of Petroleum and Natural Gas launched a "JI-VAN" scheme in 2017 to provide financial aid to non-food renewable feedstock-based biorefineries, especially for commercial-scale 2G-ethanol facilities. Also, Brazil's government and China's National Development and Reform Commission (NDRC) have taken E100 and E10 policies to blend biorefinery ethanol with gasoline or other petrol-derived fuels (Singh et al. 2022). Globally, these initiatives will lay the foundation for innovative utilization of the LCB resources for products while improving social, economic, and environmental benefits.

CAZymes comprise many enzymes with diverse functions and are widely applied in the biofuel, textile, health science, and food industries, with an impact on biorefineries (Chettri, Verma, and Verma 2020). They are also employed in fermentation, food, brewing, and baking industries, waste treatment, bioremediation, and the production of nutraceuticals (Shi, Wang, and Huang 2018). The global increasing demand for CAZymes was valued at approximately USD 3.5 billion in 2023 and is projected to be USD 6.2 billion by 2032, with a Compound Annual Growth Rate (CAGR) of 6.5% (More et al. 2024). However, due to the incompleteness and confidentiality of existing CAZymes-based biorefinery processes, they face challenges like data availability and stakeholder participation. To encounter the drawbacks and promote the establishment of new biorefineries, the IEA presents open-access biorefinery concepts for their environmental and economic feasibility. Furthermore, this opens access to generic data and information on the possible pathways for future biorefineries. This will also support stakeholders in reconsidering the specific character of a biorefinery for a comprehensible adaptation from its existing setup (Lindorfer et al. 2019). Hence, new government policies and private investment plans are essential for future sustainable biorefineries.

5.5 CONCLUSION AND FUTURE PROSPECTS

The sustainable conversion of biomass into useful products has advanced significantly with the introduction of CAZymes into biorefinery operations. Complex polysaccharides present in LCB are broken down by CAZymes, which makes it easier to produce biofuels, bioplastics, and other biochemicals. The performance of CAZymes has been enhanced by recent developments in enzyme engineering, especially through techniques like directed evolution and rational design, which allow them to operate efficiently under the harsh circumstances common in industrial applications. Furthermore, novel CAZymes from a variety of microbiological sources have been identified and characterized, thanks to the application of omics technology, expanding the range of biomass conversion techniques accessible.

Future possibilities for CAZymes in biorefinery applications appear to be very bright. It is anticipated that future studies will produce CAZymes with improved stability, activity, and specificity for a range of substrates. The use of cutting-edge techniques such as CRISPR/Cas9 will further enhance microbial strains, increasing their ability to generate these critical enzymes on a large scale. Furthermore, metagenomic research is anticipated to find new CAZymes from previously unknown habitats, expanding the range of enzymatic tools specialized to certain biomass types. Customizing enzyme cocktails for specific feedstocks will improve biomass conversion efficiency while cutting enzyme manufacturing costs. Considering the varied composition of lignocellulosic materials, this customized method is essential. As global sustainability programs gain traction, CAZymes will play an important role in promoting a circular economy that maximizes resource recovery while minimizing environmental damage.

To summarize, CAZyme engineering and optimization have considerable promise for turning waste into valuable resources while also enhancing sustainable biorefinery processes. By utilizing these breakthroughs, we may make significant strides toward lowering our reliance on fossil fuels while promoting environmental sustainability. Collaborative efforts among researchers, industry leaders, and policymakers will be critical in realizing the full benefits of CAZyme technology and driving forward developments within the biorefinery.

ACKNOWLEDGMENT

The authors would like to thank the Department of Biotechnology, National Institute of Technology Warangal, for providing the infrastructure and library facilities for procuring references and plagiarism analysis.

AUTHOR CONTRIBUTION STATEMENT

Satwika Das: Writing, original draft preparation; Anbunithi K.: Writing, original draft preparation; Rajesh R. O.: Writing, original draft preparation; Shahin Mohammed S. K.: Writing, original draft preparation; Ashish A. Prabhu: Writing, Supervision, Reviewing and Editing.

FUNDING

Not applicable.

CONFLICT OF INTEREST STATEMENT

The authors declare that they have no competing interests.

REFERENCES

Ameri, Rihab, José Luis García, Amel Bouanane Derenfed, Nathalie Pradel, Sawssan Neifar, Sonia Mhiri, Monia Mezghanni, Nadia Zaraî Jaouadi, Jorge Barriuso, and Samir Bejar. 2022. "Genome Sequence and Carbohydrate Active Enzymes (CAZymes) Repertoire of the Thermophilic Caldicoprobacter Algeriensis TH7C1T." *Microbial Cell Factories* 21 (1): 91. doi:10.1186/s12934-022-01818-0.

André, Isabelle, Gabrielle Potocki-Véronèse, Sophie Barbe, Claire Moulis, and Magali Remaud-Siméon. 2014. "CAZyme Discovery and Design for Sweet Dreams." *Current Opinion in Chemical Biology* 19 (1): 17–24. doi:10.1016/j.cbpa.2013.11.014.

Armendáriz-Ruiz, Mariana, Jorge Alberto Rodríguez-González, Rosa María Camacho-Ruíz, and Juan Carlos Mateos-Díaz. 2018. "Carbohydrate Esterases: An Overview." *Methods in Molecular Biology* 1835: 39–68. doi:10.1007/978-1-4939-8672-9_2.

Bains, Rajneesh K., Seyed Amirhossein Nasseri, Jacob F. Wardman, and Stephen G. Withers. 2024. "Advances in the Understanding and Exploitation of Carbohydrate-Active Enzymes." *Current Opinion in Chemical Biology* 80 (June): 102457. doi:10.1016/j.cbpa.2024.102457.

Benini, S. 2020. "Carbohydrate-Active Enzymes: Structure, Activity, and Reaction Products." *International Journal of Molecular Sciences* 21(8), 2727. https://doi.org/10.3390/ijms21082727

Bornscheuer, U. T., G. W. Huisman, R. J. Kazlauskas, S. Lutz, J. C. Moore, and K. Robins. 2012. "Engineering the Third Wave of Biocatalysis." *Nature* 485 (7397): 185–94. doi:10.1038/nature11117.

Botella, Carolina, Ana Diaz, Ignacio de Ory, Colin Webb, and Ana Blandino. 2007. "Xylanase and Pectinase Production by Aspergillus Awamori on Grape Pomace in Solid State Fermentation." *Process Biochemistry* 42 (1): 98–101. doi:10.1016/j.procbio.2006.06.025.

Chandukishore, T., Satwika Da s, K. Narasimhulu, and Ashish A. Prabhu. 2023. "Biovalorisation of Mixed Food Waste through Newly Isolated Thermo Tolerant Fungal Cell Factories: A Step toward Transforming Waste to Wealth." *Sustainable Chemistry and Pharmacy* 35 (August): 101230. doi:10.1016/j.scp.2023.101230.

Chettri, Dixita, Susmita Nad, Ujjal Konar, and Anil Kumar Verma. 2022. "CAZyme from Gut Microbiome for Efficient Lignocellulose Degradation and Biofuel Production." *Frontiers in Chemical Engineering* 4 (November). doi:10.3389/fceng.2022.1054242.

Chettri, Dixita, Ashwani Kumar Verma, and Anil Kumar Verma. 2020. "Innovations in CAZyme Gene Diversity and Its Modification for Biorefinery Applications." *Biotechnology Reports* 28: e00525. doi:10.1016/j.btre.2020.e00525.

Choi, In Seong, Eun Jin Cho, Jae-Hak Moon, and Hyeun-Jong Bae. 2015. "Onion Skin Waste as a Valorization Resource for the By-Products Quercetin and Biosugar." *Food Chemistry* 188: 537–42. doi:10.1016/j.foodchem.2015.05.028.

Chuaboon, Litavadee, Thanyaporn Wongnate, Pangrum Punthong, Cholpisit Kiattisewee, Narin Lawan, Chia Yi Hsu, Chun Hung Lin, Uwe T. Bornscheuer, and Pimchai Chaiyen. 2019. "One-Pot Bioconversion of l-Arabinose to l-Ribulose in an Enzymatic Cascade." *Angewandte Chemie – International Edition* 58 (8): 2428–32. doi:10.1002/anie.201814219.

Contesini, Fabiano Jares, Rasmus John Normand Frandsen, and André Damasio. 2021. "Editorial: CAZymes in Biorefinery: From Genes to Application." *Frontiers in Bioengineering and Biotechnology* 9 (February). doi:10.3389/fbioe.2021.622817.

Das, Satwika, T Chandukishore, Rangabhashiyam Selvasembian, and Ashish A. Prabhu. 2024. "Mixed Food Waste Valorization Using a Thermostable Glucoamylase Enzyme Produced by a Newly Isolated Filamentous Fungus: A Sustainable Biorefinery Approach." *Chemosphere* 352 (March): 141480. doi:10.1016/J.CHEMOSPHERE.2024.141480.

Datta, Rahul. 2024. "Enzymatic Degradation of Cellulose in Soil: A Review." *Heliyon* 10 (1): e24022. doi:10.1016/j.heliyon.2024.e24022.

de Souza, Thaiza S. P., and Haroldo Yukio Kawaguti. 2021. "Cellulases, Hemicellulases, and Pectinases: Applications in the Food and Beverage Industry." *Food and Bioprocess Technology* 14 (8): 1446–77. doi:10.1007/s11947-021-02678-z.

Drula, Elodie, Marie-Line Garron, Suzan Dogan, Vincent Lombard, Bernard Henrissat, and Nicolas Terrapon. 2022. "The Carbohydrate-Active Enzyme Database: Functions and Literature." *Nucleic Acids Research* 50 (D1): D571–77. doi:10.1093/nar/gkab1045.

Garron, Marie-Line, and Bernard Henrissat. 2019. "The Continuing Expansion of CAZymes and Their Families." *Current Opinion in Chemical Biology* 53 (December): 82–87. doi:10.1016/j.cbpa.2019.08.004.

Gong, Liang, Ajit Kumar Passari, Chunxiao Yin, Vijay Kumar Thakur, John Newbold, William Clark, Yueming Jiang, Shanmugam Kumar, and Vijai Kumar Gupta. 2024. "Sustainable Utilization of Fruit and Vegetable Waste Bioresources for Bioplastics Production." *Critical Reviews in Biotechnology* 44 (2): 236–54. doi:10.1080/07388551.2022.2157241.

Guadix-Montero, Susana, and Meenakshisundaram Sankar. 2018. "Review on Catalytic Cleavage of C–C Inter-Unit Linkages in Lignin Model Compounds: Towards Lignin Depolymerisation." *Topics in Catalysis* 61 (3): 183–98. doi:10.1007/s11244-018-0909-2.

Guo, Mingxin, and Weiping Song. 2019. "The Growing U.S. Bioeconomy: Drivers, Development and Constraints." *New Biotechnology* 49: 48–57. doi:10.1016/j.nbt.2018.08.005.

Hage, Hayat, and Marie-Noëlle Rosso. 2021. "Evolution of Fungal Carbohydrate-Active Enzyme Portfolios and Adaptation to Plant Cell-Wall Polymers." *Journal of Fungi (Basel, Switzerland)* 7 (3): 185.doi:10.3390/jof7030185.

Haldar, Dibyajyoti, and Mihir Kumar Purkait. 2020. "Lignocellulosic Conversion into Value-Added Products: A Review." *Process Biochemistry* 89: 110–33. doi:10.1016/j.procbio.2019.10.001.

Himmel, Michael E., Shi-You Ding, David K. Johnson, William S. Adney, Mark R. Nimlos, John W. Brady, and Thomas D. Foust. 2007. "Biomass Recalcitrance: Engineering Plants and Enzymes for Biofuels Production." *Science* 315 (5813): 804–7. doi:10.1126/science.1137016.

Islam, Farjana, and Narayan Roy. 2018. "Screening, Purification and Characterization of Cellulase from Cellulase Producing Bacteria in Molasses." *BMC Research Notes* 11 (1): 445. doi:10.1186/s13104-018-3558-4.

James, Jennylynd A., and Byong H. Lee. 1997. "Glucoamylases: Microbial Sources, Industrial Applications and Molecular Biology – A Review." *Journal of Food Biochemistry* 21 (6): 1–52. doi:10.1111/j.1745-4514.1997.tb00223.x.

Joshi, Namrata, Jasneet Grewal, Lukasz Drewniak, and Kumar Pranaw. 2024. "Bioprospecting CAZymes Repertoire of Aspergillus Fumigatus for Eco-Friendly Value-Added Transformations of Agro-Forest Biomass." *Biotechnology for Biofuels and Bioproducts* 17 (1): 1–13. doi:10.1186/s13068-023-02453-6.

Kuhad, Ramesh Chander, Deepa Deswal, Sonia Sharma, Abhishek Bhattacharya, Kavish Kumar Jain, Amandeep Kaur, Brett I. Pletschke, Ajay Singh, and Matti Karp. 2016. "Revisiting Cellulase Production and Redefining Current Strategies Based on Major Challenges." *Renewable and Sustainable Energy Reviews* 55 (March): 249–72. doi:10.1016/j.rser.2015.10.132.

Kumar, Adarsh, and Ram Chandra. 2020. "Ligninolytic Enzymes and Its Mechanisms for Degradation of Lignocellulosic Waste in Environment." *Heliyon* 6 (2): e03170. doi:10.1016/j.heliyon.2020.e03170.

Levasseur, Anthony, Elodie Drula, Vincent Lombard, Pedro M. Coutinho, and Bernard Henrissat. 2013. "Expansion of the Enzymatic Repertoire of the CAZy Database to Integrate Auxiliary Redox Enzymes." *Biotechnology for Biofuels* 6 (1): 41. doi:10.1186/1754-6834-6-41.

Lindorfer, J., Lettner, M., Hesser, F., Fazeni, K., Rosenfield, D., Annevelink, B., and Mandl, M. 2019. "Technical, economic and environmental assessment of biorefinery concepts: Developing a practical approach for characterisation." *IEA Bioenergy task* 42: 2281312.

Linhardt, R. J., P. M. Galliher, and C. L. Cooney. 1987. "Polysaccharide Lyases." *Applied Biochemistry and Biotechnology* 12 (2): 135–76. doi:10.1007/BF02798420.

Masuelli, M. A. (Ed.). (2022). *Pectins – The New-Old Polysaccharides*. BoD–Books on Demand., p. 166. doi:10.5772/intechopen.100247.

More, Anuradha B., V. Chandola, and Shruti Bhat. 2024. "Carbohydrate Active Enzyme Market." DATA INTELO. www.dataintelo.com

Nakamura, Aline M., Alessandro S. Nascimento, and Igor Polikarpov. 2017. "Structural Diversity of Carbohydrate Esterases." *Biotechnology Research and Innovation* 1 (1): 35–51. doi:10.1016/j.biori.2017.02.001.

Naranjo, Javier M., Carlos A. Cardona, and Juan C. Higuita. 2014. "Use of Residual Banana for Polyhydroxybutyrate (PHB) Production: Case of Study in an Integrated Biorefinery." *Waste Management* 34 (12): 2634–40. doi:10.1016/j.wasman.2014.09.007.

Nath, Soumitra. 2024. "Biotechnology and Biofuels: Paving the Way towards a Sustainable and Equitable Energy for the Future." *Discover Energy* 4 (1). doi:10.1007/s43937-024-00032-w.

Nizami, A. S., M. Rehan, M. Waqas, M. Naqvi, O. K. M. Ouda, K. Shahzad, R. Miandad, M. Z. Khan, M. Syamsiro, I. M. I. Ismail, and Deepak Pant. 2017. "Waste Biorefineries: Enabling Circular Economies in Developing Countries." *Bioresource Technology* 241: 1101–17. doi:10.1016/j.biortech.2017.05.097.

Öner, Ebru Toksoy. 2013. "Microbial Production of Extracellular Polysaccharides from Biomass BT." In Z. Fang (Ed.), *Pretreatment Techniques for Biofuels and Biorefineries.* 35–56. Berlin, Heidelberg: Springer Berlin Heidelberg. doi:10.1007/978-3-642-32735-3_2.

Sánchez-González, Ángel, Francisco J. Martín-Martínez, and J. A. Dobado. 2017. "The Role of Weak Interactions in Lignin Polymerization." *Journal of Molecular Modeling* 23 (3): 80. doi:10.1007/s00894-017-3257-4.

Shallom, Dalia, and Yuval Shoham. 2003. "Microbial Hemicellulases." *Current Opinion in Microbiology* 6 (3): 219–28. doi:10.1016/S1369-5274(03)00056-0.

Shankar, Kumar, Neelkant S. Kulkarni, Rakesh Sajjanshetty, S. K. Jayalakshmi, and Kuruba Sreeramulu. 2020. "Co-Production of Xylitol and Ethanol by the Fermentation of the Lignocellulosic Hydrolysates of Banana and Water Hyacinth Leaves by Individual Yeast Strains." *Industrial Crops and Products* 155: 112809. doi:10.1016/j.indcrop.2020.112809.

Shi, Yanan, Xuefeng Wang, and Aixiang Huang. 2018. "Proteomic Analysis and Food-Grade Enzymes of Moringa Oleifer Lam. a Lam. Flower." *International Journal of Biological Macromolecules* 115: 883–90. doi:10.1016/j.ijbiomac.2018.04.109.

Silva, Caio O. G., Raissa P. Vaz, and Edivaldo X. F. Filho. 2018. "Bringing Plant Cell Wall-Degrading Enzymes into the Lignocellulosic Biorefinery Concept." *Biofuels, Bioproducts and Biorefining* 12 (2): 277–89. doi:10.1002/bbb.1832.

Silva, Silvio S., Solange I. Mussatto, Júlio C. Santos, Diego T. Santos, and Juliana Polizel. 2007. "Cell Immobilization and Xylitol Production Using Sugarcane Bagasse as Raw Material." *Applied Biochemistry and Biotechnology* 141 (2–3): 215–27. doi:10.1007/BF02729063.

Singh, Nisha, Reeta Rani Singhania, Poonam S. Nigam, Cheng Di Dong, Anil Kumar Patel, and Munish Puri. 2022. "Global Status of Lignocellulosic Biorefinery: Challenges and Perspectives." *Bioresource Technology* 344 (PB): 126415. doi:10.1016/j.biortech.2021.126415.

Srivastava, Neha, Rekha Rawat, Harinder Singh Oberoi, and Pramod W. Ramteke. 2015. "A Review on Fuel Ethanol Production From Lignocellulosic Biomass." *International Journal of Green Energy* 12 (9): 949–60. doi:10.1080/15435075.2014.890104.

Sukumaran, Rajeev K., Reeta Rani Singhania, Gincy Marina Mathew, and Ashok Pandey. 2009. "Cellulase Production Using Biomass Feed Stock and Its Application in Lignocellulose Saccharification for Bio-Ethanol Production." *Renewable Energy* 34 (2): 421–24. doi:10.1016/j.renene.2008.05.008.

Sun, Xianyun, Ziyong Liu, Yinbo Qu, and Xuezhi Li. 2008. "The Effects of Wheat Bran Composition on the Production of Biomass-Hydrolyzing Enzymes by Penicillium Decumbens BT – Biotechnology for Fuels and Chemicals." In Adney, W. S., McMillan, J. D., Mielenz, J., and Klasson K. T., (Eds.), *Applied Biochemistry and Biotechnology,* 239–48. Totowa, NJ: Humana Press.

Sutherland, Ian W. 1995. "Polysaccharide Lyases." *FEMS Microbiology Reviews* 16 (4): 323–47. doi:10.1111/j.1574-6976.1995.tb00179.x.

The White House. 2012. "National Bioeconomy Blueprint, April 2012." *Industrial Biotechnology* 8 (3): 97–102. doi:10.1089/ind.2012.1524.

Ubando, Aristotle T., Charles B. Felix, and Wei Hsin Chen. 2020. "Biorefineries in Circular Bioeconomy: A Comprehensive Review." *Bioresource Technology* 299: 122585. doi:10.1016/j.biortech.2019.122585.

Usmani, Zeba, Minaxi Sharma, Abhishek Kumar Awasthi, Tiit Lukk, Maria G Tuohy, Liang Gong, Phuong Nguyen-Tri, Alan D. Goddard, Roslyn M. Bill, S. Chandra Nayak, Vijai Kumar Gupta. 2021. "Lignocellulosic Biorefineries: The Current State of Challenges and Strategies for Efficient Commercialization." *Renewable and Sustainable Energy Reviews* 148: 111258. doi:10.1016/j.rser.2021.111258.

van den Brink, Joost, and Ronald P. de Vries. 2011. "Fungal Enzyme Sets for Plant Polysaccharide Degradation." *Applied Microbiology and Biotechnology* 91 (6): 1477–92. doi:10.1007/s00253-011-3473-2.

Wardman, Jacob F., and Stephen G. Withers. 2024. "Carbohydrate-Active Enzyme (CAZyme) Discovery and Engineering via (Ultra)High-Throughput Screening." *RSC Chemical Biology* 5 (7): 595–616. doi:10.1039/D4CB00024B.

Yamaguchi, Aritomo, Osamu Sato, Naoki Mimura, Yoshihiko Hirosaki, Hirokazu Kobayashi, Atsushi Fukuoka, and Masayuki Shirai. 2014. "Direct Production of Sugar Alcohols from Wood Chips Using Supported Platinum Catalysts in Water." *Catalysis Communications* 54: 22–26. doi:10.1016/j.catcom.2014.05.008.

Yang, Bao, Huiling Liu, Jiali Yang, Vijai Kumar Gupta, and Yueming Jiang. 2018. "New Insights on Bioactivities and Biosynthesis of Flavonoid Glycosides." *Trends in Food Science & Technology* 79 (September): 116–24. doi:10.1016/j.tifs.2018.07.006.

6 Emerging Roles of Carbohydrate-Active Nanoenzymes in Biomedical Applications

Minal Richi, Maryam Khan, and Mohd Hasan Mujahid

ABBREVIATIONS

Ag-NPs	Silver nanoparticles
BCMG-SPION	Glycol chitosan-coated superparamagnetic iron oxide
BET	Brunauer-Emmett-Teller
CANEs	Carbohydrate active nanoenzymes
CAZymes	Carbohydrate-active enzymes
CNT	Carbon nanotubes
CVD	Chemical vapor deposition
DDS	Drug delivery system
DLS	Dynamic light scattering
DSC	Differential scanning calorimetry
EDS	Energy dispersive X-ray spectrophotometer
FDA	The Food and Drug Administration
GO	Graphene oxide
GPC	Gingerol-phytosome-chitosan
MRI	Magnetic resonance imaging
MWCNTs	Multiwalled carbon nanotubes
NCs	Nanocomposites
NIRF	Near-infrared fluorescence
NPs	Nanoparticles
ROS	Reactive oxygen species
SEM	Scanning electron microscopy
SQUID	Superconducting quantum interference device
TEM	Transmission electron microscopy
TGA	Thermal galvanometric analysis
VSM	Vibrating sample magnetometer

DOI: 10.1201/9781003561682-6

XRD	X-ray diffraction
β-TCP	β-Tricalcium phosphate

6.1 INTRODUCTION

6.1.1 Overview of Carbohydrates as Macromolecular Biomaterials

A polymer is a substance made up of multiple repeating units called monomers. Due to the variety of functional groups within their structure, polymers can be combined with both high- and low-molecular-weight materials to impart unique properties. Additionally, they can be conjugated with drugs and tailored for various biomedical applications (Abedini et al. 2018). Among biopolymers, carbohydrates are the most abundant and diverse in nature (Agwa et al. 2023). Natural carbohydrates, or saccharides, are polymers made from carbon, oxygen, and hydrogen. These are classified into monosaccharides, disaccharides, and polysaccharides (Maji 2019).

Disaccharides consist of two monosaccharide units linked together through a glycosidic bond, with hydrogen bonds providing further structural stability. Polysaccharides, on the other hand, are more complex and are formed by the bonding of more than ten monosaccharide units *via* glycosidic linkages and hydrogen bonds. Polysaccharides can be classified into two main categories: homo-polysaccharides and hetero-polysaccharides (Su et al. 2021). In nature, carbohydrates are found in a wide range of sources, including plants, animals, algae, parasites, and various microorganisms such as yeast, fungi, and bacteria. These natural carbohydrates, like many other biopolymers, are characterized by properties that make them highly desirable for a variety of applications. Some of the remarkable properties are that they are biocompatible, biodegradable, low-cost, safe, eco-friendly, hydrophilic, easily accessible, and recyclable (Maji 2019; Garron and Henrissat 2019; Abedini et al. 2018; Benalaya et al. 2024).

Carbohydrates are abundant in the cell walls of bacteria, fungi, and plants, as well as in the extracellular matrix of animals. They are essential components of large energy-storage molecules like starch and glycogen, as well as structural polysaccharides like chitin and cellulose (Khowala, Verma, and Banik 2008). Like other polymers, carbohydrates are macromolecules with various active functional groups, including carbonyl, aldehyde, and ketone groups. Many common carbohydrates are considered preferable to synthetic polymers due to their ease of tailoring. They are often classified as homogeneous and bio-adhesive, with controllable functionality and rigidity, which makes them ideal for biomedical applications (Di et al. 2022; Agwa et al. 2023). Unlike synthetic polymers, carbohydrate-based polymers exhibit intrinsic bioactivity toward cells while also allowing for the incorporation of various acquired functionalities. Simultaneously, amidogen and/or hydroxyl groups at various sites on carbohydrate molecules can be further modified to achieve desirable physical and chemical properties (Di et al. 2022). These unique characteristics distinguish carbohydrates and place them at the forefront of glycobiology. Prominent examples of carbohydrates include cellulose, chitosan, chitin, alginate, pectin, pullulan, xanthan gum, starch, and cyclodextrin (Adrian, Mihai, and Vodnar 2019).

As targeting ligands, carbohydrates can bind to and recognize specific receptors, thereby enhancing cell–matrix and cell–cell interactions (Su et al. 2021). Despite the extensive understanding of the chemical and physical properties of simple carbohydrates, the structural complexity and variability of polysaccharides continue to pose challenges. The production of carbohydrate-based polymers remains less controllable than that of synthetic polymers, limiting their widespread success in biomaterials. Ongoing scientific efforts to explore carbohydrate technologies are critical for unlocking their full potential in the formation of nanoconjugates and biocomposites. These advances are particularly relevant in the pharmaceutical industry, where carbohydrates play a crucial role in the development of effective polymeric drug delivery systems as natural macromolecules and biologically active substances (Abedini et al. 2018; Qin et al. 2024; Gopinath et al. 2018).

6.1.2 Introduction of Carbohydrate Active Nanoenzymes

Carbohydrate-active enzymes (CAZymes) are a diverse group of enzymes that catalyze the breakdown, modification, or synthesis of complex carbohydrates, including polysaccharides, oligosaccharides, and glycoconjugates. These enzymes are critical to many biological processes, such as digestion, cellular signaling, and the degradation of plant biomass. CAZymes play a key role in the cycling of carbon in ecosystems, particularly through their involvement in the breakdown of plant cell walls, which are composed of complex carbohydrates like cellulose, hemicellulose, and pectin (J. F. Wardman and Withers 2024). An innovative class of nanomaterials, known as carbohydrate-active nanoenzymes (CANEs), mimics the actions of the natural enzymes like CAZymes. They can carry out particular metabolic activities by altering saccharides or breaking down complicated sugar. Usually created at the nanoscale (1–100 nm), these nanoenzymes combine the special qualities of nanomaterials with the functional qualities of conventional enzymes (Khan et al. 2019; Peptu et al. 2014; Shen et al. 2025). However, CANEs are synthetic or modified forms of these natural enzymes that are frequently made by adding organic nanomaterials, inorganic nanoparticles (NPs), or hybrid systems to improve catalytic stability, specificity, and efficiency. Recently, CANEs have attracted a lot of interest because of their potential uses in a number of industries, including biomedicine, biotechnology, and environmental sciences (Jebali et al. 2017).

Compared to conventional enzymes, CANEs have improved reactivity and catalytic capabilities due to the NPs' small size and high surface-to-volume ratio (X. Ren et al. 2022). As a result, CANEs can demonstrate stronger substrate affinities and faster reaction rates. It is possible to design CANEs to have particular carbohydrate identification capabilities, including focusing on particular monosaccharide units or glycosidic linkages. Specialized applications, such as pharmaceutical delivery and food preparation, are made possible by this focused action. CANEs frequently show better resistance to degradation, pH tolerance, and heat stability than natural enzymes. Because of this, they are more resilient and reusable in industrial and particularly biomedical applications (Shen et al. 2025; Seddiqi et al. 2021; Kato and Ishiwa 2015).

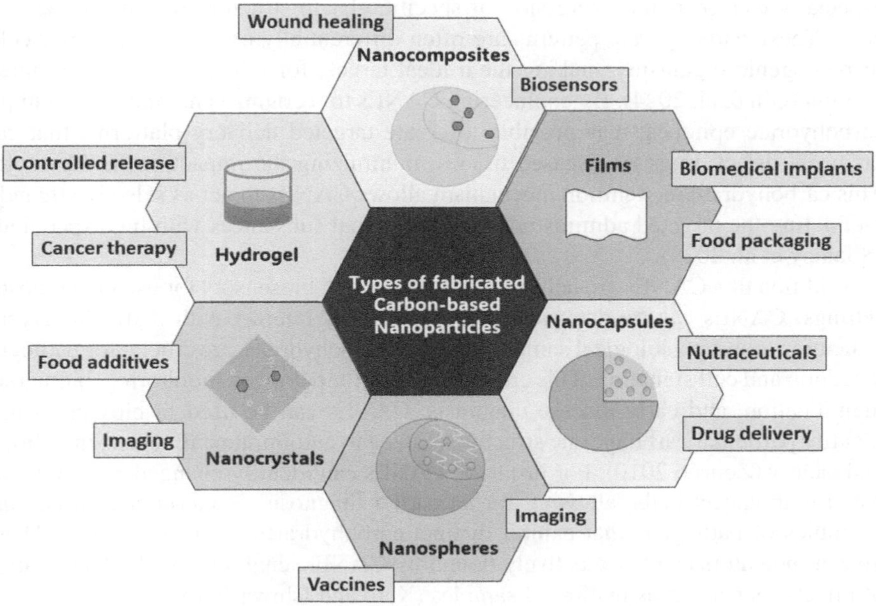

FIGURE 6.1 Illustration of different types of carbohydrate-based nanoparticles or nanocomposites used for various applications.

Combining the benefits of nanoengineering with the catalytic potential of CAZymes, CANEs are becoming a very promising tool in the biomedical sciences (Corrêa et al. 2020; J. F. Wardman and Withers 2024; Alibolandi et al. 2021). Utilizing the unique properties of nanotechnology, such as stability, reactivity, and large surface area, CANEs have a great deal of promise for developing a range of biological applications, ranging from drug delivery to diagnostic devices (Falsafi et al. 2023; B. Kang et al. 2015). Moreover, carbohydrate-based nanoparticles have recently gained attention in climate change research due to their unique properties, sustainability, and potential to address environmental challenges. Firstly, they are naturally made without the use of any chemicals; hence, on degradation, they have less probability of causing harm to the environment. Secondly, they help in carbon sequestration, as they are derived from biodegradable and carbon-neutral materials like cellulose, chitosan, or starch. Lastly, engineered CANEs can remove heavy metals and organic pollutants from water, preventing further damage to ecosystems already stressed by climate change. Different types of fabricated CANEs used for a variety of applications are illustrated in Fig. 6.1.

6.1.3 Significance of CANEs in Biomedical Sciences

The development of targeted drug delivery systems (DDS) is one of the most promising biomedical applications of CANEs. A critical feature of many disease processes, especially cancer, is the expression of specific glycan structures on the surfaces of cells. These carbohydrate patterns are often differentially expressed on tumor cells or pathogenic organisms, making them ideal targets for selective therapeutic intervention (Qin et al. 2024). By engineering CANEs to recognize and bind these unique carbohydrate epitopes, it is possible to create targeted delivery platforms that can precisely direct drugs to diseased tissues, minimizing the impact on healthy cells. This carbohydrate recognition mechanism allows CANEs to act as selective ligands, facilitating the targeted administration of medicinal substances with high specificity (Silantev et al. 2023).

Additionally, CANEs are helpful in the creation of biosensors for use in diagnostic settings. CANEs can be designed to recognize and interact with particular glycan structures found in biological samples because carbohydrates are crucial for pathogen detection and cell signaling. This creates chances for pathogen monitoring, biomarker identification, and early disease diagnosis. CANEs can be used in biosensors that identify particular carbohydrate structures linked to autoimmune illnesses, infections, and cancer (Zourob 2010). For instance, CANEs can identify changed glycosylation patterns in cancer cells, allowing for an earlier diagnosis. Bacteria and viruses are examples of pathogens that exhibit distinct carbohydrates on their surfaces. These glycan indicators can be selectively bound by CANEs, enabling quick identification of infectious pathogens in clinical samples (Kato and Ishiwa 2015).

Additionally, CANEs can be included in biomaterials intended for tissue regeneration or wound healing. Their capacity to interact with polysaccharides such as chitin or hyaluronic acid promotes better cellular motility, adhesion, and proliferation—all of which are essential for tissue regeneration and repair. By altering the extracellular matrix of tissues, CANEs can improve the regeneration and healing of cartilage, bone, skin, and other tissues. Materials based on carbohydrates are often biocompatible, which lowers the possibility of unfavorable immune reactions and encourages tissue integration (Kato and Ishiwa 2015; Ruoying Wang et al. 2024; Alibolandi et al. 2021; J. F. Wardman and Withers 2024).

The literature reported in recent years has mainly focused either on the applications of carbohydrate carriers or NPs alone. This chapter emphasizes the advantageous properties of both carbohydrates and nanotechnology to address a range of medical challenges. The novelty lies in the unique ability of CANEs to harness the inherent properties of carbohydrates like biocompatibility, biodegradability, and specificity and combine them with the extraordinary properties of NPs. The nanocomposites (NCs) thus formed can be reused with higher activity and stability and have a greater potential to improve drug delivery, diagnostic imaging, and enhanced tissue generation. These innovations drive the growing importance of CANEs in the development of next-generation therapeutics and diagnostic solutions. This chapter starts with the introduction of CAZymes and CANEs along with their biomedical significance. The next section describes the classification of CANEs and how they can be synthesized and characterized by various techniques. The four types of carbohydrate enzymes that have been focused in this article are chitosan, alginate, cellulose, and starch-based nanocarriers, and their role in diagnostics, drug delivery, and tissue engineering has been described in detail. At the end, challenges and future prospects have been discussed.

6.2 FUNDAMENTALS OF CANES

Carbohydrate-active enzymes (CAZymes) are extremely important enzymes used in biomedical applications. They are highly used for the detection of diseases and treatment of various disorders. Some carbohydrates have an inherent propensity to bind through particular receptors to specific kinds of cells and offer the possibility of targeted delivery systems based on receptor-mediated endocytosis, which is a process involving invagination of the plasma membrane and subsequent formation of vesicles to deliver agents deep into the tissue of a tumor. Besides, they can be simply grafted onto the surface of drug carriers through chemical modification and deliver an in-depth analysis of carbohydrate-modified NPs toward improving active targeted combinatorial chemo/phototherapy (Cavalcanti and Santana 2019; Alonso, Rendueles, and Díaz 2013). The activity of CAZymes is entropically controlled by factors responsible for the availability of carbohydrates and hence the regulation of metabolic processes (Kartal et al. 2011). CAZymes can thereby make improvements in various fields, such as carbon sequestration, waste management, and biofuels (Hage and Rosso 2021; Bains et al. 2024). One of the principal uses of CAZymes lies in mitigating climate change. CAZymes, especially glycosyltransferases and glycosidases, are involved in the conversion of lignocellulosic biomass into fermentable sugars, which will eventually be converted into biofuels. Nanozymes describe nanomaterials exhibiting inherent enzyme-like activity, catalyzing the conversion of substrates through kinetics and processes similar to those involved in the operation of real enzymes in *in vivo* settings, while the term "nanozyme" was first discovered by Wei and Wang in 2013 (An et al. 2023; Wei and Wang 2013). Nanozymes represent one class of nanomaterials characterized by multivalent catalytic sites and an average particle size under 100 nm, with much higher environmental adaptability and extended lifespan than their chemical counterparts. Enzymes in nanoform mimic one of the most prominent approaches for circumventing some inherent disadvantages of natural enzymes (H. Liu et al. 2022; Qiao et al. 2023). These properties include nanoscale dimensions that can be easily manipulated; surface charge and surface chemistry to make it feasible to facilitate; and customization and optimization of NP-enzyme complexes for applications ranging from biomedical to diagnostic testing (Ahmadi et al. 2023).

6.2.1 Classification of CANEs

6.2.1.1 Alginate-Based Nanoparticles

Alginate, a water-soluble polymer made from alginic acid, is made up of 1-4 monomeric units connected by α-D-mannuronate and β-L-guluronate at varying lengths in the chain (Fig. 6.2a). It can be produced by microorganisms and is extracted from brown algae. Alginate-based CANEs are a promising field of study. Alginate's properties can be utilized to immobilize enzymes, enhancing catalytic efficacy. These nanoenzymes have greater stability and reusability as well as operating efficiency, making them suitable for several applications in biotechnology fields (Weng et al. 2023). Because of their excellent properties like hydrophilicity, biological adaptability, mucoadhesive properties, bioavailability, eco-friendliness, and

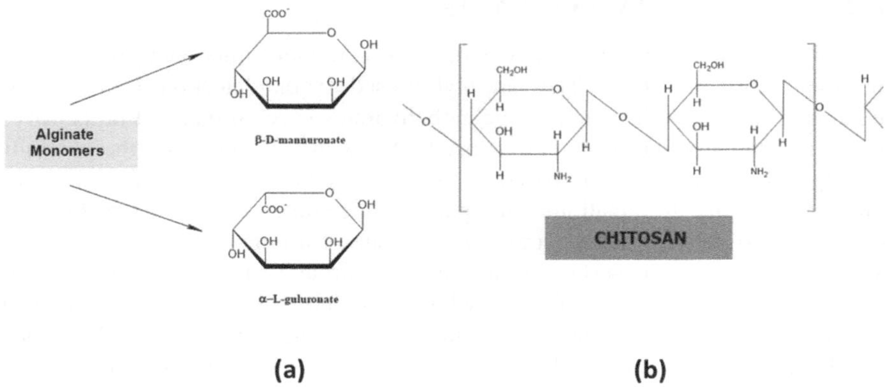

FIGURE 6.2 (a) Diagrammatic representation of alginate monomers existing into two structural forms; (b) Schematic representation of chitosan polymers with repeating sugar residues.

cost-effectiveness, alginates have been widely studied. Biopolymers from alginate form micro- and nano-systems for the controlled and targeted drug delivery in cancer treatment, where anticancer drugs are encapsulated to enhance bioavailability, prolong release, optimize pharmacokinetics, and facilitate bio-clearance (Iravani and Varma 2022). Alginate-based nanocarriers are promising delivery vehicles for bioactive substances. In this regard, in the list of promising wall materials, calcium alginate and sodium alginate should be underlined, mainly due to their non-toxicity and cost-efficiency as well as a very simple technological process of production, biocompatibility, and biodegradability (Karim et al. 2022). In summary, immobilizing the enzymes by alginate microspheres significantly minimizes the loss in activities after immobilization. For example, glucose oxidase coated with particular multilayer nanofilms retained an activity level of more than 84% for three months after the coating and remained active throughout (Srivastava et al. 2005). As demonstrated by Liu et al., the superparamagnetic sodium alginate NPs have been produced for the covalent immobilization of enzymes for the effective reusability of enzymes like Candida rugosa lipase (Xiao Liu et al. 2012). Another study by Obireddy et al. involved alteration of gel-based nanocarriers with "Favipiravir" as a model of a poorly water-soluble and pharmaceutically relevant molecule for the investigation of the effect of three inorganic nanoparticulate modifiers. These consisted of silver NPs (Ag-NPs), montmorillonite clay, and iron-based magnetic NPs, and the effect on encapsulation and sustained release of poorly soluble drugs was also assessed (Obireddy et al. 2023). *In vitro* studies have shown remarkable efficacy in the customized and prolonged release of both hydrophobic and hydrophilic pharmaceuticals. Micro-alginate can protect the stomach and deliver the drug to the colon as a colon-targeted release of orally administered medications. Alginate NPs and microgels may be advantageous in improving stability and drug bioavailability. Alginate-based wound dressings have been widely used. Although alginate systems have proven to have promising results in *in vivo* animal studies, further research on clinical safety and efficacy is needed before they can be used on humans for pharmacological approval (Lai et al. 2024). These alginate nanozymes can reduce greenhouse gas emissions efficiently and also encourage carbon sequestration, which will eventually help fight climate change. Enhancing the stability and efficiency of enzymes such as carbonic anhydrase, alginate-based nanozymes can sequester carbon by transforming CO_2 into stable carbonates (R. Yadav et al. 2011).

6.2.1.2 Chitosan-Based Nanoparticles

The most notable contribution of NPs in nanomedicine is their ability to enhance the bioavailability of the drug. The increasing role of chitosan-based NPs as flexible tools in DDS is well documented by their biocompatibility and non-toxicity (Rizeq et al. 2019). The chitosan molecules possess an active hydroxyl as well as an amino group within them (Fig. 6.2b). Due to this, these allow further association of other functional groups through molecular or chemical modification (cycloaddition, sulfation, and thiolation), hence expanding the utilization of these molecules (Gao and Wu 2022). The Food and Drug Administration (FDA) has approved chitosan as safe for use in a variety of medicinal applications. It can be used to deliver a wide range of therapies, including anticancer medicines, proteins, and vaccines, and this can be achieved in various modes, such as oral and nasal administration (Iacob et al. 2021; Boroumand et al. 2021). The positive charge of chitosan makes it more active in association with negatively charged cell membranes, thereby increasing cellular uptake *via* multiple mechanism routes. The mechanism of using the proton sponge effect leads ChNPs to avoid endosomal degradation and ensure efficient drug release inside the cells. The modification of the chitosan surface results in enhanced specificity of targeting action, which in turn increases the bioavailability of the delivered pharmaceuticals (Sharifi-Rad et al. 2021; Aibani et al. 2021). Moreover, the polycationic properties of chitosan endow the compound with bactericidal activity. Chitosan exerts its antibacterial effect by tying up negatively charged lipopolysaccharides and surface proteins on the membranes of microbial cells with its positively charged NH_2 groups. This binding causes the lysis of cellular components and eventually leads to cell destruction (Ruiz-Rico, Sancenón, and Barat 2023). There is evidence regarding the antifungal activity of chitosan against certain species of fungi, besides its well-established antibacterial activities. The most obvious evidence, based on the research, is that chitosan can be absorbed into the hyphae and will inhibit the enzymes that have a role to play in its development (Basseri et al. 2019). Lin et al. (2020) successfully synthesized histatin1-modified thiolated chitosan hydronanogels, which demonstrated significant advantages in promoting wound healing compared to free hydrogels. These hydronanogels showed great potential for use as wound dressings, offering several benefits for skin repair. Moreover, the hydronanogels demonstrated the ability to release the drug (histatin1) at the target site under physiological conditions, ensuring that the active therapeutic agents were delivered precisely where needed. This targeted release system minimized the potential for systemic side effects and ensured a controlled release, which is especially beneficial for treating chronic or difficult-to-heal wounds (Lin et al. 2020).

Chitosan nanofibers can be used as scaffolds that support the development and maturation of osteoblasts, thus enhancing both the number and quality of trabecular bone and making healing easier (Ho et al. 2014). A study developed liposome-loaded chitosan and poly(ε-caprolactone) nanofibers and explored their possibility of controlling the release of paclitaxel in breast cancer treatment. Nanofibers showed a sustained release of paclitaxel in a period of 30 days coupled with enhanced viability of cells and death of human breast cancer cells. *In vivo* studies exhibited the modulation of tumor mass (Hasanbegloo et al. 2023). Drug release from ChPs involves mechanisms such as the swelling of the polymer, diffusion of the adsorbed drug, diffusion of the drug through the polymeric matrix, erosion or degradation of the polymer, and a combination of erosion and degradation. Other processes include drug diffusion through the matrix's polymeric structure (R. Singh and Lillard Jr 2009; S. Liu, Yang, and Ho 2018). Although ChNPs have enormous drug delivery benefits, it is important to continue with research to fully understand their interactions within biological systems and implications toward improving their therapeutic applications.

Chitosan, a natural biopolymer derived from chitin, has gained attention in agricultural applications due to its biodegradability, biocompatibility, and ability to form NCs that can encapsulate bioactive compounds. The controlled release of nutrients and bioactive compounds, enabled by chitosan NCs, offers significant potential for improving agricultural productivity, particularly under challenging environmental conditions. This approach is crucial for enhancing plant growth and yield, as it ensures that essential nutrients are delivered in a controlled and sustained manner, optimizing their availability to plants over time. When integrated with various nutrients or pesticides, chitosan-based NCs can provide targeted delivery, ensuring that these substances are released at the right time and place to maximize their effect on plant health and growth (Subramanian et al. 2020; Hidangmayum and Dwivedi 2022).

6.2.1.3 Cellulose-Based Nanoparticles

Cellulose is a naturally occurring polysaccharide that has received tremendous attention due to its abundance, biodegradability, and cost-effectiveness (Sampath et al. 2016; Gunathilake et al. 2018). Cellulose structure consists of repeating units of glucose linked via β-1,4-glycosidic bonds as shown in Fig. 6.3(a); the three-hydroxyl group present on each glucose unit forms hydrogen bonds between the adjacent units, resulting in the formation of a stable crystalline structure also known as the microfibrils.

Cellulose-based NPs have gained popularity as promising agents in biomedical applications due to their potential for biocompatibility, biodegradability, and uniqueness. These NPs can be synthesized using various approaches, including ionic liquid solutions, which improve the efficiency of drug delivery and minimize cytotoxicity levels. Their applications are varied, such as drug DDS, tissue engineering, and wound healing, therefore showing how versatile these NPs can be for healthcare purposes (Fuster et al. 2023). CNPs act as an effective DDS that can attain targeted and regulated release of medicinal substances. Improved efficacy along with minimized toxicity stems from better treatment efficiency and reduced side effects of the drugs. This versatility finds its basis due to its enormous surface area; hence, it is suitable for most therapeutic applications (Khorsandi et al. 2024).

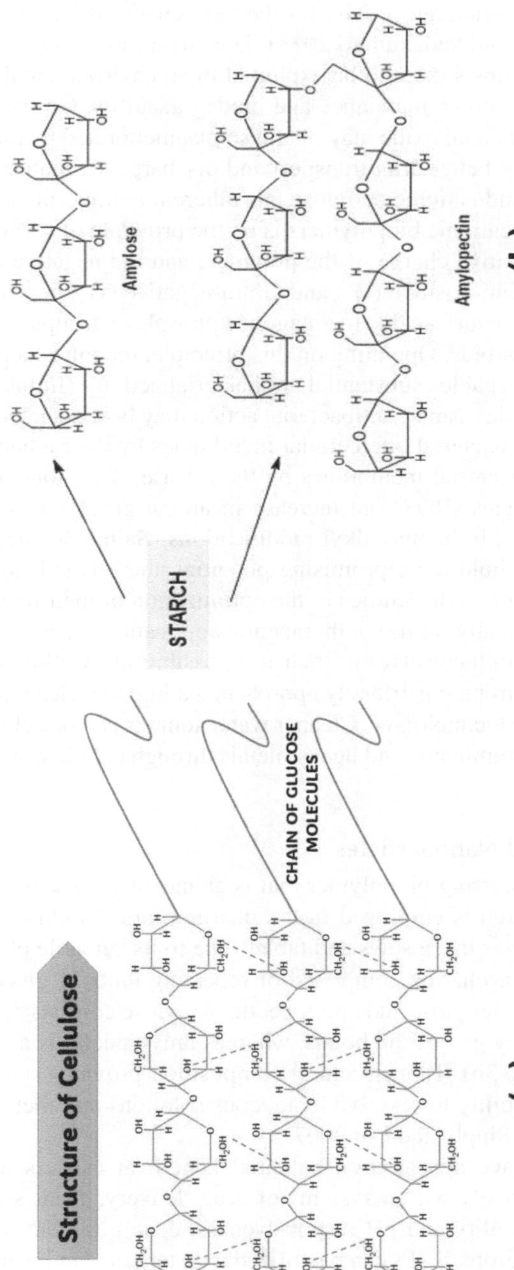

FIGURE 6.3 (a) Illustration of cellulose made up of chain of glucose molecules; (b) Structural forms of starch existing in linear and branched chains.

Another group of researchers demonstrated that cellulosic membrane wound dressings containing cellulose nanofibers possessed highly desirable mechanical features and showed antibacterial activity for their effectiveness in the treatment process (Senthil, Kavukcu, and Vedakumari 2023). The integration of cellulose materials within cancer therapy allows them to be exploited in an environmentally responsive manner through their inherent abundance and biodegradability. On the contrary, the integration of a metal or metal oxide may obtain some ameliorated therapeutic results because it may allow for better drug transport and discharge mechanisms (Patel and Geed 2024). Silver-bound/cationic cellulose has inherent antimicrobial activity. The antibacterial activity of cationic biopolymers is on the principle of great interactional affinity between the positive charge of the polymers and the negative charge of the bacterial cells. Both Gram-positive (G^+) and Gram-negative (G^-) bacterial cells bear a negative charge on their surface due to antagonist phosphate groups existing in peptidoglycan and phospholipids. Operating on this principle, the introduction of a positive charge to cellulose enables substantial antibacterial activity (Imtiaz et al. 2020). Cationic cellulose's mechanism of antibacterial action may be associated with several functions: destabilizing bacterial intercellular membranes by the exchange of Ca^{2+} or Mg^{2+} ions, disturbing bacterial membranes by the release of K^+ ions, the induction of reactive oxygen species (ROS), an increase in amine groups, or enhancements in lipophilicity produced by amino-alkyl modifications (Saini, Belgacem, and Bras 2017). Although CNPs hold quite promising potential, there is still room for overcoming some of the inherent difficulties in the optimization of their manufacture and functionalization specifically for use in therapeutic applications. This calls for further research to properly avail themselves of their full possibilities. Cellulose nanozymes can be the potential environment-friendly approach, leading to a cleaner environment if utilized in sustainable technologies. Cleaner water sources can be achieved through efficient removal of contaminants and heavy metals through cellulose NPs (Abda and Konwarh 2021).

6.2.1.4 Starch-Based Nanoparticles

Starch is a naturally occurring biopolymer that is abundantly present in rice, wheat, corn, and potatoes. Starch is employed in the pharmaceutical industry as a binder, disintegrant, and stabilizer in capsules and tablets due to its versatile physiochemical properties. It is a polysaccharide comprised of repeating units of glucose with two structural forms, namely amylose and amylopectin. Amylose consists of linear chains of α-D-glucose linked by glycosidic bonds, whereas amylopectin is a branched biopolymer (refer to Fig. 6.3b). This structural composition provides starch its unique properties, such as its ability to dissolve in aqueous solutions and metabolize within the human body (Kaur, Singh, and Liu 2007).

Starch-based NPs have appeared as potential beneficial carriers in biomedical applications, more precisely within systems of drug delivery. Their special properties include biocompatibility and pH responsiveness, enabling them to encapsulate therapeutic agents; therefore, SNPs can be utilized with targeted and regulated release mechanisms. SNPs can be designed to identify and attach to certain cancer cells, hence improving the accuracy of medication administration (Pei et al. 2024). The release of medications from SNPs can occur by many methods, including diffusion, swelling, and surface erosion, all regulated by the pH of the environment around them (Yu et al. 2021). The techniques that are used to synthesize SNPs have a significant impact on the features of these molecules. One example of this is the micro-emulsion technique, which allows for the adjustment of particle size and dispersion, both of which are necessary for the effectiveness of SNPs in medication administration (Estrada-Ramos et al. 2023). The flexibility of starch-based NC hydrogels is being demonstrated by the fact that they are now being developed for use in wound treatment and tissue regeneration purposes (da Silva, Scheibel, and Soares 2023). Despite this, dimensions of nanoscale materials are very crucial with the wide features and applications that they possess. The best size of NPs to be used for successful endocytosis is between 25 and 50 nm, while the best size for non-cationic NPs is between 6 and 34 nm (Qiu et al. 2019).

Controlled release is one of the essential components of pharmaceutical delivery systems. Starch-based hydrogels represent versatile approaches for drug delivery in a controlled form. The kinetics of drug release from such hydrogels can be controlled by changing many variables, including composition, crosslinking density, drug-loading methods, and environmental triggers. Understanding processes and kinetics in the controlled release is fundamental in engineering starch-based hydrogels for functional drug delivery systems (C.-S. Lee and Hwang 2023). Although SNPs have significant advantages in drug delivery, challenges still exist in finding an effective approach toward the synthesis and ensuring consistency in performance in the clinical setting. More research is needed to enhance their efficacy and safety profile in human systems. It was also found that hybrid nanomaterials based on starch can be highly efficient in removing organic contaminants as well as heavy metals from soil, water, or air. A hybrid nanomaterial based on starch is superior compared to traditional materials because of a large surface area and reactivity, enabling the effective process of adsorption or breakdown of harmful chemicals (Gamage, Punniamoorthy, and Madhujith 2021).

CAZymes have been broadly classified into three major functional groups that include the synthesis, breakdown, or modification of carbohydrates. The applications of such enzymes in a wide range of industries make them strong selective catalysts following millions of years of evolution, astonishingly very adaptive and readily evolved in the laboratory (J. Wardman and Withers 2024). Nanozymes are nanomaterials that have been characterized to show high enzymatic activity and, thus, are envisaged as better alternatives for natural enzymes. Nano-entities have some unique properties, including the combination of enhanced catalytic activity with outstanding stability and simple and cost-effective production methods. Nanozyme employment is feasible for various applications, including biosensing, immunoassays, disease detection and treatment, and environmental management (Gomaa 2022). CAZymes are present in all domains of life and serve critical roles in the metabolism and physiological processes of living cells. Many CAZymes are built with a modular architecture that includes multiple catalytic domains, such as carbohydrate-binding modules (CBMs) and linker domains, which often define the specificity of their substrates (Windels et al. 2024). CANEs play several roles within multiple biological processes, including glycan production, modification, and degradation. These nanozymes, particularly carbon nanomaterial-based variations, have properties that are similar to those of enzymes, which has further augmented their use in bio-applications (Dhiman et al. 2022). Elucidation of the functions of particular glycans demands the availability of identical and well-characterized glycans. Because of their intrinsic difficulty, obtaining homogeneous glycans for structure-function studies has been challenging and has complicated their chemical synthesis. A steadily more enabling strategy is thus required to combine synthetic and enzymatic methodologies using CAZymes (Bains et al. 2024. Nanozymes activated by carbohydrates, typically called CANEs, are an area that has lately drawn much attention due to their several applications in biomedicine and catalysis. In addition to being a tool for studying carbohydrate–lectin interactions, CANEs exhibit tunable properties and thus promise applications in drug delivery and molecular imaging (Marradi, Martin-Lomas, and Penades 2010).

6.2.2 Synthesis and Characterization of CANEs

6.2.2.1 Synthesis of CANEs

Catalytic NPs are acknowledged to possess distinct characteristics in comparison to normal enzymes. Nanoenzymes have various advantages such as stability, multiple utility, and low cost in general (Stasyuk et al. 2020). Some of the most common synthesizing routes for CANEs are outlined below.

6.2.2.1.1 Bottom-Up Approach

The bottom-up approach for nanoparticle synthesis is building NPs from smaller compounds or atoms. The process involves slowly constructing individual atoms or molecules into NPs using self-assembly, chemical reactions, or nucleation processes, as shown in Fig. 6.4(a). This method allows exact control over the NPs' dimensions, morphology, and content (Saleh 2022) . Furthermore, the bottom-up approach would be more relevant and logical for fabricating nanostructured systems in the biomedical field since it controls process-related size, composition, shape, and uniformity through an efficient growth process (Sen et al. 2024).

Chemical vapor deposition method

Chemical vapor deposition (CVD) can make very thin films of material deposited onto a substrate by chemical processes involving precursors in the vapor phase. The substrate, while exposed to one or more volatile precursors, will disintegrate into fragments on its surface and deposit a solid substance. This technology enables accurate control over the nanozyme composition, structure, as well as dimensions. In this regard, it allows for the design of targeted catalytic functions for the nanozymes. For the synthesis of nanozymes, CVD is widely used because it provides homogeneous coatings, high-purity materials, and large-scale manufacturing (Hachem et al. 2022; Sun et al. 2021). CVD has the potential for the synthesis of very homogeneous and finely controlled nanostructures and so may form an interesting direction for carbohydrate-activated nanozymes. GO (graphene oxide) or rGO (reduced graphene oxide) synthesized via CVD can also be functionalized by attaching carbohydrates

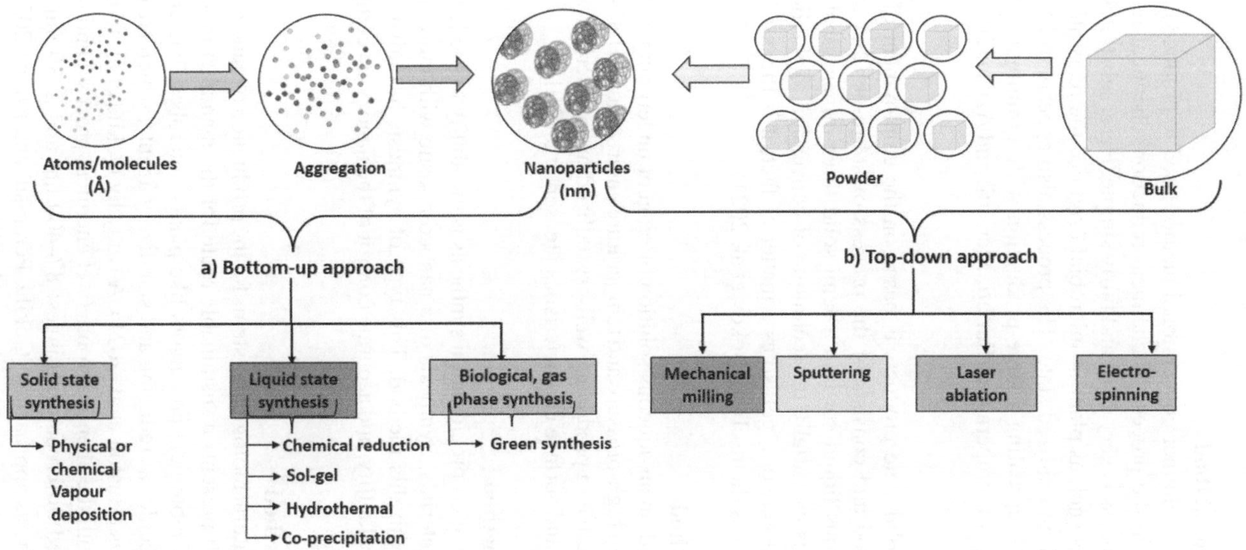

FIGURE 6.4 Schematic illustration of different methods for the synthesis of NPs using (a) Bottom-up and (b) Top-down approaches. The figure is created using a free version of BioRender software (www.biorender.com).

to enhance biological efficiency. Glucose-functionalized graphene nanozymes show good glucose oxidase-like activity and, therefore, are competent for glucose detection and diabetes monitoring (Yoon et al. 2024; Gonçales et al. 2016).

Chemical reduction method

Chemical reduction is another conventional method used to produce nanozymes. As the name suggests, in the process of chemical reduction, chemical-reducing agents are utilized. This process is very swift and fairly simple. Here, substrates can be either from natural sources such as plant or microbial origin or the reagent/chemicals that would reduce under an oxidized state. The process also appears to be economically friendly and easy for upscaling a large production with a minimum requirement of high-pressure energy or temperature (Suriati, Mariatti, and Azizan 2014)

Sol-gel method

The formation of solid-state products is based on the employment of a wet chemical synthesis method and consists of the process of sol-to-gel that transforms the colloidal, homogenous liquid sol into a semi-solid three-dimensional material that appears gel-like. It is essentially the technique of conversion of the liquid solution, or sol, into the solid network, gel, so as a matter of fact, it is the condition wherein nanozymes can express themselves (Bokov et al. 2021).

Hydrothermal method

This method is based on an aqueous solution of reagent undergoing a chemical reaction within a confined high-pressure and temperature environment. Hydrothermal synthesis is highly effective in producing well-crystallized nanomaterials with regulated forms, dimensions, and surface characteristics (Jensen et al. 2014).

Co-precipitation method

The main advantage of co-precipitation synthesis is the ability to produce pure, clean, and uniform nanomaterials. Synthesis of a nanocomposite with more than one component is feasible with this method. This type of synthesis has advantages such as ease of operation, scalability, and accurate control of the nanozyme composition (Wei and Wang 2008).

Green synthesis method

The application of natural biological systems for the production of nanozymes is termed as green synthesis. It presents a sustainable route for the manufacture of nanozymes involving the use of biological precursors like plants, microbes (bacteria, yeast, and fungus) and biomolecules as reducing and stabilizing agents, which are dependent on low-energy processes (Huston et al. 2021). A study by Gokila et al. found that Fe_3O_4 synthesized using both the chemical route (wet chemical process) and the green synthesis route exhibited different size ranges: 87–400 nm and 2–80 nm, respectively, using different synthesis approaches (Gokila, Perarasu, and Rufina 2021). The green synthesis method was sustainable and eco-friendly since the phytochemicals of the cinnamon were used as reducing and stabilizing agents to prepare Ag-NPs from the cinnamon extract. The synthesized NPs proved to be a colorimetric nanoenzyme sensor for peroxidase and xanthine oxidase-like activities. Green synthesis also minimizes the extent of environmental contamination by omitting dangerous chemicals that are often present in traditional chemical procedures. Procedures require less energy and mostly occur at room temperature so that carbon emissions can be minimized (Nazir and Amin 2021; Kumaraguruparaswami et al. 2021). The nanozymes manufactured by green synthesis are pretty effective in cleaning contaminated soil and water, bringing the ecological balance closer. Eco-friendly solutions as well as sustainable practices in several diverse industries are fostered through green nanotechnology integration (Gour and Jain 2019).

6.2.2.1.2 Top-Down Approach

The top-down approach for NP synthesis involves the fragmentation of bigger materials into smaller particles to achieve nanoscale dimensions of NPs, as illustrated in Fig. 6.4(b) (Asha and Narain 2020). Self-assembly is a good approach for organic NP synthesis. Because of its precision and ease of use, particle replication in non-wetting templates (PRINT) was recently described as a promising technique for the preparation of defined organic NPs. PRINT is a top-down imprint lithography technique that can produce monodisperse and non-spherical particles with well-controlled form, composition, and surface characteristics. Sizes range from 20 nm to more than 100 μm (Pradhan and Saha 2022; Chakraborty and Patni 2014).

Mechanical milling method

Mechanical milling is an uncomplicated and economical method for synthesizing nanozymes. The formation of nanomaterials in a simple, cost-effective manner with high yield has been an ongoing problem since the inception of nanoscience. The process entails milling precursor materials into NPs *via* high-energy ball mills. The milling process reduces particle size and increases surface energy, hence enhancing catalytic qualities that emulate natural enzyme activity (T. P. Yadav, Yadav, and Singh 2012).

Sputtering method

In the production of nanozymes, the sputtering technique involves depositing NPs onto a liquid substrate, thus bringing allowance for fine control in regard to their characteristics. This method uses vacuum sputter deposition on low-volatile liquids, allowing control over particle development and attributes such as size, shape, and composition (Nguyen, Deng, and Yonezawa 2022).

Laser ablation method

The word "ablation" refers to eliminating outermost atoms by a multiphoton excitation, namely thermal evaporation (Shahid et al. 2022). It is a high-energy technique for producing NPs, including nanozymes, also in a liquid medium using laser pulses. This method permits control over features such as dimensions and morphology of the NPs, which are of great importance for catalytic action (Huang and Lai

2018). The process initiates with the laser focusing on a target material dissolved in a liquid, which leads to material exfoliation followed by plasma formation. It is through this plasma, which expands, that the NPs can be formed easily by shock waves because the cool plasma releases energy into the liquid (Sadrolhosseini et al. 2019). A femtosecond laser, characterized by its strong pulse, is a prevalent source of radiation utilized for laser ablation in tissue engineering applications (Tran and Wen 2014).

Electrospinning method

Electrospinning is one of the most significant processes in the production of NPs, notably nanofibers from different materials. In electrospinning, a charged jet of polymer solution creates fibers as the solvent evaporates. This method is reported to be reliable, flexible, and highly manipulable in terms of the shape of fibers produced (Reddy et al. 2016). This is achieved through the help of a high-voltage electrostatic field generation and attraction of the polymer solution into a finer shape (Tran and Wen 2014). Nanofibers, thus produced, may vary in diameter from some nanometers to micrometers, thus being applied in a wide variety of applications. Some of the potential benefits of electrospun fibers are an extremely high surface-to-volume ratio, substantial porosity, and superior mechanical properties. Thus, they are applied in an extensive range of functional applications such as tissue engineering, environmental filtration, wound care, and energy solutions or even for the protection of textiles and DDS (Angel et al. 2022).

6.2.2.2 Characterization of Carbohydrate-Active Nanoenzymes

Characterizing nanoenzymes or NPs is essential for comprehending their distinctive features and enhancing their applicability in diverse domains. The characterization of NPs is usually done for two purposes: the first one deals with the determination of the size and shape of specific NPs, and the second one determines the elemental crystal structure, composition, dimensions, and morphology of a multitude of other physical features (Mourdikoudis, Pallares, and Thanh 2018). Fig. 6.5 includes a representation of diverse techniques that may be utilized to characterize the various kinds of NPs (Mujahid et al. 2022). For the morphological and internal analysis of the NPs, techniques such as scanning electron microscopy (SEM) and transmission electron microscopy (TEM) offer comprehensive records of surface morphology, interior structural insights, chemical characterization, and particle sizes (Vladár and Hodoroaba 2020; S. Zhang and Wang 2023).

DLS (Dynamic Light Scattering) is the most extensively used method for NP size distribution. It calculates all the information about the size distribution of particles by assessing the oscillations in light scattering produced through the Brownian motion of NPs within the solution. NTA (Nanoparticle Tracking Analysis) is a method used to calculate the size and distribution of NPs in a liquid solution (Babick 2020). In NTA, particles are illuminated using a laser, and the scattered light resulting from it is captured. Under the real-time observation of a microscope and camera, the scattered light will be detected so that each nanoparticle's size may be analyzed after undergoing Brownian motion (Van Der Pol et al. 2014).

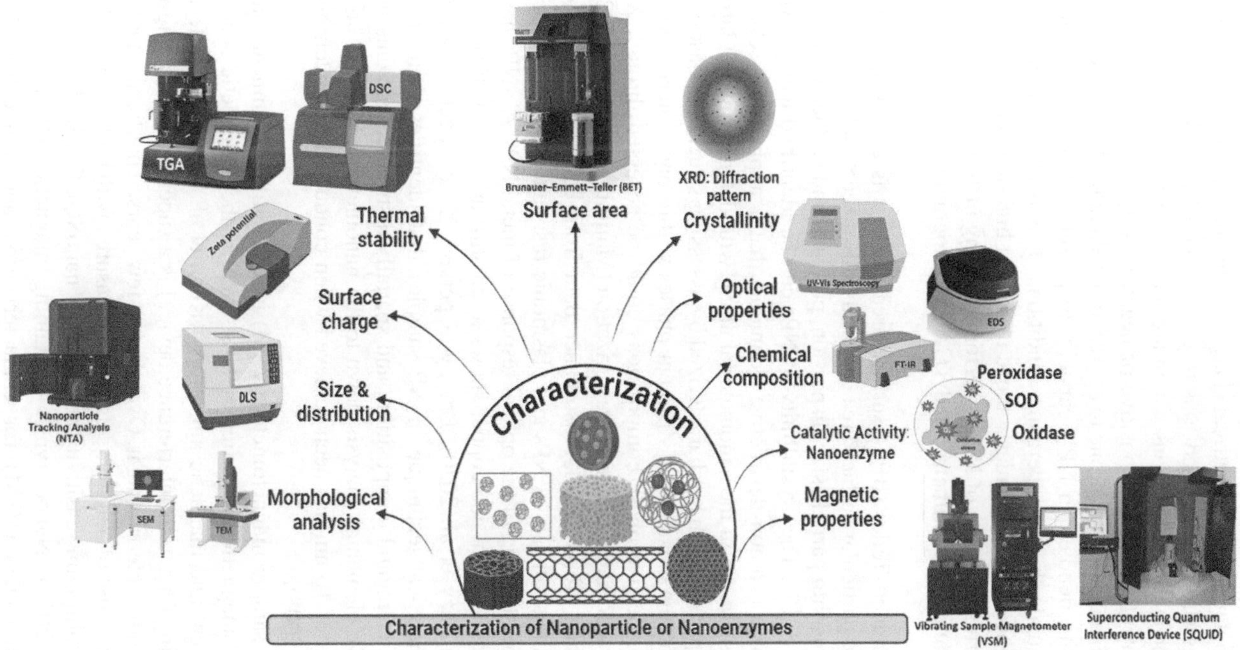

FIGURE 6.5 Schematic depiction of several methodologies employed in the characterization of nanoparticles or nanoenzymes. The figure is created using a free version of BioRender software (www.biorender.com).

Zeta potential can also be calculated by the DLS technique. The Zeta potential of NPs indicates that the suspended ions in the solution interact with the NPs when they are suspended in a liquid, thus creating an electrical double layer around the NP called electrokinetic potential. The electrical charge around the NP has two layers: the inner layer comprises strongly attached ions to the surface of the NP, and the outer one consists of oppositely charged ions which are more dispersed (Bhattacharjee 2016). A high zeta potential, whether positive or negative, signifies that the NPs are thoroughly disseminated and less prone to aggregation (Shrestha, Wang, and Dutta 2020). Thermal stability denotes an NP's capacity to preserve its chemical and physical characteristics under high-temperature conditions. This is an essential characteristic for several applications, particularly those related to high-temperature conditions or processes. Thermal stability of NPs is performed using instrumentation such as DSC (differential scanning calorimetry) and TGA (thermogravimetric analysis) (Nurazzi et al. 2022).

Brunauer-Emmett-Teller (BET) surface area analysis is a prevalent method for determining the particular surface area of porous substances, including NPs. It offers essential insights into particle size dispersion, porosity, and surface properties (Shaji and Zachariah 2017). The crystallinity of NPs can be identified using XRD (X-ray diffraction) which is a non-destructive technique, that is applied to assess the structure and phases present in the nanomaterial along with determining the lattice plane position. Nanoparticles can be characterized by assessing structural phases and phase purity. Based on the existence of different phases within one material system, their contents and respective phase compositions could be accessed through XRD as a part of steps toward nanoparticle characterization (Majithia and Barretto 2023; Reed-Gore, Moser, and Weiss 2018). Absorbance data from UV-visible spectroscopy is used to prove the synthesis of NPs and plasmonic resonance. The resonance is critical for the understanding of the optical response of metal NPs. Ag-NPs have shown the maximum at 420 nm, and copper shows a red shift at 560 nm (Krasochko et al. 2020; Alim-Al-Razy et al. 2020). Energy-dispersive X-ray spectroscopy (EDS) is an essential analytical technique in NP samples that identifies their composition concerning the elemental composition, and spatially distributed contents within the NPs. Qualitative elemental analysis down to the nanometric scale with an approximate resolution of 10 nm has been achieved when conducted in tandem with SEM (Herzing et al. 2008).

A vibrating sample magnetometer (VSM) is a scientific instrument that is used to measure the magnetic properties of NPs, such as their magnetic susceptibility, magnetic moment and magnetic anisotropy (Khan et al. 2017). In short, a SQUID (Superconducting Quantum Interference device) is a superconducting ring with two Josephson junctions interrupting it. Over the last few years, it has been established that the magnetic response of nano-objects can be measured with high accuracy by using a SQUID with a minimum sensitivity area, namely nanoSQUID (Shukla 2022; Granata and Vettoliere 2016). Nanoenzymes containing magnetic NPs require vibrating sample magnetometers (VSM) for characterization and application. VSM was utilized in the development of magnetic nano-reagents that can detect tumor markers, thus proving its versatility in biomedical applications. Experiments have proven that iron-oxide nanoparticles (IONPs) exhibit enhanced enzymatic activity upon magnetic stimulation, and VSM can measure the magnetic field precisely to understand the mechanism of enhancement (Chieh, Huang, and Shi 2015; Y. He et al. 2022). Since magnetic fields may be observed at the scale of single spins, nanoSQUIDs represent good devices for assessing the magnetic properties of nanoenzymes. NanoSQUIDs open up possibilities for observing the biological processes; among them, the most important ones are mediation transport and drug delivery processes used in anticancer treatments (Granata, Silvestrini, and Vettoliere 2020).

6.3 BIOMEDICAL APPLICATIONS OF CARBOHYDRATE-ACTIVE NANOENZYMES

CANEs have emerged as a driving force in various scientific disciplines, particularly in the field of biomedicine (Verma et al. 2020). Carbohydrate biopolymers provide unique physical and biological properties enabling the development of sustainable, biocompatible, and biodegradable NPs compared to current interventions (Jacob et al. 2018; Birrenbach and Speiser 1976; Couvreur et al. 2006). For example, traditionally, polymethacrylate and polyacrylamide were key synthetic polymers that were used for normal systemic administration. However, prolonged usage of these non-biodegradable polymers exhibits chronic toxic impact on human health and the environment (Jacob et al. 2018).

Carbohydrate-based NPs can provide an efficient alternative in imaging, biosensing, antimicrobial, regenerative medicine, and drug delivery systems for controlled, sustained, targeted release (Jacob et al. 2018). Moreover, these characteristics make CANEs highly desirable in various applications, particularly in the field of biomedical science, where such properties are crucial for safety, efficacy, and environmental impact.

6.3.1 Potential of CANEs in Diagnostics

Carbohydrate-based nanomaterials offer unique physiochemical properties that are excellent for diagnostic applications in the field of biomedical research. Since carbohydrates play a major role in mediating biomolecules and biological interactions, the fabrication of CANEs for biorecognition opens a wide range of platforms for cellular imaging, especially in sensing and imaging applications (Hao et al. 2016). Carbohydrates are often coupled with NPs or NCs to detect and visualize specific biological targets. In contrast to antibodies and enzymes, CANEs display stability, functionality, and specificity in various cellular environments (Adak, Li, and Lin 2015). This section explores the various potential of CANEs as imaging and biosensing tools in biomedical research.

6.3.1.1 Biosensors

CANEs-based biosensors leverage a broad range of properties of biopolymers. They offer a unique advantage of biomolecule immobilization and selective and sensitive detection of analytes in food safety, environmental, and biomedical applications.

Chitosan conjugated with metal oxides NC are commonly used for the development of biosensors. One such example is the HRP (horseradish peroxidase) biosensor based on Al_2O_3 and chitosan. It shows excellent sensitivity and selectivity in the detection of phenolic compounds, particularly hydroquinone, making it a promising tool for both environmental and biomedical applications (Xiaojuan Liu et al. 2011). To develop biosensors for measurement of G6P level in blood serum samples with glucose-6-phosphate dehydrogenase (G6PD) enzyme, a study was conducted to compare biosensor models with polypyrrole (PPy) conjugated with chitosan-Fe_3O_4 NC attached to G6PD (BS-2) in comparison to the PPy-G6PD conjugated biosensor (BS-1) (Sahin et al. 2019). The two biosensors were analyzed based on a blood serum interference study with major components such as urea, glucose, and cystine against G6P detection. BS-2 demonstrated superior sensitivity and selectivity against blood components with a linear range of 0.025–0.05 mM as compared to the range of BS-1 (0.025–0.25 mM), indicating its potential in measuring the level of G6P even at lower concentrations. In addition, chitosan was applied in the development of a nanoscale biosensor by the unique fabrication of multiwalled carbon nanotubes (MWCNT), platinum NP, and a chitosan-silica matrix enhanced with immobilized enzyme glucose oxidase (X. Kang et al. 2008). The unique combination offers rapid quantification of glucose with low interference, allowing precise, stable, and reproducible glucose monitoring and measurement in human serum samples. In another study, alginate cryogel coupled with $NiFe_2O_4$ NPs also enhanced glucose detection as they strengthened enzyme immobilization and electron transfer with a low detection limit of 0.32 mM, making it a sensitive and efficient glucose biomarker (Fatoni et al. 2021).

Few studies have recently shown interest in the development of NC using layer-by-layer (LbL) assembly design to form biosensors with versatile properties. A study by Nangare et al. shows that the conjugate of platinum-chitosan-sodium alginate (Ch-NA-Pt) LbL NC enhances detection by cationic and anionic interaction due to the presence of both chitosan and alginate, respectively. Ch-NA-Pt biosensor successfully detected BACE-1 (β-site amyloid precursor protein cleaving enzyme-1), which is a key biomarker in Alzheimer's disease. Moreover, selective and real-time detection in complicated BACE-1 specimens showed their practical usefulness in early detection of Alzheimer's disease. Similarly, Wu et al. (2007) created a glucose biosensor using a six-layer configuration LbL assembly of chitosan-gold-glucose oxidase. Such a configuration exhibited long-term stability, higher catalytic activity, rapid detection of glucose in 8 seconds, and a broad range of 0.5 to 16 mM with a minimum detection limit of 7.0 μM.

Biosensors based on cellulose are more cost-effective compared to current commercial methods in the detection of circulating tumor cells (CTC). These sensors were assembled by coating a cellulose acetate membrane with melamine and furfural polymerized NP (Zhao et al. 2024). The rough nanoscale surface of these composites facilitated accurate isolation of CTCs from the blood samples, allowing quantitative, robust detection *via* colorimetric covalent organic framework-based nanozymes.

6.3.1.2 Imaging

Carbohydrate-based NPs and NCs in imaging systems operating as carriers for contrast agents or fluorescent biomarkers facilitate precise visualization of biological dynamics, early-onset disease detection, and advanced diagnosis in the fields of neurology, oncology, and pathogenic diseases. This section mainly emphasizes imagining a CANE carrier of contrasting agents for tumor visualization. Gadolinium (Gd^{3+}) chelates are commonly used contrast agents in magnetic resonance imaging (MRI) due to their ability to modify the magnetic properties of tissues, which enhances the image quality (Asadollahzade et al. 2022). Despite that, Gd^{3+} chelate lacks stability at low pH and may dissociate from the chelate, causing systemic toxicity. Additionally, metal leakage, aggregation, and complex synthesis are some of the issues that have prompted researchers for alternative approaches in MRI (Asadollahzade et al. 2022).

Among alternative solutions, chitosan-based nanogels provide a matrix that is covalently stabilized by Mn-t-CDTA (manganese-trans-1,2-diaminocyclohexane-N, N, N', N'-tetra acetic acid), and Mn^{3+} can also function as a cross-linker and contrasting agent (Carniato et al. 2023). The addition of chitosan nanogels improves relaxation rate and increases stability and efficient tumor accumulation in preclinical studies. Alginate-based hydrogels can also incorporate Mn ions or Mn-NP, which has aided manganese-enhanced MRI in precision, efficacy, and monitoring of cell transplantation and stem cell delivery (Araszkiewicz et al. 2022).

In addition, Joshi and his group have developed nanohybrids with chitosan-encapsulated doxorubicin (DOX) and iron oxides. This NC consisted of inherent magnetic properties and was used in both imaging and treatment of cancer (Joshi et al. 2023). The nanohybrids were characterized by TEM, XRD, and FTIR; sustained effective drug delivery; and specifically enhanced tumor detection in MRI. In another study, chitosan complexed with radioisotopes such as ^{64}Cu, ^{68}Ga, ^{90}Y, ^{153}Sm, ^{166}Ho, and ^{177}Lu by ionotropic gelation produced radiolabeled NP (Gaikwad et al. 2021). Studies in the C57BL/6 mice model showed that chitosan-radioisotope remains attached to the NPs without significant degradation during imaging and treatment.

In a study by Chung et al., an NC probe was developed using glycol chitosan-coated superparamagnetic iron oxide (BCMG-SPION). Dual-mode imaging with MRI and NIRF (Near-Infrared Fluorescence) recognized tumor progression and matrix metalloproteinase-2 activity (used for emission of NIRF signals). The emission of NIRF signals peaked at 48 hours post-injection with BCMG-SPION. Animal studies revealed 34% enhanced contrast in comparison to normal muscle tissue. The hybrid probes increased image contrast while being specific in the identification of the tumor. NIRF imaging of extracted organs from BCMG-SPION-treated models presented eight times higher fluorescent signal emission than non-cancerous tissue (Chung et al. 2019).

Time is of crucial essence in the detection of cancer. Current conventional imaging techniques are limited in their sensitivity and resolution in detecting cancer in its initial stages (Pulumati et al. 2023). These studies highlight the potential of innovative approaches that can greatly contribute to current limitations in the detection and imaging of cancer diagnostics. Nevertheless, further evaluations are required to assess the long-term stability, biodistribution, and cytotoxicity of these NPs and NCs.

6.3.2 CANEs-Based Drug Delivery Systems

Carbohydrate- or polysaccharide-based NPs play a multidisciplinary role in the DDS. In general, polysaccharides consist of repeating units of high molecular weight monosaccharides (Yang and Zhang 2009). This enables them to have an increased surface area-to-volume ratio, facilitating surface modifications (Yang and Zhang 2009; Jacob et al. 2018). This characteristic of polysaccharide has been exploited in the development of the drug delivery system (Jacob et al. 2018).

The challenges associated with non-functionalized NPs are limited bioavailability, inadequate solubility, enzymatic degradation hindering intestinal absorption, lack of specificity to the target site, reduced therapeutic potency, and adverse effects (Kukoyi 2016). However, alteration to the surface defends it against rapid opsonization and elimination by the mononuclear phagocyte system, making it a more suitable therapeutic option (Nitta and Numata 2013).

Four extensively studied polysaccharide-based NPs in DDS are chitosan, alginate, cellulose, and starch, as they are excellent for delivering agents such as proteins, peptides, and nucleic acids (Calvo et al. 1997; Wang et al. 1997). These versatile properties make CANEs a promising tool for enhancing drug delivery systems and minimizing the use of chemical interventions that affect both human health and the environment. Table 6.1 summarizes the recent drugs used in DDS utilizing carbohydrate-based carriers.

6.3.2.1 Chitosan-Based DDS

Chitosan is a derivative of chitin, abundantly found in the shells of marine crustaceans such as crab, shrimp, lobster, and krill. Chitin is functionalized as chitosan by alkaline partial deacetylation as shown in Fig 6.2. The term 'partially deacetylated α-chitin' refers to the removal of a few acetyl groups, which creates a structure that is insoluble in aqueous solutions and inorganic solvents (Lueßen et al. 1994). Chitosan is composed of glucosamine and N-acetylglucosamine units, making it hydrophilic in nature (Lueßen et al. 1994). This facilitates protonation under acidic conditions, making the polymer soluble in mildly acidic solutions. This is a crucial feature in the preparation of hydrogel or gels with other polyanions, increasing the duration of exposure and enhancing the absorption efficacy of the drug (Bystricky et al. 1991; Xu et al. 2018).

For example, curcumin is a highly potent bioactive compound known for its poor aqueous solubility and bioavailability profile (Bhawana et al. 2011). However, encapsulating curcumin with chitosan-carboxymethyl nanocellulose formulation significantly improved drug delivery kinetics and prolonged drug release against *Pseudomonas aeruginosa* (Samrot et al. 2018). Chitosan nanohybrid is evidenced to improve the biocompatibility and better targeted release of drugs, which can be visualized in chitosan-vanillin hybrid with calcium ferrite NPs that significantly enhanced biocompatibility and controlled the rate of curcumin release in both gastric fluids (pH 1.2) and body fluids (pH 7.4) environments, respectively (Kamaraj et al. 2018). The drug release profile at varying magnetic fields indicated an increased rate of drug release and specificity to the target site. In a similar study, curcumin was efficiently encapsulated with chitosan-Tween 20 hybrid through the method of spray drying, which extended the burst release up to 2 hours, constructing a sustained drug release system (O'Toole et al. 2012).

Gingerol is a phytochemical known for its antibacterial properties. However, its poor bioavailability and low aqueous solubility reduce the efficacy of the compound (Kim et al. 2005; Singh et al. 2018). To mitigate this challenge, chitosan was combined with

TABLE 6.1
List of Drugs Embedded in Fabricated Carbohydrate-Based NPs or NCs

Biopolymer	Modified Material	Drug/Model Drug	References
1. List of chitosan biopolymers in DDS			
N-maleoyl chitosan	β-Cyclodextrin	Ketoprofen	Hou et al. (2017)
Chitosan	2-Chloro-N, N-diethylethylamine hydrochloride	Quercetin	de Oliveira Pedro et al. (2018)
Chitosan	Succinic anhydride, glutaric anhydride	Paclitaxel and docetaxel	Skorik et al. (2017)
Chitosan	Glycidyltrimethyl ammonium chloride, gelatin	Dopamine	Y. Ren et al. (2017)
Chitosan	Poly(DL-lactide-co-glycolide)	Donepezil	Takeuchi et al. (2017)
Catechol modified-Chitosan	Genipin	Sulfasalazine	Jinke Xu et al. (2017)
Chitosan	Calcium carbonate	Methotrexate	J. Wang et al. (2018)
Chitosan	Gold	Curcumin	K. M. Rao et al. (2018)
Chitosan		Benzydamine	Jiang, Cai, and Li (2017)
2. List of alginate biopolymers in DDS			
Alginate	Glutamic acid	Zidovudine	Joshy et al. (2018)
Alginate	Ethylene glycol tetra acetic acetic acid	Doxorubicin-loaded liposomes	Shtenberg et al. (2018)
Sodium alginate	Agar	Indomethacin	Yin, Wang, and Wang (2018)
Sodium alginate		Retinoic acid	F. Wang, He, and Chen (2018)
Sodium alginate	Graphene oxide, Protamine sulphate	Doxorubicin	Xie et al. (2018)
Sodium alginate	Graphene oxide	5-Fluorouracil	B. Zhang et al. (2017)
Alginate	Brushite	Ibuprofen	Dabiri et al. (2017)
Sodium alginate	Calcium chloride	Ketoprofen	Bonilla et al. (2018)
Alginic acid	Poly(ethylene glycol) bis(amine)	Doxorubicin	Tawfik et al. (2018)
3. List of cellulose biopolymers in DDS			
Carboxymethyl cellulose	Aminated fumed graphene	Doxorubicin	Z. Rao et al. (2018)
Cellulose	Poly(methyl methacrylate)	Betulinic acid	Dai and Si (2017)

(continued)

TABLE 6.1 (Continued)
List of Drugs Embedded in Fabricated Carbohydrate-Based NPs or NCs

Biopolymer	Modified Material	Drug/Model Drug	References
Microcrystalline Cellulose	Cetyltrimethylammonium bromide	Luteolin and luteoloside	Qing et al. (2016)
Carboxymethyl cellulose	Calcium cation, chitosan and graphene oxide	5-Fluorouracil	Ren Wang et al. (2017)
Carboxymethyl cellulose	Graphene quantum dot	Doxorubicin	Javanbakht and Namazi (2018)
Passion fruit peel cellulose		Tetracycline	Wijaya et al. (2017)
Hydroxyl ethyl cellulose	Ferrous-ferric oxide	Doxorubicin	Bekaroğlu, İşçi, and İşçi (2017)
	4. List of Starch biopolymers in DDS		
Starch	Acetic anhydride	Doxorubicin	Xiao et al. (2016)
Starch	Graphene oxides	Hydroxycamptothecin	K. Liu et al. (2015)
Starch	Polyvinyl alcohol	Chlorpheniramine maleate	Jaiturong et al. (2018)
Aminated starch	Iron oxide, glutaraldehyde, genipin and citric acid	Curcumin	Saikia et al. (2016)
Starch	Ferrous-ferric oxide, poly(ethylene phthalate)	Tungstophosphoric acid	Hamidian and Tavakoli (2016)
Carboxymethyl starch	Lecithin	Mesalamine	Mihaela Friciu et al. (2013)
mPEGylated starch	3,3'-Dithiodipropionic acid	Doxorubicin	C. Wu et al. (2016)
Starch acetate	Polyethylene glycol, gelatin	Cisplatin	Raj and Prabha (2016)

Source: Extracted from Jacob et al. (2018).

phytosome NPs to form a gingerol-phytosome-chitosan (GPC) complex (Singh et al. 2018). Cell culture studies established the improved biocompatibility of GPC-NC, along with sustained release of gingerol. In support of these initial findings, preclinical analysis also indicated that the GPC complex improved bioavailability, prolonged oral absorption in hematological samples of rabbit, and enhanced antibacterial activity against *Staphylococcus aureus* and *Escherichia coli*. Chitosan can also be utilized as nanocrystals functionalized with antimicrobial peptides, enabling robust permeation of bacterial cell walls specifically targeting antibiotic-resistant bacteria (Hrapovic et al. 2022).

6.3.2.2 Alginate-Based DDS

Alginate is a naturally occurring anionic biopolymer found in the family *Phaeophyceae*, also known as brown algae (Smidsrod et al. 1990). The major component of its structure is composed of repeating units of α-L-guluronate and β-D-mannonate residues, linked by glycosidic bonds (refer to Fig. 6.3) (Fischer and Dörfel 1955). Both α-L-guluronate and β-D-mannonate residues contain a carboxyl group, which at a physiological pH of 7.4 facilitates deprotonation, resulting in a negatively charged, anionic polysaccharide alginate (Lee and Mooney 2012). This key feature enables alginate to form gel or hydrogel with divalent cations without requiring chemical and extreme interventions, making it an ideal candidate for encapsulation of drugs and cells in DDS (George and Abraham 2006).

Early experiments conducted in 2007 for DDS of nifedipine focused on the development of microspheres by conjugating methylcellulose and sodium alginate that carried "nifedipine" (Ramesh et al. 2007). The encapsulation with carbohydrate-based microspheres extended drug release lasting up to 12 hours. Subsequently, alginate-$CaCO_3$-DNA NC embedded with chemotherapeutic agent DOX and cytotoxic gene achieved successful dual delivery of both gene therapy and chemotherapy (Zhao et al. 2012). The alginate NC significantly improved gene transfection, effectively elevating cell inhibition while selectively reducing the cell viability. Alginate combined with glutamic acid functions as an NC carrier for the anti-HIV drug "Zidovudine," establishing gradual sustained drug release in a stimulated human physiological environment with phosphate-buffered saline at pH 7.4 (Joshy et al. 2018). Chitosan-alginate NCs have also become a popular encapsulating polymer for antibiotic drugs like 'vancomycin,' improving the drug delivery mechanism to the targeted site (Unagolla et al. 2018).

In another study by Ding et al., Zein-Alginate nanogel loaded with curcumin displayed promising results in increasing encapsulation efficiency, enhanced stability, and controlled release in a stimulated environment. The unique fabrication and characterization of Zein NPs with alginate created a hybrid NC that facilitated calcium-induced gelation. Thus, the presence of calcium enhanced the photostability and thermal stability of curcumin along with gradual and sustained release, increasing its bioavailability (Ding et al. 2023).

6.3.2.3 Cellulose-Based DDS

The structure of cellulose makes it an ideal candidate for a wide range of applications in DDS (Fig. 6.4). Cellulose NPs are made by introducing various structural modifications through the process of nanoprecipitation or by acid or alkali hydrolysis, giving versatile surface properties and colloidal stability (Löbmann and Svagan 2017). Cotton or jute fibers are fabricated or modified to form cellulose NPs that can create a controlled and sustained drug delivery system (Orasugh et al. 2018; Chin et al. 2018). Jute NC fabricated by conjugation of cellulose nanofibrils with hydroxypropylmethycellulose functions as a carrier of macro- or small drugs for sustained drug delivery (Orasugh et al. 2018). Nanocapsules can be formulated through green and sonochemical methods for cancer therapy by utilizing carboxymethylcellulose-based nanocapsules functionalized with folic acid on the surface (S. He et al. 2021). The presence of folate on the nanocapsules surface mitigates precise binding on folate receptors that are overexpressed in a tumor cell membrane and mediates receptor-based endocytosis. In addition, cellulose is also used to create nanocrystalline cellulose using a combination of enzymatic hydrolysis and sonication (Karimian et al. 2019). Using a similar method, You et al. (2022) developed a multifunctional cellulose nanocrystal system that acts as a carrier of NC conjugate Cu^{2+} ions and Au NPs embedded with DOX, enhancing ROS generation and promoting Fenton reaction in cancer therapy.

In a recent study by Ramzan et al., hydroxypropyl cellulose (HPC) was optimized for delivery of delafloxacin. The characterization of the system showed that cellulose derivative plays a pivotal part in enhancing solubility, stability, and drug release rate by creating amorphous dispersion and stabilizing delafloxacin. The presence of HPC enhanced the bioavailability of the drug, making it an effective formulation for oral administration (Ramzan, Yousaf, and Khan 2024).

6.3.2.4 Starch-Based DDS

Modification of ampholytic starch involves the addition of functional groups such as carboxymethyl and aminoethyl to the polysaccharide chain for controlled and sustained drug release. In a study conducted with aspirin, metformin, and mesalamine encapsulated with ampholytic starch derivatives, self-assembled matrices assisted controlled drug release for the first 2 hours in a stimulated gastric environment while maintaining the drug diffusion up to 24 hours in stimulated intestinal fluid (Sakeer et al. 2018).

In another study by Mallakpour et al., a nanohybrid complex of starch/chitosan/CNT was developed as a hydrophobic drug model. This study involved functionalization of MWCNTs by attaching D-glucose to the structure, which enhanced the interaction between the MWCNT and the starch matrix, forming amphiphilic esters. The hybrid NCs successfully carried the drug 'zolpidem' in a hydrophobic delivery system. Such a fabrication method of MWCNTs is preferred, as they are rapidly formed, economical, ecofriendly, and high-quality NCs (Mallakpour and Khodadadzadeh 2018).

Starch NCs are also used in cancer therapeutics specifically targeting tumor cells. In a study conducted, redox-sensitive DDS utilized hydroxyethyl-starch-doxorubicin (HES-SS-DOX) conjugate to release DOX to targeted tumor cells (Hu et al. 2016). Cell line investigations revealed that HES-SS-DOX selectively delivered DOX, initiating apoptosis while reducing the cytotoxicity effect of DOX on healthy tissues, thus creating a safe and effective cancer chemotherapy. To summarize, the redox-sensitive DDS reduces the use of synthetic and potentially hazardous materials while offering sustainable targeted drug delivery. A subsequent study of starch as a DOX carrier developed by coupling hydrazidomethyl starch and a magnetic core claimed more than a fourfold increase in the rate of drug release from the carrier to the cancer cell environment (Nowak et al. 2024).

6.3.3 Efficacy of Carbohydrate Biopolymers in Tissue Engineering

Tissue engineering is an interdisciplinary field bringing together biology, medicine, and engineering and is focused on developing functional tissues as a substitute to replace or repair damaged tissues. In recent years, carbohydrate-based NPs have evolved as a promising tool in tissue engineering; the commonly used biopolymers include chitosan, dextran, hyaluronic acid, and starch (Gomez-Florit et al. 2020). The key features of CANEs in tissue engineering comprise the ability of surface modification, mimicking the extracellular matrix, sustained delivery of growth factors, encapsulation of cells, non-cytotoxicity, biocompatibility, and biodegradability (Malafaya, Silva, and Reis 2007; Mano et al. 2007; Gomez-Florit et al. 2020). CANEs are biodegradable, unlike metal and synthetic polymers, which can persist in the environment for a long period and contribute to emerging issues of metal toxicity and microplastics (Gomez-Florit et al. 2020; Mano et al. 2007). Carbohydrate biopolymers demonstrate controlled enzymatic degradability, which is considered optimal for tissue repair as it enhances the healing process, eliminating the need for surgery (Lichte et al. 2011; Sabir et al. 2009). As was previously discussed in DDS, CANEs effectively transport growth factors, peptides, and extracellular matrix components in addition to medications, thereby fostering a milieu that promotes cell adhesion, proliferation, and differentiation (Malafaya et al. 2007). Matching the mechanical characteristics of the target tissue as well as the implant is crucial in tissue engineering. Because of these biopolymers' structural makeup, researchers can alter the size, shape, and rigidity of NPs based on the target tissue; as a result, CANEs are becoming a cutting-edge method in tissue engineering. (Butscher et al. 2011).

6.3.3.1 Bone Tissue Engineering

Bone tissue damage has always been a challenging issue in medicine. According to recent studies, tissue engineering scaffolds with unfunctional, single-unit structure and composition fail to sufficiently repair complex bone defects (Lichte et al. 2011). Despite rapid development in tissue engineering, CANEs are being considered over traditional replacement surgery for their biodegradability, healing, and regenerative properties.

An earlier study conducted by Cai et al. demonstrated that NCs made from oxidizing alginate, gelatin, and tricalcium phosphate demonstrated promising biocompatibility with osteoblasts along with the capability of controlled drug release promoting bone repair and bone regeneration (Cai et al. 2007). Another study combined starch NPs with naturally derived biodegradable thermoplastic and β-tricalcium phosphate (β-TCP) to develop an efficient NC for bone tissue engineering application (Taherimehr, Bagheri, and Taherimehr 2021). Addition of thermoplastic starch with β-TCP significantly strengthens the mechanical properties, improving the bioactivity by nucleating hydroxyapatite (an important mineral for formation of bone) in stimulated body fluid, supporting bone tissue formation. Further investigations showed desirable degradability and swelling in BSF. Cytotoxic analysis validated that these NCs are non-toxic to the cells (Taherimehr, Bagheri, and Taherimehr 2021). In a recent study by Tang et al., innovative models of (3D) chitosan-based composite scaffolds embedded with zinc oxide NPs that function as an efficient tissue scaffold for bone repair have been presented (Tang et al. 2025). The study concluded that chitosan scaffolds facilitated controlled release of zinc oxide NPs, promoting mechanical stability and cell adhesion of these scaffolds.

6.3.3.2 Cartilage Regeneration

Cartilage is a flexible connective tissue found in various parts of the body. It is rich in collagen and proteoglycans. Cartilage lacks blood supply due to the absence of nerve fiber and relies heavily on the surrounding tissue for oxygen and nutrients;

thus, damage to the cartilage tissue can be detrimental and difficult to repair (Flierl et al. 2013).

A study developed an innovative dual release system by combining chitosan NPs and electrospun fiber scaffolds embedded with growth factor Nell-1 molecules (C. Wang et al. 2017). Chitosan hybrid scaffolds promoted sustained drug release, prolonging the duration of drug diffusion and bioactivity compared to scaffolds that are directly embedded with growth factors. Further evaluation in a controlled environment indicated successful delivery of the Nell-1 molecule promoting differentiation of human bone mesenchymal stem cells to cartilage. Similar investigations by Li and his group evaluated the delivery of transforming growth factors (TGF-β1) and bone morphogenetic protein-2 (BMP-2) carried by hybrid chitosan NP and silk fibroin hydrogel scaffold (Li et al. 2021). Both *ex vivo* and *in vivo* results indicated efficient delivery of TGFβ1 and BMP-2, as the results showed enhanced chondrogenic differentiation and sustained and gradual diffusion of growth factors, promoting stem cells to develop into cartilage. Hence, the combined NC approach holds a promising potential in improving cartilage injuries and degeneration.

6.3.3.3 Wound Healing

Wound healing is a complex biological process that involves the interplay of cells, growth factors, and extracellular components to restore the integrity and function of the tissue. Developing CANE-based approaches allows delivery of bioactive molecules, promotes tissue regeneration, provides a moist condition for regeneration, and exhibits antimicrobial properties.

Wound healing can be an extremely slow and infection-prone process for a diabetic patient. However, hydrogels made with NC can aid the process; this is explicitly seen in a study that developed a novel conjugate by bringing together fibroblast growth factor-gold-chitosan-hyaluronic acid (FGF-Au@CS/HA) (Xin Liu et al. 2024). The results from biological evaluation revealed enhanced cell viability, proliferation, migration, and antibacterial efficacy. In similar research for diabetic wound healing, chitosan-β-glycerophosphate hydrogel was embedded with a formulation of chlorogenic acid-indocyanine green-zinc ions (CIZ). The hydrogel acts as a smart photothermal drug delivery system, enhancing antioxidation, antimicrobial activity, and anti-inflammatory effects and promoting angiogenesis in diabetic foot ulcers (Wu et al. 2025). Ferulic acid is a potent antioxidant and antidiabetic drug. However, due to it being administered orally, the bioavailability is compromised. A solution to this has been developed by implanting ferulic acid into chitosan-based NPs for its topical administration; the formulation favors sustained release, enhanced mechanical properties and, most importantly, improved bioavailability, becoming a promising candidate in diabetic wound healing (Bhardwaj et al. 2024).

Cellulose-chitosan NC is studied for versatile application in tissue regeneration. Coupling iron oxide NP with cellulose-chitosan enhanced its superparamagnetic properties, which improves the cellulose-chitosan scaffold swelling ratio of 2445% ± 23.34%, maintaining excellent tensile strength (5.08 MPa) and absorption of fluids with very low cytotoxic adverse effects (Rezazadeh et al. 2024).

6.4 CHALLENGES AND LIMITATIONS OF CANES

CANEs display excellent biocompatibility. However, the interaction of NPs in the body can elicit cytotoxicity and lead to adverse effects such as immune response and inflammation. While carbohydrates are mostly stable, their biodegradability can challenge the stability of the NPs and NCs in a cellular environment, especially when exposed to changes in the pH. Further studies need to evaluate the stability of these NCs at various temperatures, dosage, and their limiting shelf life in practical applications. Carbohydrates are abundantly present in nature; however, scalability of carbohydrates to NPs and NCs can be challenging for commercial production. Another challenge is consistency and uniformity in the size of these NPs, hence making large-scale manufacturing problematic.

Although CANEs are easily manufactured and functionalized according to the target molecule, they often face challenges scaling the drug because of the minuscule size of NPs, which frequently leads to drug carriers having a lower load capacity. The receptor's target specificity may be diminished by NPs' low ligand–receptor interaction, which is primarily observed in tissue engineering. High specificity and affinity to biological receptors may eventually become difficult, as a result. Furthermore, challenges might be faced due to variable tumor environments that might impede NP cellular penetration, medication delivery, and imaging in cancer cells. In spite of all odds, continuous research and development is paving the road to overcome them, making these techniques more adaptable in medicinal applications.

6.5 CONCLUSION AND FUTURE PROSPECTS

CANEs are naturally occurring biopolymers that come from renewable resources such as algae, crustaceans, and plants and are a promising, versatile class of nanomaterials with a wide range of potential applications, ranging from biomedical to environmental uses. Such biopolymeric NPs offer extremely small sizes, which expose a large surface area for interaction of biomolecules and functionalization, enabling better biocompatibility and biodegradability. By combining the principles of green technology, the catalytic efficiency of natural enzymes, and the unique properties of nanomaterials, CANEs offer a powerful platform for advancing glycobiology, biotechnology, and drug development. Studies reviewed in this article suggest that CANEs display excellent morphology, topography, and crystallinity, making them applicable in versatile fields. The key factor of these NPs is that they can be combined with other biomolecules to produce NCs of metals, proteins, nucleic acid, and other carbohydrates, which further improve their characteristics.

Due to their extraordinary properties, CANE can be employed in water treatment, cosmetics, and the food industry. However, they are emerging as an innovative diagnostic and therapeutic approach in the field of biomedical research. In the literature highlighted in this chapter, particularly for DDS, CANEs aided controlled and sustained drug release, further enhancing the drug target specificity, improving therapeutic efficacy, and reducing adverse effects. In tissue engineering, these polymers promote cell adhesion, proliferation, and regeneration for damaged and degenerating tissues. Moreover, CANEs also play an important role as carriers in the chemotherapeutic approach. While imaging with NCs is still in its infancy, the studies show great potential in detecting initial stages of cancer. CANEs can also mitigate an unpredictable tumor environment, increasing the drug penetration in the tumor cell; this could drastically affect the mortality rate of cancer patients and show better clinical outcomes. Nevertheless, further research and large-scale preclinical and clinical studies are required to fabricate NPs with better pharmacokinetic, drug release kinetics, and nanoparticle functionalization.

With ongoing advancements in nanotechnology, biopolymers, and personalized medicine, these CANEs are envisioned to become more efficient, target-specific, economic, and scalable. Further advancements in tissue engineering will be able to develop complex functioning tissue, aiding the ongoing crisis in regenerative medicine for organ donation and organ transplantation. Continued research in their synthesis, targeting capabilities, and biocompatibility will pave the way for their broader application in various scientific and industrial fields.

ACKNOWLEDGEMENT

The authors would like to sincerely thank their family and friends for their unwavering support throughout the preparation of this chapter and their affiliated organizations for providing professional assistance necessary for the completion of this task.

AUTHOR CONTRIBUTION STATEMENT

MR, MHM, MK: Writing, original draft preparation, and editing. MK: Supervision, reviewing, and finalizing the manuscript.

FUNDING

This research did not receive any type of funding from public, commercial, or not-for-profit agencies.

CONFLICT OF INTEREST STATEMENT

The authors declare that there is no potential conflict of interest.

REFERENCES

Abda, Ebrahim M., and Rocktotpal Konwarh. 2021. "Harnessing the Sustainable Bioresource, Cellulose at the Nanoscale for Multifarious Environmental Applications." In J. Aravind, M. kamaraj, M. Prashanthi Devi, and S. Rajakumar (Eds.), *Strategies and Tools for Pollutant Mitigation: Avenues to a Cleaner Environment*, 65–91.Springer.

Abedini, Fatemeh, Mohammad Ebrahimi, Abbas Hemmati Roozbehani, Abraham J. Domb, and Hossein Hosseinkhani. 2018. "Overview on Natural Hydrophilic Polysaccharide Polymers in Drug Delivery." *Polymers for Advanced Technologies* 29 (10): 2564–73. https://doi.org/10.1002/pat.4375.

Adak, Avijit K., Ben-Yuan Li, and Chun-Cheng Lin. 2015. "Advances in Multifunctional Glycosylated Nanomaterials: Preparation and Applications in Glycoscience." *Carbohydrate Research* 405 (March): 2–12. https://doi.org/10.1016/j.carres.2014.07.026.

Adrian, Gheorghe, Mihaela Mihai, and Dan Cristian Vodnar. 2019. "The Use of Chitosan, Alginate, and Pectin in the Biomedical and Food Sector—Biocompatibility, Bioadhesiveness, and Biodegradability." *Polymers* 11 (11): 1837.

Agwa, Mona M., Heba Elmotasem, Hassan Elsayed, Abdallah S. Abdelsattar, Ahmed M. Omer, Doaa T. Gebreel, Mohamed S. Mohy-Eldin, and Moustafa M.G. Fouda. 2023. "Carbohydrate Ligands-Directed Active Tumor Targeting of Combinatorial Chemotherapy/Phototherapy-Based Nanomedicine: A Review." *International Journal of Biological Macromolecules* 239 (March): 124294. https://doi.org/10.1016/j.ijbiomac.2023.124294.

Ahmadi, Sepideh, Kamal Rahimizadeh, Ali Shafiee, Navid Rabiee, and Siavash Iravani. 2023. "Nanozymes and Their Emerging Applications in Biomedicine." *Process Biochemistry* 131: 154–74.

Aibani, Noorjahan, Raj Rai, Parth Patel, Grace Cuddihy, and Ellen K. Wasan. 2021. "Chitosan Nanoparticles at the Biological Interface: Implications for Drug Delivery." *Pharmaceutics* 13 (10): 1686.

Alibolandi, Mona, Marzieh Mohammadi, Reihaneh Teimouri, Maryam Babaei, Jebraeil Movaffagh, and Mohammad Ramezani. 2021. Carbohydrate-Based Materials in Tissue Engineering Applications. In Sougata Jana and Subrata Jana (Eds.), *Nanoengineering of Biomaterials: Drug Delivery & Biomedical Applications, 2 Volumes*. Wiley-VCH. https://doi.org/10.1002/9783527832095.ch18.

Alim-Al-Razy, Md, G. M. Asik Bayazid, Raihan Ur Rahman, Rajib Bosu, and Shahla Samim Shamma. 2020. "Silver Nanoparticle Synthesis, UV-Vis Spectroscopy to Find Particle Size and Measure Resistance of Colloidal Solution." *Journal of Physics: Conference Series* (IOP Publishing) 1706: 12020.

Alonso, Saúl, Manuel Rendueles, and Mario Díaz. 2013. "Bio-Production of Lactobionic Acid: Current Status, Applications and Future Prospects." *Biotechnology Advances* 31 (8): 1275–91.

An, Mengying, Meng-Qi He, Caishi Lin, Yuebo Wu, Yongjian Ai, Hongbo Xin, and Qionglin Liang. 2023. "Recent Progress of Nanozymes with Different Spatial Dimensions for Bioanalysis." *Materials Today Nano* 22: 100330.

Angel, Nicole, Songnan Li, Feng Yan, and Lingyan Kong. 2022. "Recent Advances in Electrospinning of Nanofibers from Bio-Based Carbohydrate Polymers and Their Applications." *Trends in Food Science & Technology* 120: 308–24.

Araszkiewicz, Antonina M., Eduarda P. Oliveira, Terje Svendsen, Katarzyna Drela, Piotr Rogujski, Izabela Malysz-Cymborska, … & L. Stanaszek. 2022. "Manganese-Labeled Alginate Hydrogels for Image-Guided Cell Transplantation." *International Journal of Molecular Sciences* 23 (5): 2465. https://doi.org/10.3390/ijms23052465.

Asadollahzade, Elnaz, Fereshteh Ghadiri, Zahra Ebadi, and Abdorreza Naser Moghadasi. 2022. "The Benefits and Side Effects of Gadolinium-Based Contrast Agents in Multiple Sclerosis Patients." *Revista Da Associacao Medica Brasileira (1992)* 68 (8): 979–81. https://doi.org/10.1590/1806-9282.20220643.

Asha, Anika Benozir, and Ravin Narain. 2020. "Nanomaterials Properties." In Ravi Narain (Ed.), *Polymer Science and Nanotechnology*, 343–59. Elsevier.

Babick, Frank. 2020. "Dynamic Light Scattering (DLS)." In V.D. Hodoroaba, W.E.S. Unger, and A.G. Shard (Eds.), *Characterization of Nanoparticles*, 137–72. Elsevier.

Bains, Rajneesh K., Seyed Amirhossein Nasseri, Jacob F. Wardman, and Stephen G. Withers. 2024. "Advances in the Understanding and Exploitation of Carbohydrate-Active Enzymes." *Current Opinion in Chemical Biology* 80: 102457.

Basseri, Hamidreza, Ronak Bakhtiyari, Sayed Jamal Hashemi, Mojgan Baniardelani, Hadi Shahraki, and Laila Hosainpour. 2019. "Antibacterial/Antifungal Activity of Extracted Chitosan from American Cockroach (Dictyoptera: Blattidae) and German Cockroach (Blattodea: Blattellidae)." *Journal of Medical Entomology* 56 (5): 1208–14.

Bekaroğlu, Maide Gökçe, Yavuz İşçi, and Sevim İşçi. 2017. "Colloidal Properties and in Vitro Evaluation of Hydroxy Ethyl Cellulose Coated Iron Oxide Particles for Targeted Drug Delivery." *Materials Science and Engineering: C* 78 (September): 847–53. https://doi.org/10.1016/j.msec.2017.04.030.

Benalaya, Ikbel, Gilberto Alves, João Lopes, and Luís R. Silva. 2024. "A Review of Natural Polysaccharides: Sources, Characteristics, Properties, Food, and Pharmaceutical Applications." *International Journal of Molecular Sciences* 25 (2). https://doi.org/10.3390/ijms25021322.

Bhardwaj, Harish, and Rajendra Kumar Jangde. 2024. "Development and Characterization of Ferulic Acid-Loaded Chitosan Nanoparticle Embedded- Hydrogel for Diabetic Wound Delivery." *European Journal of Pharmaceutics and Biopharmaceutics* 201 (August): 114371. https://doi.org/10.1016/j.ejpb.2024.114371.

Bhattacharjee, Sourav. 2016. "DLS and Zeta Potential–What They Are and What They Are Not?" *Journal of Controlled Release* 235: 337–51.

Bhawana, Rupesh Kumar Basniwal, Harpreet Singh Buttar, V. K. Jain, and Nidhi Jain. 2011. "Curcumin Nanoparticles: Preparation, Characterization, and Antimicrobial Study." *Journal of Agricultural and Food Chemistry* 59 (5): 2056–61. https://doi.org/10.1021/jf104402t.

Birrenbach, G., and P. P. Speiser. 1976. "Polymerized Micelles and Their Use as Adjuvants in Immunology." *Journal of Pharmaceutical Sciences* 65 (12): 1763–66. https://doi.org/10.1002/jps.2600651217.

Bokov, Dmitry, Abduladheem Turki Jalil, Supat Chupradit, Wanich Suksatan, Mohammad Javed Ansari, Iman H. Shewael, Gabdrakhman H. Valiev, and Ehsan Kianfar. 2021. "Nanomaterial by Sol-gel Method: Synthesis and Application." *Advances in Materials Science and Engineering* 2021 (1): 5102014.

Bonilla, Pablo, Eva M. Arias, Conxita Solans, and María José García-Celma. 2018. "Influence of Crosslinked Alginate on Drug Release from Highly Concentrated Emulsions." *Colloids and Surfaces A: Physicochemical and Engineering Aspects* 536 (January): 148–55. https://doi.org/10.1016/j.colsurfa.2017.07.026.

Boroumand, Homa, Fereshteh Badie, Samaneh Mazaheri, Zeynab Sadat Seyedi, Javid Sadri Nahand, Majid Nejati, Hossein Bannazadeh Baghi, Mohammad Abbasi-Kolli, Bita Badehnoosh, and Maryam Ghandali. 2021. "Chitosan-Based Nanoparticles against Viral Infections." *Frontiers in Cellular and Infection Microbiology* 11: 643953.

Butscher, A., M. Bohner, S. Hofmann, L. Gauckler, and R. Müller. 2011. "Structural and Material Approaches to Bone Tissue Engineering in Powder-Based Three-Dimensional Printing." *Acta Biomaterialia* 7 (3): 907–20. https://doi.org/10.1016/j.actbio.2010.09.039.

Bystricky, S., A. Malovikova, and T. Sticzay. 1991. "Interaction of Acidic Polysaccharides with Polylysine Enantiomers. Conformation Probe in Solution." *Carbohydrate Polymers* 15 (3): 299–308. https://doi.org/10.1016/0144-8617(91)90044-D.

Cai, K., J. Zhang, L. Deng, L. Yang, Y. Hu, C. Chen, L. Xue, and L. Wang. 2007. "Physical and Biological Properties of a Novel Hydrogel Composite Based on Oxidized Alginate, Gelatin and Tricalcium Phosphate for Bone Tissue Engineering." *Advanced Engineering Materials* 9 (12): 1082–88. https://doi.org/10.1002/adem.200700222.

Calvo, P., C. Remunan-Lopez, J. L. Vila-Jato, and M. J. Alonso. 1997. "Novel Hydrophilic Chitosan-Polyethylene Oxide Nanoparticles as Protein Carriers." *Journal of Applied*

Polymer Science 63 (1): 125–32. https://doi.org/10.1002/(SICI)1097-4628(19970103)63:1<125::AID-APP13>3.0.CO;2-4.

Carniato, Fabio, Marco Ricci, Lorenzo Tei, Francesca Garello, Chiara Furlan, Enzo Terreno, Enrico Ravera, Giacomo Parigi, Claudio Luchinat, and Mauro Botta. 2023. "Novel Nanogels Loaded with Mn(II) Chelates as Effective and Biologically Stable MRI Probes." *Small* 19 (42). https://doi.org/10.1002/smll.202302868.

Cavalcanti, André D. D., and Maria H. A. Santana. 2019. "Structural and Surface Properties Control the Recovery and Purity of Bio-Hyaluronic Acid upon Precipitation with Isopropyl Alcohol." *Colloids and Surfaces A: Physicochemical and Engineering Aspects* 573 : 112–18.

Chakraborty, Amlan, and Pranav Patni. 2014. "A Review on Microfabricated Engineered Particle Systems for Drug Delivery-PRINT." *International Journal of Advanced Information Science and Technology* 3 (1): 62–66.

Chieh, Jen-Jie, Kai-Wen Huang, and Jin-Cheng Shi. 2015. "Sub-Tesla-Field Magnetization of Vibrated Magnetic Nanoreagents for Screening Tumor Markers." *Applied Physics Letters* 106 (7): 073703-1–073703-4.

Chin, Suk Fun, Fiona Beragai Jimmy, and Suh Cem Pang. 2018. "Size Controlled Fabrication of Cellulose Nanoparticles for Drug Delivery Applications." *Journal of Drug Delivery Science and Technology* 43 (February): 262–66. https://doi.org/10.1016/j.jddst.2017.10.021.

Chung, In Jae, Seong Ik Jeon, Eui-Joon Cha, Youngro Byun, Ick Chan Kwon, Young Il Kim, Kwangmeyung Kim, and Cheol-Hee Ahn. 2019. "Activatable NIRF/MRI Dual Imaging Probe Using Bio-Inspired Coating of Glycol Chitosan on Superparamagnetic Iron Oxide Nanoparticles." *Journal of Industrial and Engineering Chemistry* 76 (August): 403–9. https://doi.org/10.1016/j.jiec.2019.04.006.

Corrêa, Camila L., Glaucia E. O. Midorikawa, Edivaldo Ximenes Ferreira Filho, Eliane Ferreira Noronha, Gabriel S. C. Alves, Roberto Coiti Togawa, Orzenil Bonfim Silva-Junior, Marcos Mota do Carmo Costa, Priscila Grynberg, and Robert N. G. Miller. 2020. "Transcriptome Profiling-Based Analysis of Carbohydrate-Active Enzymes in Aspergillus Terreus Involved in Plant Biomass Degradation." *Frontiers in Bioengineering and Biotechnology* 8 (October): 1–20. https://doi.org/10.3389/fbioe.2020.564527.

Couvreur, Patrick, Ruxandra Gref, Karine Andrieux, and Claude Paul Malvy. 2006. "Nanotechnologies for Drug Delivery: Application to Cancer and Autoimmune Diseases." *Progress in Solid State Chemistry* 34 (2–4): 231–35.

da Silva, Laiane Souza, Jóice Maria Scheibel, and Rosane Michele Soares. 2023. "Starch-Based Nanocomposite Hydrogels for Biomedical Applications." In *Functional Nanocomposite Hydrogels*, 311–38. Elsevier.

Dabiri, Seyed Mohammad Hossein, Alberto Lagazzo, Fabrizio Barberis, Amirreza Shayganpour, Elisabetta Finocchio, and Laura Pastorino. 2017. "New In-Situ Synthetized Hydrogel Composite Based on Alginate and Brushite as a Potential PH Sensitive Drug Delivery System." *Carbohydrate Polymers* 177 (December): 324–33. https://doi.org/10.1016/j.carbpol.2017.08.046.

Dai, Lin, and Chuan-Ling Si. 2017. "Cellulose-Graft-Poly(Methyl Methacrylate) Nanoparticles with High Biocompatibility for Hydrophobic Anti-Cancer Drug Delivery." *Materials Letters* 207 (November): 213–16. https://doi.org/10.1016/j.matlet.2017.07.090.

de Oliveira Pedro, Rafael, Francisco M. Goycoolea, Susana Pereira, Carla C. Schmitt, and Miguel G. Neumann. 2018. "Synergistic Effect of Quercetin and PH-Responsive DEAE-Chitosan Carriers as Drug Delivery System for Breast Cancer Treatment." *International Journal of Biological Macromolecules* 106 (January): 579–86. https://doi.org/10.1016/j.ijbiomac.2017.08.056.

Dhiman, Nisha, Somnath Ghosh, Yogendra Kumar Mishra, and Kumud Malika Tripathi. 2022. "Prospects of Nano-Carbons as Emerging Catalysts for Enzyme-Mimetic Applications." *Materials Advances* 3 (7): 3101–22.

Di, Xiangjie, Xiao Liang, Chao Shen, Yuwen Pei, Bin Wu, and Zhiyao He. 2022. "Carbohydrates Used in Polymeric Systems for Drug Delivery: From Structures to Applications." *Pharmaceutics* 14 (4). https://doi.org/10.3390/pharmaceutics14040739.

Ding, Ran, Minghao Zhang, Qiaomei Zhu, Yuanyuan Qu, Xin Jia, and Lijun Yin. 2023. "Curcumin Loaded Zein-Alginate Nanogels with 'Core-Shell' Structure: Formation, Characterization and Simulated Digestion." *International Journal of Biological Macromolecules* 251 (November): 126201. https://doi.org/10.1016/j.ijbiomac.2023.126201.

Estrada-Ramos, Marco Antonio, José Manuel Rojas-García, Salvador López-Morales, Sonia Martínez-Gallegos, Ernesto Rivera, and Javier Illescas. 2023. "Synthesis and Characterization of Starch Nanoparticles Obtained through the Micro-Emulsion Method." *MRS Advances* 8 (3): 83–88.

Falsafi, Seid Reza, Fuat Topuz, Dagmara Bajer, Zahra Mohebi, Maryam Shafieiuon, Hajar Heydari, Shruti Rawal, et al. 2023. "Metal Nanoparticles and Carbohydrate Polymers Team up to Improve Biomedical Outcomes." *Biomedicine and Pharmacotherapy* 168 (October): 115695. https://doi.org/10.1016/j.biopha.2023.115695.

Fatoni, Amin, Aziz Wijonarko, Mekar Dwi Anggraeni, Dadan Hermawan, Hartiwi Diastuti, and Zusfahair. 2021. "Alginate NiFe2O4 Nanoparticles Cryogel for Electrochemical Glucose Biosensor Development." *Gels* 7 (4): 272. https://doi.org/10.3390/gels7040272.

Fischer, F. G., and Helmut Dörfel. 1955. "Die Polyuronsäuren Der Braunalgen (Kohlenhydrate Der Algen I)." *Hoppe-Seyler´s Zeitschrift Für Physiologische Chemie* 302 (Jahresband): 186–203. https://doi.org/10.1515/bchm2.1955.302.1-2.186.

Flierl, Michael A., Wade R. Smith, Cyril Mauffrey, Kaan Irgit, Allison E. Williams, Erin Ross, Gabrielle Peacher, David J. Hak, and Philip F. Stahel. 2013. "Outcomes and Complication Rates of Different Bone Grafting Modalities in Long Bone Fracture Nonunions: A Retrospective Cohort Study in 182 Patients." *Journal of Orthopaedic Surgery and Research* 8 (September): 33. https://doi.org/10.1186/1749-799X-8-33.

Friciu, Maria Mihaela, Tien Canh Le, Pompilia Ispas-Szabo, and Mircea Alexandru Mateescu. 2013. "Carboxymethyl Starch and Lecithin Complex as Matrix for Targeted Drug Delivery: I. Monolithic Mesalamine Forms for Colon Delivery." *European Journal of Pharmaceutics and Biopharmaceutics* 85 (3): 521–30. https://doi.org/10.1016/j.ejpb.2013.03.007.

Fuster, Marta G., Imane Moulefera, M. Noelia Muñoz, Mercedes G. Montalbán, and Gloria Víllora. 2023. "Synthesis of Cellulose Nanoparticles from Ionic Liquid Solutions for Biomedical Applications." *Polymers* 15 (2): 382.

Gaikwad, Ganesh, Nanda Rohra, Chandan Kumar, Sachin Jadhav, Haladhar Dev Sarma, Lalit Borade, Sudipta Chakraborty, et al. 2021. "A Facile Strategy for Synthesis of a Broad Palette of Intrinsically Radiolabeled Chitosan Nanoparticles for Potential Use in Cancer Theranostics." *Journal of Drug Delivery Science and Technology* 63 (June): 102485. https://doi.org/10.1016/j.jddst.2021.102485.

Gamage, Ashoka, Thiviya Punniamoorthy, and Terrence Madhujith. 2021. "Starch-Based Hybrid Nanomaterials for Environmental Remediation." In Martins Ochubiojo Emeje (Ed.), *Starch-Evolution and Recent Advances*, Volume 101697, pp. 1–16. Intechopen.

Gao, Yanyun, and Yuling Wu. 2022. "Recent Advances of Chitosan-Based Nanoparticles for Biomedical and Biotechnological Applications." *International Journal of Biological Macromolecules* 203: 379–88.

Garron, Marie Line, and Bernard Henrissat. 2019. "The Continuing Expansion of CAZymes and Their Families." *Current Opinion in Chemical Biology* 53: 82–87. https://doi.org/10.1016/j.cbpa.2019.08.004.

George, Meera, and T. Emilia Abraham. 2006. "Polyionic Hydrocolloids for the Intestinal Delivery of Protein Drugs: Alginate and Chitosan — A Review." *Journal of Controlled Release* 114 (1): 1–14. https://doi.org/10.1016/j.jconrel.2006.04.017.

Gokila, V., V. T. Perarasu, and R. Rufina. 2021. "Qualitative Comparison of Chemical and Green Synthesized Fe 3 O 4 Nanoparticles." *Advances in Nano Research* 10 (1): 71–76.

Gomaa, Eman Zakaria. 2022. "Nanozymes: A Promising Horizon for Medical and Environmental Applications." *Journal of Cluster Science* 33 (4): 1275–97.

Gomez-Florit, Manuel, Alberto Pardo, Rui M. A. Domingues, Ana L. Graça, Pedro S. Babo, Rui L. Reis, and Manuela E. Gomes. 2020. "Natural-Based Hydrogels for Tissue Engineering Applications." *Molecules* 25 (24): 5858. https://doi.org/10.3390/molecules25245858.

Gonçales, Vinicius R., Rafael N. P. Colombo, Marco A. O. S. Minadeo, Elaine Y. Matsubara, J. Maurício Rosolen, and Susana I. Córdoba de Torresi. 2016. "Three-Dimensional Graphene/Carbon Nanotubes Hybrid Composites for Exploring Interaction between Glucose Oxidase and Carbon Based Electrodes." *Journal of Electroanalytical Chemistry* 775: 235–42.

Gopinath, V., S. Saravanan, A. R. Al-Maleki, Manickam Ramesh, and Jamuna Vadivelu. 2018. "A Review of Natural Polysaccharides for Drug Delivery Applications: Special Focus on Cellulose, Starch and Glycogen." *Biomedicine and Pharmacotherapy* 107 (July): 96–108. https://doi.org/10.1016/j.biopha.2018.07.136.

Gour, Aman, and Narendra Kumar Jain. 2019. "Advances in Green Synthesis of Nanoparticles." *Artificial Cells, Nanomedicine, and Biotechnology* 47 (1): 844–51.

Granata, Carmine, Paolo Silvestrini, and Antonio Vettoliere. 2020. "Nano Superconducting Quantum Interference Device." In *21st Century Nanoscience–A Handbook*, 10–14. CRC Press.

Granata, Carmine, and Antonio Vettoliere. 2016. "Nano Superconducting Quantum Interference Device: A Powerful Tool for Nanoscale Investigations." *Physics Reports* 614: 1–69.

Gunathilake, Thennakoon M. Sampath Udeni, Yern Chee Ching, Cheng Hock Chuah, Hazlee Azil Illias, Kuan Yong Ching, Ramesh Singh, and Liou Nai-Shang. 2018. "Influence of a Nonionic Surfactant on Curcumin Delivery of Nanocellulose Reinforced Chitosan Hydrogel." *International Journal of Biological Macromolecules* 118 (October): 1055–64. https://doi.org/10.1016/j.ijbiomac.2018.06.147.

Hachem, Kadda, Mohammad Javed Ansari, Raed Obaid Saleh, Hamzah H. Kzar, Moaed E. Al-Gazally, Usama S. Altimari, Shaymaa Abed Hussein, Halah T. Mohammed, Ali Thaeer Hammid, and Ehsan Kianfar. 2022. "Methods of Chemical Synthesis in the Synthesis of Nanomaterial and Nanoparticles by the Chemical Deposition Method: A Review." *BioNanoScience* 12 (3): 1032–57.

Hage, Hayat, and Marie-Noëlle Rosso. 2021. "Evolution of Fungal Carbohydrate-Active Enzyme Portfolios and Adaptation to Plant Cell-Wall Polymers." *Journal of Fungi* 7 (3): 185.

Hamidian, Hooshang, and Tayebeh Tavakoli. 2016. "Preparation of a New Fe 3 O 4 /Starch-g -Polyester Nanocomposite Hydrogel and a Study on Swelling and Drug Delivery Properties." *Carbohydrate Polymers* 144 (June): 140–48. https://doi.org/10.1016/j.carbpol.2016.02.048.

Hao, Nanjing, Kitjanit Neranon, Olof Ramström, and Mingdi Yan. 2016. "Glyconanomaterials for Biosensing Applications." *Biosensors and Bioelectronics* 76 (February): 113–30. https://doi.org/10.1016/j.bios.2015.07.031.

Hasanbegloo, Kimiya, Solmaz Banihashem, Babak Faraji Dizaji, Sara Bybordi, Nika Farrokh-Eslamlou, Parvaneh Ghaderi-shekhi Abadi, Fariborz Sharifian Jazi, and Mohammad Irani. 2023. "Paclitaxel-Loaded Liposome-Incorporated Chitosan (Core)/

Poly (ε-Caprolactone)/Chitosan (Shell) Nanofibers for the Treatment of Breast Cancer." *International Journal of Biological Macromolecules* 230: 123380.

He, Shihao, Shuangling Zhong, Qingye Meng, Yu Fang, Yueming Dou, Yan Gao, and Xuejun Cui. 2021. "Sonochemical Preparation of Folate-Decorated Reductive-Responsive Carboxymethylcellulose-Based Nanocapsules for Targeted Drug Delivery." *Carbohydrate Polymers* 266 (August): 118174. https://doi.org/10.1016/j.carbpol.2021.118174.

He, Yuan, Xiaoyong Chen, Ye Zhang, Yanyun Wang, Mengyao Cui, Galong Li, Xiaoli Liu, and Haiming Fan. 2022. "Magnetoresponsive Nanozyme: Magnetic Stimulation on the Nanozyme Activity of Iron Oxide Nanoparticles." *Science China Life Sciences* 65 (1): 184–92.

Herzing, Andrew A., Masashi Watanabe, Jennifer K. Edwards, Marco Conte, Zi-Rong Tang, Graham J. Hutchings, and Christopher J. Kiely. 2008. "Energy Dispersive X-Ray Spectroscopy of Bimetallic Nanoparticles in an Aberration Corrected Scanning Transmission Electron Microscope." *Faraday Discussions* 138: 337–51.

Hidangmayum, Akash, and Padmanabh Dwivedi. 2022. "Chitosan Based Nanoformulation for Sustainable Agriculture with Special Reference to Abiotic Stress: A Review." *Journal of Polymers and the Environment* 30 (4): 1264–83.

Ho, Ming-Hua, Mei-Hsiu Liao, Yi-Ling Lin, Chien-Hao Lai, Pei-I Lin, and Ruei-Ming Chen. 2014. "Improving Effects of Chitosan Nanofiber Scaffolds on Osteoblast Proliferation and Maturation." *International Journal of Nanomedicine* 9: 4293–4304.

Hou, Xinyu, Wenjuan Zhang, Muye He, Yiben Lu, Kaiyan Lou, and Feng Gao. 2017. "Preparation and Characterization of β -Cyclodextrin Grafted N-Maleoyl Chitosan Nanoparticles for Drug Delivery." *Asian Journal of Pharmaceutical Sciences* 12 (6): 558–68. https://doi.org/10.1016/j.ajps.2017.07.007.

Hrapovic, Sabahudin, Camilo F. Martinez-Farina, Jessie Sui, Jean-Danick Lavertu, and Usha D. Hemraz. 2022. "Design of Chitosan Nanocrystals Decorated with Amino Acids and Peptides." *Carbohydrate Polymers* 298 (December): 120108. https://doi.org/10.1016/j.carbpol.2022.120108.

Hu, Hang, Yihui Li, Qing Zhou, Yanxiao Ao, Chan Yu, Ying Wan, Huibi Xu, Zifu Li, and Xiangliang Yang. 2016. "Redox-Sensitive Hydroxyethyl Starch–Doxorubicin Conjugate for Tumor Targeted Drug Delivery." *ACS Applied Materials & Interfaces* 8 (45): 30833–44. https://doi.org/10.1021/acsami.6b11932.

Huang, Hongkun, and Jiancheng Lai. 2018. "Mechanism Study of Nanomaterial Synthesis by Pulsed Laser Ablation in Liquid." In Rongshi Xiao, Minghui Hong, and Jian Liu (Eds.), *Advanced Laser Processing and Manufacturing II*, 10813: 92–98. SPIE.

Huston, Matthew, Melissa DeBella, Maria DiBella, and Anisha Gupta. 2021. "Green Synthesis of Nanomaterials." *Nanomaterials* 11 (8): 2130.

Iacob, Andreea Teodora, Florentina Geanina Lupascu, Maria Apotrosoaei, Ioana Mirela Vasincu, Roxana Georgiana Tauser, Dan Lupascu, Simona Eliza Giusca, Irina-Draga Caruntu, and Lenuta Profire. 2021. "Recent Biomedical Approaches for Chitosan Based Materials as Drug Delivery Nanocarriers." *Pharmaceutics* 13 (4): 587.

Imtiaz, Yusha, Beza Tuga, Christopher W. Smith, Alexander Rabideau, Long Nguyen, Yali Liu, Sabahudin Hrapovic, Karina Ckless, and Rajesh Sunasee. 2020. "Synthesis and Cytotoxicity Studies of Wood-Based Cationic Cellulose Nanocrystals as Potential Immunomodulators." *Nanomaterials* 10 (8): 1603.

Iravani, Siavash, and Rajender S. Varma. 2022. "Alginate-Based Micro-and Nanosystems for Targeted Cancer Therapy." *Marine Drugs* 20 (10): 598.

Jacob, Joby, Józef T. Haponiuk, Sabu Thomas, and Sreeraj Gopi. 2018. "Biopolymer Based Nanomaterials in Drug Delivery Systems: A Review." *Materials Today Chemistry* 9 (September): 43–55. https://doi.org/10.1016/j.mtchem.2018.05.002.

Jaiturong, Patthanakorn, Busaban Sirithunyalug, Sukum Eitsayeam, Chawalinee Asawahame, Pratchaya Tipduangta, and Jakkapan Sirithunyalug. 2018. "Preparation of Glutinous Rice Starch/Polyvinyl Alcohol Copolymer Electrospun Fibers for Using as a Drug Delivery Carrier." *Asian Journal of Pharmaceutical Sciences* 13 (3): 239–47. https://doi.org/10.1016/j.ajps.2017.08.008.

Javanbakht, Siamak, and Hassan Namazi. 2018. "Doxorubicin Loaded Carboxymethyl Cellulose/Graphene Quantum Dot Nanocomposite Hydrogel Films as a Potential Anticancer Drug Delivery System." *Materials Science and Engineering: C* 87 (June): 50–59. https://doi.org/10.1016/j.msec.2018.02.010.

Jebali, Ali, Elham Khajeh Nayeri, Sima Roohana, Shiva Aghaei, Maede Ghaffari, Karim Daliri, and Garcia Fuente. 2017. "Nano-Carbohydrates: Synthesis and Application in Genetics, Biotechnology, and Medicine." *Advances in Colloid and Interface Science* 240: 1–14. https://doi.org/10.1016/j.cis.2016.11.002.

Jensen, Kirsten M. Ø., Christoffer Tyrsted, Martin Bremholm, and Bo B. Iversen. 2014. "In Situ Studies of Solvothermal Synthesis of Energy Materials." *ChemSusChem* 7 (6): 1594–1611.

Jiang, Wang-Zhan, Yang Cai, and Hao-Ying Li. 2017. "Chitosan-Based Spray-Dried Mucoadhesive Microspheres for Sustained Oromucosal Drug Delivery." *Powder Technology* 312 (May): 124–32. https://doi.org/10.1016/j.powtec.2017.02.021.

Joshi, Bhavana, Uma Shankar, Supriya Vishwakarma, Amit Kumar, Ashok Kumar, Deepti Joshi, and Abhijeet Joshi. 2023. "Multifunctional Ultrasmall Theranostic Nanohybrids Developed by Ultrasonic Atomizer for Drug Delivery and Magnetic Resonance Imaging." *ACS Applied Bio Materials* 6 (5): 1943–52. https://doi.org/10.1021/acsabm.3c00151.

Joshy, K. S., M. Alex Susan, S. Snigdha, Kalarikkal Nandakumar, A. Pothen Laly, and Thomas Sabu. 2018. "Encapsulation of Zidovudine in PF-68 Coated Alginate Conjugate Nanoparticles for Anti-HIV Drug Delivery." *International Journal of Biological Macromolecules* 107 (February): 929–37. https://doi.org/10.1016/j.ijbiomac.2017.09.078.

Kamaraj, Sriram, Uma Maheswari Palanisamy, Meera Sheriffa Begum Kadhar Mohamed, Arthanareeswaran Gangasalam, Gover Antoniraj Maria, and Ruckmani Kandasamy. 2018. "Curcumin Drug Delivery by Vanillin-Chitosan Coated with Calcium Ferrite Hybrid Nanoparticles as Carrier." *European Journal of Pharmaceutical Sciences* 116 (April): 48–60. https://doi.org/10.1016/j.ejps.2018.01.023.

Kang, Biao, Till Opatz, Katharina Landfester, and Frederik R. Wurm. 2015. "Carbohydrate Nanocarriers in Biomedical Applications: Functionalization and Construction." *Chemical Society Reviews* 44 (22): 8301–25. https://doi.org/10.1039/c5cs00092k.

Kang, Xinhuang, Zhibin Mai, Xiaoyong Zou, Peixiang Cai, and Jinyuan Mo. 2008. "Glucose Biosensors Based on Platinum Nanoparticles-Deposited Carbon Nanotubes in Sol–Gel Chitosan/Silica Hybrid." *Talanta* 74 (4): 879–86. https://doi.org/10.1016/j.talanta.2007.07.019.

Karim, Aiman, Abdur Rehman, Jianguo Feng, Asia Noreen, Elham Assadpour, Mohammad Saeed Kharazmi, Zhang Lianfu, and Seid Mahdi Jafari. 2022. "Alginate-Based Nanocarriers for the Delivery and Controlled-Release of Bioactive Compounds." *Advances in Colloid and Interface Science* 307: 102744.

Karimian, Ansar, Hadi Parsian, Maryam Majidinia, Mahdi Rahimi, Seyed Mostafa Mir, Hossein Samadi Kafil, Vahid Shafiei-Irannejad, Maryam Kheyrollah, Hassan Ostadi, and Bahman Yousefi. 2019. "Nanocrystalline Cellulose: Preparation, Physicochemical Properties, and Applications in Drug Delivery Systems." *International Journal of Biological Macromolecules* 133 (July): 850–59. https://doi.org/10.1016/j.ijbiomac.2019.04.117.

Kartal, Önder, Sebastian Mahlow, Alexander Skupin, and Oliver Ebenhöh. 2011. "Carbohydrate-active Enzymes Exemplify Entropic Principles in Metabolism." *Molecular Systems Biology* 7 (1): 542.

Kato, Kentaro, and Akiko Ishiwa. 2015. "The Role of Carbohydrates in Infection Strategies of Enteric Pathogens." *Tropical Medicine and Health* 43 (1): 41–52. https://doi.org/10.2149/tmh.2014-25.

Kaur, Lovedeep, Jaspreet Singh, and Qiang Liu. 2007. "Starch – A Potential Biomaterial for Biomedical Applications." In M. Reza Mozafari (Ed.), *Nanomaterials and Nanosystems for Biomedical Applications*, 83–98. Springer Netherlands. https://doi.org/10.1007/978-1-4020-6289-6_5.

Khan, I., K. Saeed, and I. Khan. 2019. "Nanoparticles: Properties, Applications and Toxicities." *Arabian Journal of Chemistry* 12 (7): 908–31.

Khan, U.S., A. Rahim, N. Khan, N. Muhammad, F. Rehman, K. Ahmad, and J. Iqbal. 2017. "Aging Study of the Powdered Magnetite Nanoparticles." *Materials Chemistry and Physics* 189: 86–89.

Khorsandi, Danial, Serena Jenson, Atefeh Zarepour, Arezoo Khosravi, Navid Rabiee, Siavash Iravani, and Ali Zarrabi. 2024. "Catalytic and Biomedical Applications of Nanocelluloses: A Review of Recent Developments." *International Journal of Biological Macromolecules* 268: 131829.

Khowala, Suman, Deepak Verma, and Samudra P. Banik. 2008. "Carbohydrtaes." In Mamata Kankara, N.C. Sharma, Prakash C. Sharma, B.L Somani, and P.C. Misra (Eds.), *Biomolecules (Introduction,Structure and Functions)*. National Science Digital Library(NSDL).

Kim, Eok-Cheon, Jeong-Ki Min, Tae-Yoon Kim, Shin-Jeong Lee, Hyun-Ok Yang, Sanghwa Han, Young-Myeong Kim, and Young-Guen Kwon. 2005. "[6]-Gingerol, a Pungent Ingredient of Ginger, Inhibits Angiogenesis in Vitro and in Vivo." *Biochemical and Biophysical Research Communications* 335 (2): 300–308. https://doi.org/10.1016/j.bbrc.2005.07.076.

Krasochko, Petr, Rudolf Korochkin, S Gvozdev, and M Ponaskov. 2020. "Optical Properties of Colloidal Solutions of Metal Nanoparticles." *Scientific Horizons* 23 (10): 47–53. https://doi.org/10.48077/scihor.23(10).2020.47-53.

Kukoyi, Adeshola Raheem. 2016. "Economic Impacts of Natural Polymers." In Ololade Olatunji (Ed.), *Natural Polymers*, 339–62. Springer International Publishing. https://doi.org/10.1007/978-3-319-26414-1_11.

Kumaraguruparaswami, Manonmani, Senthilkumar Kandasamy, Naveen Kumar Manickam, Balaji Dhandapani, Gokilam Mohankumar, and Sangeetha Arunachalam. 2021. "Green Technologies for the Biosynthesis of Nanoparticles and Their Applications for Environmental Sustainability." In R. Praveen Kumar, and B. Bharatiraja (Eds.), *Nanomaterials*, 65–78. Elsevier.

Lai, Joanne, Abul Kalam Azad, Wan Mohd Azizi Wan Sulaiman, Vinoth Kumarasamy, Vetriselvan Subramaniyan, and Salah Abdalrazak Alshehade. 2024. "Alginate-Based Encapsulation Fabrication Technique for Drug Delivery: An Updated Review of Particle Type, Formulation Technique, Pharmaceutical Ingredient, and Targeted Delivery System." *Pharmaceutics* 16 (3): 370.

Lee, Chung-Sung, and Hee Sook Hwang. 2023. "Starch-Based Hydrogels as a Drug Delivery System in Biomedical Applications." *Gels* 9 (12): 951.

Lee, Kuen Yong, and David J. Mooney. 2012. "Alginate: Properties and Biomedical Applications." *Progress in Polymer Science* 37 (1): 106–26. https://doi.org/10.1016/j.progpolymsci.2011.06.003.

Li, Yuan, Yanping Liu, and Qiang Guo. 2021. "Silk Fibroin Hydrogel Scaffolds Incorporated with Chitosan Nanoparticles Repair Articular Cartilage Defects by Regulating TGF-B1 and BMP-2." *Arthritis Research & Therapy* 23 (1): 50. https://doi.org/10.1186/s13075-020-02382-x.

Lichte, P., H. C. Pape, T. Pufe, P. Kobbe, and H. Fischer. 2011. "Scaffolds for Bone Healing: Concepts, Materials and Evidence." *Injury* 42 (6): 569–73. https://doi.org/10.1016/j.injury.2011.03.033.

Lin, Zhen, Riwang Li, Yi Liu, Yaowu Zhao, Ningjian Ao, Jing Wang, Lihua Li, and Gang Wu. 2020. "Histatin1-Modified Thiolated Chitosan Hydrogels Enhance Wound Healing by Accelerating Cell Adhesion, Migration and Angiogenesis." *Carbohydrate Polymers* 230: 115710.

Liu, Haoyu, Ying Xu, Hui Geng, Yongdong Chen, and Xiaohu Dai. 2022. "Contributions of MOF-808 to Methane Production from Anaerobic Digestion of Waste Activated Sludge." *Water Research* 220: 118653.

Liu, Kunping, Yimin Wang, Huiming Li, and Yixiang Duan. 2015. "A Facile One-Pot Synthesis of Starch Functionalized Graphene as Nano-Carrier for PH Sensitive and Starch-Mediated Drug Delivery." *Colloids and Surfaces B: Biointerfaces* 128 (April): 86–93. https://doi.org/10.1016/j.colsurfb.2015.02.010.

Liu, Shanshan, Shili Yang, and Paul C. Ho. 2018. "Intranasal Administration of Carbamazepine-Loaded Carboxymethyl Chitosan Nanoparticles for Drug Delivery to the Brain." *Asian Journal of Pharmaceutical Sciences* 13 (1): 72–81.

Liu, Xiao, Xia Chen, Yanfeng Li, Yanjun Cui, Hao Zhu, and Weiwei Zhu. 2012. "Preparation of Superparamagnetic Sodium Alginate Nanoparticles for Covalent Immobilization of Candida Rugosa Lipase." *Journal of Nanoparticle Research* 14: 1–7.

Liu, Xiaojuan, Liqiang Luo, Yaping Ding, and Yanhong Xu. 2011. "Amperometric Biosensors Based on Aluminananoparticles-Chitosan-Horseradish Peroxidase Nanobiocomposites for the Determination of Phenolic Compounds." *The Analyst* 136 (4): 696–701. https://doi.org/10.1039/C0AN00752H.

Liu, Xin, Shengwei Peng, Yongju Pei, Yuanyuan Huo, Yadi Zong, Jianwei Ren, and Jing Zhao. 2024. "Facile Fabrication of Chitosan/Hyaluronic Acid Hydrogel-Based Wound Closure Material Co-Loaded with Gold Nanoparticles and Fibroblast Growth Factor to Improve Anti-Microbial and Healing Efficiency in Diabetic Wound Healing and Nursing Care." *Regenerative Therapy* 26 (June): 1018–29. https://doi.org/10.1016/j.reth.2024.10.003.

Löbmann, Korbinian, and Anna J. Svagan. 2017. "Cellulose Nanofibers as Excipient for the Delivery of Poorly Soluble Drugs." *International Journal of Pharmaceutics* 533 (1): 285–97. https://doi.org/10.1016/j.ijpharm.2017.09.064.

Lueßen, H. L., C.-M. Lehr, C.-O. Rentel, A. B. J. Noach, A. G. de Boer, J. C. Verhoef, and H. E. Junginger. 1994. "Bioadhesive Polymers for the Peroral Delivery of Peptide Drugs." *Journal of Controlled Release* 29 (3): 329–38. https://doi.org/10.1016/0168-3659(94)90078-7.

Maji, Biswajit. 2019. Introduction to Natural Polysaccharides. In Saurabh Maiti (Ed.), *Functional Polysaccharides for Biomedical Applications*, 1–20. Elsevier Ltd. https://doi.org/10.1016/B978-0-08-102555-0.00001-7.

Majithia, Maithili, and Delicia A. Barretto. 2023. "Biocompatible Green-Synthesized Nanomaterials for Therapeutic Applications." In Pranay Pradeep Morajkar and Milind Mohan Nayak (Eds.), *Advances in Nano and Biochemistry*, 285–367. Elsevier.

Malafaya, Patrícia B., Gabriela A. Silva, and Rui L. Reis. 2007. "Natural–Origin Polymers as Carriers and Scaffolds for Biomolecules and Cell Delivery in Tissue Engineering Applications." *Advanced Drug Delivery Reviews* 59 (4–5): 207–33. https://doi.org/10.1016/j.addr.2007.03.012.

Mallakpour, Shadpour, and Leila khodadadzadeh. 2018. "Ultrasonic-Assisted Fabrication of Starch/MWCNT-Glucose Nanocomposites for Drug Delivery." *Ultrasonics Sonochemistry* 40 (January): 402–9. https://doi.org/10.1016/j.ultsonch.2017.07.033.

Mano, J. F., G. A. Silva, H. S. Azevedo, P. B. Malafaya, R. A. Sousa, S. S. Silva, L. F. Boesel, et al. 2007. "Natural Origin Biodegradable Systems in Tissue Engineering and Regenerative Medicine: Present Status and Some Moving Trends." *Journal of The Royal Society Interface* 4 (17): 999–1030. https://doi.org/10.1098/rsif.2007.0220.

Marradi, Marco, Manuel Martin-Lomas, and Soledad Penades. 2010. "Glyconanoparticles: Polyvalent Tools to Study Carbohydrate-Based Interactions." *Advances in Carbohydrate Chemistry and Biochemistry* 64: 211–90.

Mourdikoudis, Stefanos, Roger M. Pallares, and Nguyen T. K. Thanh. 2018. "Characterization Techniques for Nanoparticles: Comparison and Complementarity upon Studying Nanoparticle Properties." *Nanoscale* 10 (27): 12871–934.

Mujahid, Mohd Hasan, Tarun Kumar Upadhyay, Fahad Khan, Pratibha Pandey, Moon Nyeo Park, Amit Baran Sharangi, Mohd Saeed, Vijay J. Upadhye, and Bonglee Kim. 2022. "Metallic and Metal Oxide-Derived Nanohybrid as a Tool for Biomedical Applications." *Biomedicine & Pharmacotherapy* 155: 113791.

Nazir, Ruqeya, and Insha Amin. 2021. "Role of Green Nanotechnology in Alleviating Climate Change." In *Microbiomes and the Global Climate Change*, 365–74.

Nguyen, Mai Thanh, Lianlian Deng, and Tetsu Yonezawa. 2022. "Control of Nanoparticles Synthesized via Vacuum Sputter Deposition onto Liquids: A Review." *Soft Matter* 18 (1): 19–47.

Nitta, Sachiko, and Keiji Numata. 2013. "Biopolymer-Based Nanoparticles for Drug/Gene Delivery and Tissue Engineering." *International Journal of Molecular Sciences* 14 (1): 1629–54. https://doi.org/10.3390/ijms14011629.

Nowak, Paweł, Anna Ilnicka, and Marta Ziegler-Borowska. 2024. "Hydrazidomethyl Starch as a PH-Sensitive Coating for Magnetic Core in Tailored Magnetic Nanoparticles with Selective Doxorubicin Release." *International Journal of Biological Macromolecules*, November, 137716. https://doi.org/10.1016/j.ijbiomac.2024.137716.

Nurazzi, N. M., N. Abdullah, M. N. F. Norrrahim, S. H. Kamarudin, S. Ahmad, S. S. Shazleen, M. Rayung, M. R. M. Asyraf, R. A. Ilyas, and M. Kuzmin. 2022. "Thermogravimetric Analysis (TGA) and Differential Scanning Calorimetry (DSC) of PLA/Cellulose Composites." In Jyotishkumar Parameswaranpillai, Suchart Siengchin, Nisa V. Salim, Jinu Jacob George, and Aishwarya Poulose (Eds.), *Polylactic Acid-Based Nanocellulose and Cellulose Composites*, 145–64. CRC Press.

Obireddy, Sreekanth Reddy, Shirisha Bellala, Madhavi Chintha, Akkulanna Sake, Subha Marata Chinna Subbarao, and Wing-Fu Lai. 2023. "Synthesis and Properties of Alginate-Based Nanoparticles Incorporated with Different Inorganic Nanoparticulate Modifiers for Enhanced Encapsulation and Controlled Release of Favipiravir." *Arabian Journal of Chemistry* 16 (7): 104751.

Orasugh, Jonathan Tersur, Nayan Ranjan Saha, Dipak Rana, Gunjan Sarkar, Md. Masud Rahaman Mollick, Atiskumar Chattoapadhyay, Bhairab Chandra Mitra, Dibyendu Mondal, Swapan Kumar Ghosh, and Dipankar Chattopadhyay. 2018. "Jute Cellulose Nano-Fibrils/Hydroxypropylmethylcellulose Nanocomposite: A Novel Material with Potential for Application in Packaging and Transdermal Drug Delivery System." *Industrial Crops and Products* 112 (February): 633–43. https://doi.org/10.1016/j.indcrop.2017.12.069.

O'Toole, Martin G., Richard M. Henderson, Patricia A. Soucy, Brigitte H. Fasciotto, Patrick J. Hoblitzell, Robert S. Keynton, William D. Ehringer, and Andrea S. Gobin. 2012. "Curcumin Encapsulation in Submicrometer Spray-Dried Chitosan/Tween 20 Particles." *Biomacromolecules* 13 (8): 2309–14. https://doi.org/10.1021/bm300564v.

Patel, Pitambar, and Sachin Rameshrao Geed. 2024. "Recent Advancements in the Application of Nanomaterial in Modern Drug Delivery and Future Perspective." In Maulin

P. Shah, Navaneet Bharadawaj, and Lakhan Kumar (Eds.), *Biogenic Nanomaterials for Environmental Sustainability: Principles, Practices, and Opportunities*, 319–51. Springer.

Pei, JinJin, Yuqiang Yan, Selvaraj Jayaraman, Ponnulakshmi Rajagopal, Prabhu Manickam Natarajan, Vidhya Rekha Umapathy, Sridevi Gopathy, Jeane Rebecca Roy, Janaki Coimbatore Sadagopan, and Dwarakesh Thalamati. 2024. "A Review on Advancements in the Application of Starch-Based Nanomaterials in Biomedicine: Precision Drug Delivery and Cancer Therapy." *International Journal of Biological Macromolecules* 265: 130746.

Peptu, C.A., L. Ochiuz, L. Alupei, C. Peptu, and M. Popa. 2014. "Carbohydrate Based Nanoparticles for Drug Delivery Across Biological Barriers." *Journal of Biomedical Nanotechnology* 10 (9): 2107–48.

Pradhan, Subhashree Subhasmita, and Sampa Saha. 2022. "Advances in Design and Applications of Polymer Brush Modified Anisotropic Particles." *Advances in Colloid and Interface Science* 300: 102580.

Pulumati, Akhil, Anika Pulumati, Bilikere S. Dwarakanath, Amit Verma, and Rao V. L. Papineni. 2023. "Technological Advancements in Cancer Diagnostics: Improvements and Limitations." *Cancer Reports* 6 (2). https://doi.org/10.1002/cnr2.1764.

Qiao, Yu, Chang Sun, Juan Jian, Tianyu Zhou, Xiangxin Xue, Jinghui Shi, Guangbo Che, and Guangfu Liao. 2023. "Efficient Removal of Organic Pollution via Photocatalytic Degradation over a TiO2@ HKUST-1 Yolk-Shell Nanoreactor." *Journal of Molecular Liquids* 385: 122383.

Qin, Henan, Yibin Teng, Rui Dai, Aman Wang, and Jiwei Liu. 2024. "Glycan-Based Scaffolds and Nanoparticles as Drug Delivery System in Cancer Therapy." *Frontiers in Immunology* 15 (May): 1–15. https://doi.org/10.3389/fimmu.2024.1395187.

Qing, Weixia, Yong Wang, Youyou Wang, Dongbao Zhao, Xiuhua Liu, and Jinhua Zhu. 2016. "The Modified Nanocrystalline Cellulose for Hydrophobic Drug Delivery." *Applied Surface Science* 366 (March): 404–9. https://doi.org/10.1016/j.apsusc.2016.01.133.

Qiu, Chao, Yao Hu, Zhengyu Jin, David Julian McClements, Yang Qin, Xueming Xu, and Jinpeng Wang. 2019. "A Review of Green Techniques for the Synthesis of Size-Controlled Starch-Based Nanoparticles and Their Applications as Nanodelivery Systems." *Trends in Food Science & Technology* 92: 138–51.

Raj, V., and G. Prabha. 2016. "Synthesis, Characterization and in Vitro Drug Release of Cisplatin Loaded Cassava Starch Acetate–PEG/Gelatin Nanocomposites." *Journal of the Association of Arab Universities for Basic and Applied Sciences* 21 (1): 10–16. https://doi.org/10.1016/j.jaubas.2015.08.001.

Ramesh Babu, V., Malladi Sairam, Kallappa M. Hosamani, and Tejraj M. Aminabhavi. 2007. "Preparation of Sodium Alginate–Methylcellulose Blend Microspheres for Controlled Release of Nifedipine." *Carbohydrate Polymers* 69 (2): 241–50. https://doi.org/10.1016/j.carbpol.2006.09.027.

Ramzan, Muhammad, Abid Mehmood Yousaf, and Ikram Ullah Khan. 2024. "Fabrication and In-Vitro Characterization of Delafloxacin-Laden Electrosprayed Hydroxypropyl Cellulose Nanoparticulated Solid Dispersions for Enhanced Aqueous Solubility and Release Rate of the Drug." *International Journal of Polymeric Materials and Polymeric Biomaterials* 73 (16): 1367–73. https://doi.org/10.1080/00914037.2023.2289521.

Rao, Kummara Madhusudana, Anuj Kumar, Maduru Suneetha, and Sung Soo Han. 2018. "PH and Near-Infrared Active; Chitosan-Coated Halloysite Nanotubes Loaded with Curcumin-Au Hybrid Nanoparticles for Cancer Drug Delivery." *International Journal of Biological Macromolecules* 112 (June): 119–25. https://doi.org/10.1016/j.ijbiomac.2018.01.163.

Rao, Ziqie, Hongyu Ge, Liangling Liu, Chen Zhu, Lian Min, Meng Liu, Lihong Fan, and Dan Li. 2018. "Carboxymethyl Cellulose Modified Graphene Oxide as PH-Sensitive Drug Delivery System." *International Journal of Biological Macromolecules* 107 (February): 1184–92. https://doi.org/10.1016/j.ijbiomac.2017.09.096.

Reddy, A. Babul, G. Siva Mohan Reddy, Veluri Sivanjineyulu, J. Jayaramudu, Kokkarachedu Varaprasad, and Emmanuel Rotimi Sadiku. 2016. "Hydrophobic/Hydrophilic Nanostructured Polymer Blends." In Sabu Thomas, Robert Shanks, and Sarthachandran Chandrasekharakurup (Eds.), *Design and Applications of Nanostructured Polymer Blends and Nanocomposite Systems*, 385–411. Elsevier.

Reed-Gore, Erin R., Robert D. Moser, and Charles Arthur Weiss. 2018. "Characterization of Nanosized Crystallites Using X-Ray Diffraction (XRD): Standard Operating Procedure Series: Characterization (C)." Report number:ERDC/GSL SR-18-3. US Army Engineer Research and Development Center (ERDC).

Ren, Xiangyi, Dongxu Chen, Yan Wang, Huifang Li, Yabing Zhang, Hongying Chen, Xi Li, and Minfeng Huo. 2022. "Nanozymes-Recent Development and Biomedical Applications." *Journal of Nanobiotechnology* 20 (1): 1–18. https://doi.org/10.1186/s12951-022-01295-y.

Ren, Yizhuo, Xin Zhao, Xiaofeng Liang, Peter X. Ma, and Baolin Guo. 2017. "Injectable Hydrogel Based on Quaternized Chitosan, Gelatin and Dopamine as Localized Drug Delivery System to Treat Parkinson's Disease." *International Journal of Biological Macromolecules* 105 (December): 1079–87. https://doi.org/10.1016/j.ijbiomac.2017.07.130.

Rezazadeh, Nahid, Effat Alizadeh, Somaieh Soltani, Soodabeh Davaran, and Neda Esfandiari. 2024. "Synthesis and Characterization of a Magnetic Bacterial Cellulose-Chitosan Nanocomposite and Evaluation of Its Applicability for Osteogenesis." *BioImpacts* 14 (6): 30159. https://doi.org/10.34172/bi.2024.30159.

Rizeq, Balsam R., Nadin N. Younes, Kashif Rasool, and Gheyath K. Nasrallah. 2019. "Synthesis, Bioapplications, and Toxicity Evaluation of Chitosan-Based Nanoparticles." *International Journal of Molecular Sciences* 20 (22): 5776.

Ruiz-Rico, María, Félix Sancenón, and José M. Barat. 2023. "Evaluation of the in Vitro and in Situ Antimicrobial Properties of Chitosan-Functionalised Silica Materials." *LWT* 173: 114373.

Sabir, Muhammad Iqbal, Xiaoxue Xu, and Li Li. 2009. "A Review on Biodegradable Polymeric Materials for Bone Tissue Engineering Applications." *Journal of Materials Science* 44 (21): 5713–24. https://doi.org/10.1007/s10853-009-3770-7.

Sadrolhosseini, Amir Reza, Mohd Adzir Mahdi, Farideh Alizadeh, and Suraya Abdul Rashid. 2019. "Laser Ablation Technique for Synthesis of Metal Nanoparticle in Liquid." In *Laser Technology and Its Applications*, 63–83. IntechOpen.

Sahin, Selmihan, Ismail Ozmen, Gizem Yildirim Bastemur, and Sabriye Percin Ozkorucuklu. 2019. "Development of Voltammetric Glucose-6-Phosphate Biosensors Based on the Immobilization of Glucose-6-Phosphate Dehydrogenase on Polypyrrole- and Chitosan-Coated Fe3O4 Nanoparticles/Polypyrrole Nanocomposite Films." *Applied Biochemistry and Biotechnology* 188 (4): 1145–57. https://doi.org/10.1007/s12010-019-02979-2.

Saikia, Chinmayee, Monoj K. Das, Anand Ramteke, and Tarun K. Maji. 2016. "Effect of Crosslinker on Drug Delivery Properties of Curcumin Loaded Starch Coated Iron Oxide Nanoparticles." *International Journal of Biological Macromolecules* 93 (December): 1121–32. https://doi.org/10.1016/j.ijbiomac.2016.09.043.

Saini, Seema, Mohamed Naceur Belgacem, and Julien Bras. 2017. "Effect of Variable Aminoalkyl Chains on Chemical Grafting of Cellulose Nanofiber and Their Antimicrobial Activity." *Materials Science and Engineering: C* 75: 760–68.

Sakeer, Khalil, Pompilia Ispas-Szabo, Nassim Benyerbah, and Mircea Alexandru Mateescu. 2018. "Ampholytic Starch Excipients for High Loaded Drug Formulations: Mechanistic Insights." *International Journal of Pharmaceutics* 535 (1–2): 201–16. https://doi.org/10.1016/j.ijpharm.2017.11.005.

Saleh, Tawfik Abdo. 2022. *Surface Science of Adsorbents and Nanoadsorbents: Properties and Applications in Environmental Remediation*. Academic Press.

Sampath, Udeni, Yern Ching, Cheng Chuah, Johari Sabariah, and Pai-Chen Lin. 2016. "Fabrication of Porous Materials from Natural/Synthetic Biopolymers and Their Composites." *Materials* 9 (12): 991. https://doi.org/10.3390/ma9120991.

Samrot, Antony V., Ujjala Burman, Sheryl Ann Philip, and Kumar Chandrasekaran. 2018. "Synthesis of Curcumin Loaded Polymeric Nanoparticles from Crab Shell Derived Chitosan for Drug Delivery." *Informatics in Medicine Unlocked* 10: 159–82. https://doi.org/10.1016/j.imu.2017.12.010.

Seddiqi, Hadi, Erfan Oliaei, Hengameh Honarkar, Jianfeng Jin, Lester C. Geonzon, Rommel G. Bacabac, and Jenneke Klein-Nulend. 2021. Cellulose and Its Derivatives: Towards Biomedical Applications. *Cellulose* 28: 1893–1931. https://doi.org/10.1007/s10570-020-03674-w.

Sen, Aparajita, Jyoti Oswalia, Sneha Yadav, Meenakshi Vachher, and Arti Nigam. 2024. "Recent Trends in Nanozyme Research and Their Potential Therapeutic Applications." *Current Research in Biotechnology* 7: 100205.

Senthil, Rethinam, Serdar Batıkan Kavukcu, and Weslen S Vedakumari. 2023. "Cellulose Based Biopolymer Nanoscaffold: A Possible Biomedical Applications." *International Journal of Biological Macromolecules* 246: 125656.

Shahid, Mehmood, Suresh Sagadevan, Waqar Ahmed, Yiqiang Zhan, and Pakorn Opaprakasit. 2022. "Metal Oxides for Optoelectronic and Photonic Applications: A General Introduction." In Suresh Sagadevan, Jiban Podder, and Farauq Mohammad (Eds.), *Metal Oxides for Optoelectronics and Optics-Based Medical Applications*, 3–31. Elsevier.

Shaji, Ashin, and Ajesh K. Zachariah. 2017. "Surface Area Analysis of Nanomaterials." In Sabu Thomas, Raju Thomas, Ajesh K. Zachariah, and Raghvendra Kumar Mishra (Eds.), *Thermal and Rheological Measurement Techniques for Nanomaterials Characterization*, 197–231. Elsevier.

Sharifi-Rad, Javad, Cristina Quispe, Monica Butnariu, Lia Sanda Rotariu, Oksana Sytar, Simona Sestito, Simona Rapposelli, Muhammad Akram, Mehwish Iqbal, and Akash Krishna. 2021. "Chitosan Nanoparticles as a Promising Tool in Nanomedicine with Particular Emphasis on Oncological Treatment." *Cancer Cell International* 21 (1): 318.

Shen, Bowen, Zhaoxu Yan, Yuxian Wang, Liying Zhu, Qianru Zhao, and Ling Jiang. 2025. "Nanozyme-Chitosan-Aerogel Immobilized Enzyme-Driven Biocatalytic Cascade for Therapeutic Engineering of Diabetic Wounds." *Carbohydrate Polymers* 347 (August 2024): 122690. https://doi.org/10.1016/j.carbpol.2024.122690.

Shrestha, Sweta, Bo Wang, and Prabir Dutta. 2020. "Nanoparticle Processing: Understanding and Controlling Aggregation." *Advances in Colloid and Interface Science* 279: 102162.

Shtenberg, Yarden, Mor Goldfeder, Hodaya Prinz, Janna Shainsky, Yasmine Ghantous, Imad Abu El-Naaj, Avi Schroeder, and Havazelet Bianco-Peled. 2018. "Mucoadhesive Alginate Pastes with Embedded Liposomes for Local Oral Drug Delivery." *International Journal of Biological Macromolecules* 111 (May): 62–69. https://doi.org/10.1016/j.ijbiomac.2017.12.137.

Shukla, Vineeta. 2022. "Introduction of Vibrating Sample Magnetometer for Magnetic Characterization." In Sabu Thomas, Ajay Vasudeo Rane, and Nandakumar Kalarikkal

(Eds.), *Handbook of Magnetic Hybrid Nanoalloys and Their Nanocomposites*, 483–505. Springer.

Silant'ev, Vladimir E., Mikhail E. Shmelev, Andrei S. Belousov, Aleksandra A. Patlay, Roman A. Shatilov, Vladislav M. Farniev, and Vadim V. Kumeiko. 2023. "How to Develop Drug Delivery System Based on Carbohydrate Nanoparticles Targeted to Brain Tumors." *Polymers* 15 (11): 1–27. https://doi.org/10.3390/polym15112516.

Singh, Rajesh, and James W. Lillard Jr. 2009. "Nanoparticle-Based Targeted Drug Delivery." *Experimental and Molecular Pathology* 86 (3): 215–23.

Singh, Rudra Pratap, H. V. Gangadharappa, and K. Mruthunjaya. 2018. "Phytosome Complexed with Chitosan for Gingerol Delivery in the Treatment of Respiratory Infection: In Vitro and in Vivo Evaluation." *European Journal of Pharmaceutical Sciences* 122 (September): 214–29. https://doi.org/10.1016/j.ejps.2018.06.028.

Skorik, Yury A., Anton A. Golyshev, Andreii S. Kritchenkov, Ekaterina R. Gasilova, Daria N. Poshina, Amal J. Sivaram, and Rangasamy Jayakumar. 2017. "Development of Drug Delivery Systems for Taxanes Using Ionic Gelation of Carboxyacyl Derivatives of Chitosan." *Carbohydrate Polymers* 162 (April): 49–55. https://doi.org/10.1016/j.carbpol.2017.01.025.

Smidsrod, O., and G. Skjakbrk. 1990. "Alginate as Immobilization Matrix for Cells." *Trends in Biotechnology* 8: 71–78. https://doi.org/10.1016/0167-7799(90)90139-O.

Srivastava, Rohit, Jonathan Q. Brown, Huiguang Zhu, and Michael J. McShane. 2005. "Stable Encapsulation of Active Enzyme by Application of Multilayer Nanofilm Coatings to Alginate Microspheres." *Macromolecular Bioscience* 5 (8): 717–27.

Stasyuk, Nataliya, Oleh Smutok, Olha Demkiv, Tetiana Prokopiv, Galina Gayda, Marina Nisnevitch, and Mykhailo Gonchar. 2020. "Synthesis, Catalytic Properties and Application in Biosensorics of Nanozymes and Electronanocatalysts: A Review." *Sensors* 20 (16): 4509.

Su, Lu, Yingle Feng, Kongchang Wei, Xuyang Xu, Rongying Liu, and Guosong Chen. 2021. "Carbohydrate-Based Macromolecular Biomaterials." *Chemical Reviews* 121 (18): 10950–29. https://doi.org/10.1021/acs.chemrev.0c01338.

Subramanian, K. S., V. Karthika, M. Praghadeesh, and A. Lakshmanan. 2020. "Nanotechnology for Mitigation of Global Warming Impacts." In V. Venkatraman, Shachi Shah, and Ram Prasad (Eds.), *Global Climate Change: Resilient and Smart Agriculture*, 315–36. Springer Nature

Sun, Luzhao, Guowen Yuan, Libo Gao, Jieun Yang, Manish Chhowalla, Meysam Heydari Gharahcheshmeh, Karen K. Gleason, Yong Seok Choi, Byung Hee Hong, and Zhongfan Liu. 2021. "Chemical Vapour Deposition." *Nature Reviews Methods Primers* 1 (1): 5.

Suriati, G., M. Mariatti, and A. Azizan. 2014. "Synthesis of Silver Nanoparticles by Chemical Reduction Method: Effect of Reducing Agent and Surfactant Concentration." *International Journal of Automotive and Mechanical Engineering* 10: 1920–27.

Taherimehr, Marzieh, Reza Bagheri, and Masoumeh Taherimehr. 2021. "In-Vitro Evaluation of Thermoplastic Starch/Beta-Tricalcium Phosphate Nano-Biocomposite in Bone Tissue Engineering." *Ceramics International* 47 (11): 15458–63. https://doi.org/10.1016/j.ceramint.2021.02.111.

Takeuchi, Issei, Tomoyoshi Takeshita, Takaaki Suzuki, and Kimiko Makino. 2017. "Iontophoretic Transdermal Delivery Using Chitosan-Coated PLGA Nanoparticles for Positively Charged Drugs." *Colloids and Surfaces B: Biointerfaces* 160 (December): 520–26. https://doi.org/10.1016/j.colsurfb.2017.10.011.

Tang, Wenmin, Peng Pan, Tiantian Chen, Jian Wang, Xiangnan Cui, Wentao Liu, and Liang Kan. 2025. "3D Chitosan Scaffolds Loaded with ZnO Nanoparticles for Bone Tissue Engineering." *Colloids and Surfaces B: Biointerfaces* 245 (January): 114199. https://doi.org/10.1016/j.colsurfb.2024.114199.

Tawfik, Salah M., Mirkomil Sharipov, Bui The Huy, Zayakhuu Gerelkhuu, Dana Biechele-Speziale, and Yong-Ill Lee. 2018. "Naturally Modified Nonionic Alginate Functionalized Upconversion Nanoparticles for the Highly Efficient Targeted PH-Responsive Drug Delivery and Enhancement of NIR-Imaging." *Journal of Industrial and Engineering Chemistry* 57 (January): 424–35. https://doi.org/10.1016/j.jiec.2017.08.051.

Tran, V., and X. Wen. 2014. "Rapid Prototyping Technologies for Tissue Regeneration." In *Rapid Prototyping of Biomaterials*, 97–155.

Unagolla, Janitha M., and Ambalangodage C. Jayasuriya. 2018. "Drug Transport Mechanisms and in Vitro Release Kinetics of Vancomycin Encapsulated Chitosan-Alginate Polyelectrolyte Microparticles as a Controlled Drug Delivery System." *European Journal of Pharmaceutical Sciences* 114 (March): 199–209. https://doi.org/10.1016/j.ejps.2017.12.012.

Van Der Pol, Edwin, Frank A. W. Coumans, Auguste Sturk, Rienk Nieuwland, and Ton G. Van Leeuwen. 2014. "Refractive Index Determination of Nanoparticles in Suspension Using Nanoparticle Tracking Analysis." *Nano Letters* 14 (11): 6195–201.

Verma, Madan L., B. S. Dhanya, Sukriti, Varsha Rani, Meenu Thakur, J. Jeslin, and Rekha Kushwaha. 2020. "Carbohydrate and Protein Based Biopolymeric Nanoparticles: Current Status and Biotechnological Applications." *International Journal of Biological Macromolecules* 154 (July): 390–412. https://doi.org/10.1016/j.ijbiomac.2020.03.105.

Vladár, András E., and Vasile-Dan Hodoroaba. 2020. "Characterization of Nanoparticles by Scanning Electron Microscopy." In Vasile-Dan Hodoroaba, Wolfgang Unger, and Alexander Shard (Eds.), *Characterization of Nanoparticles*, 7–27. Elsevier.

Wang, Chongyang, Wenxiu Hou, Xuran Guo, Juehong Li, Tu Hu, Manle Qiu, Shen Liu, Xiumei Mo, and Xudong Liu. 2017. "Two-Phase Electrospinning to Incorporate Growth Factors Loaded Chitosan Nanoparticles into Electrospun Fibrous Scaffolds for Bioactivity Retention and Cartilage Regeneration." *Materials Science and Engineering: C* 79 (October): 507–15. https://doi.org/10.1016/j.msec.2017.05.075.

Wang, Fengxiang, Shouzhi He, and Bin Chen. 2018. "Retinoic Acid-Loaded Alginate Microspheres as a Slow Release Drug Delivery Carrier for Intravitreal Treatment." *Biomedicine & Pharmacotherapy* 97 (January): 722–28. https://doi.org/10.1016/j.biopha.2017.10.109.

Wang, Jingwei, Yong Kong, Feng Liu, Dan Shou, Yongxin Tao, and Yong Qin. 2018. "Construction of PH-Responsive Drug Delivery Platform with Calcium Carbonate Microspheres Induced by Chitosan Gels." *Ceramics International* 44 (7): 7902–7. https://doi.org/10.1016/j.ceramint.2018.01.227.

Wang, Nuo, and Xue Shen Wu. 1997. "Preparation and Characterization of Agarose Hydrogel Nanoparticles for Protein and Peptide Drug Delivery." *Pharmaceutical Development and Technology* 2 (2): 135–42. https://doi.org/10.3109/10837459709022618.

Wang, Ren, Dan Shou, Ouyang Lv, Yong Kong, Linhong Deng, and Jian Shen. 2017. "PH-Controlled Drug Delivery with Hybrid Aerogel of Chitosan, Carboxymethyl Cellulose and Graphene Oxide as the Carrier." *International Journal of Biological Macromolecules* 103 (October): 248–53. https://doi.org/10.1016/j.ijbiomac.2017.05.064.

Wang, Ruoying, Shixin Xu, Miaomiao Zhang, Wei Feng, Chengwei Wang, Xuefeng Qiu, Jierui Li, and Wen Zhao. 2024. "Multifunctional Chitosan-Based Hydrogels Loaded with Iridium Nanoenzymes for Skin Wound Repair." *Carbohydrate Polymers* 342 (February): 122325. https://doi.org/10.1016/j.carbpol.2024.122325.

Wardman, Jacob F., and Stephen G. Withers. 2024. "Carbohydrate-Active Enzyme (CAZyme) Discovery and Engineering via (Ultra)High-Throughput Screening." *RSC Chemical Biology* 5 (7): 595–616. https://doi.org/10.1039/d4cb00024b.

Wei, Hui, and Erkang Wang. 2008. "Fe3O4 Magnetic Nanoparticles as Peroxidase Mimetics and Their Applications in H2O2 and Glucose Detection." *Analytical Chemistry* 80 (6): 2250–54.

Wei, H., and E. Wang. 2013. "Nanomaterials with Enzyme-like Characteristics (Nanozymes): Next-Generation Artificial Enzymes." *Chemical Society Reviews* 42 (14): 6060–93.

Weng, Yilun, Guangze Yang, Yang Li, Letao Xu, Xiaojing Chen, Hao Song, and Chun-Xia Zhao. 2023. "Alginate-Based Materials for Enzyme Encapsulation." *Advances in Colloid and Interface Science* 318: 102957.

Wijaya, Christian J., Stephanie N. Saputra, Felycia E. Soetaredjo, Jindrayani N. Putro, Chun X. Lin, Alfin Kurniawan, Yi-Hsu Ju, and Suryadi Ismadji. 2017. "Cellulose Nanocrystals from Passion Fruit Peels Waste as Antibiotic Drug Carrier." *Carbohydrate Polymers* 175 (November): 370–76. https://doi.org/10.1016/j.carbpol.2017.08.004.

Windels, Alex, Jorick Franceus, Jürgen Pleiss, and Tom Desmet. 2024. "CANDy: Automated Analysis of Domain Architectures in Carbohydrate-Active Enzymes." *Plos One* 19 (7): e0306410.

Wu, Bao-Yan, Shi-Hua Hou, Feng Yin, Jing Li, Zi-Xia Zhao, Jia-Dong Huang, and Qiang Chen. 2007. "Amperometric Glucose Biosensor Based on Layer-by-Layer Assembly of Multilayer Films Composed of Chitosan, Gold Nanoparticles and Glucose Oxidase Modified Pt Electrode." *Biosensors and Bioelectronics* 22 (6): 838–44. https://doi.org/10.1016/j.bios.2006.03.009.

Wu, Can, Jinlong Yang, Xiubin Xu, Chunmei Gao, Shaoyu Lü, and Mingzhu Liu. 2016. "Redox-Responsive Core-Cross-Linked MPEGylated Starch Micelles as Nanocarriers for Intracellular Anticancer Drug Release." *European Polymer Journal* 83 (October): 230–43. https://doi.org/10.1016/j.eurpolymj.2016.08.018.

Wu, Zhiwei, Weiwei Wu, Chi Zhang, Wenbiao Zhang, Yang Li, Tao Ding, Zhennan Fang, Juehua Jing, Xiaoyan He, and Fei Huang. 2025. "Enhanced Diabetic Foot Ulcer Treatment with a Chitosan-Based Thermosensitive Hydrogel Loaded Self-Assembled Multi-Functional Nanoparticles for Antibacterial and Angiogenic Effects." *Carbohydrate Polymers* 347 (January): 122740. https://doi.org/10.1016/j.carbpol.2024.122740.

Xiao, Huaxi, Tao Yang, Qinlu Lin, Gao-Qiang Liu, Lin Zhang, Fengxiang Yu, and Yuejiao Chen. 2016. "Acetylated Starch Nanocrystals: Preparation and Antitumor Drug Delivery Study." *International Journal of Biological Macromolecules* 89 (August): 456–64. https://doi.org/10.1016/j.ijbiomac.2016.04.037.

Xie, Meng, Feng Zhang, Lijiao Liu, Yanan Zhang, Yeping Li, Huaming Li, and Jimin Xie. 2018. "Surface Modification of Graphene Oxide Nanosheets by Protamine Sulfate/Sodium Alginate for Anti-Cancer Drug Delivery Application." *Applied Surface Science* 440 (May): 853–60. https://doi.org/10.1016/j.apsusc.2018.01.175.

Xu, Jinke, Mifong Tam, Sepideh Samaei, Sophie Lerouge, Jake Barralet, Mary M. Stevenson, and Marta Cerruti. 2017. "Mucoadhesive Chitosan Hydrogels as Rectal Drug Delivery Vessels to Treat Ulcerative Colitis." *Acta Biomaterialia* 48 (January): 247–57. https://doi.org/10.1016/j.actbio.2016.10.026.

Xu, Junfei, Shan Liu, Guangxue Chen, Ting Chen, Tao Song, Jing Wu, Congcan Shi, Minghui He, and Junfei Tian. 2018. "Engineering Biocompatible Hydrogels from Bicomponent Natural Nanofibers for Anticancer Drug Delivery." *Journal of Agricultural and Food Chemistry* 66 (4): 935–42. https://doi.org/10.1021/acs.jafc.7b04210.

Yadav, Renu, Nitin Labhsetwar, Swati Kotwal, and Sadhana Rayalu. 2011. "Single Enzyme Nanoparticle for Biomimetic CO 2 Sequestration." *Journal of Nanoparticle Research* 13: 263–71.

Yadav, Thakur Prasad, R. Manohar Yadav, and D. Pratap Singh. 2012. "Mechanical Milling: A Top down Approach for the Synthesis of Nanomaterials and Nanocomposites." *Nanoscience and Nanotechnology* 2 (3): 22–48.

Yang, Liqun, and Li-Ming Zhang. 2009. "Chemical Structural and Chain Conformational Characterization of Some Bioactive Polysaccharides Isolated from Natural Sources." *Carbohydrate Polymers* 76 (3): 349–61. https://doi.org/10.1016/j.carbpol.2008.12.015.

Yin, Zi-Chen, Yan-Ling Wang, and Kun Wang. 2018. "A PH-Responsive Composite Hydrogel Beads Based on Agar and Alginate for Oral Drug Delivery." *Journal of Drug Delivery Science and Technology* 43 (February): 12–18. https://doi.org/10.1016/j.jddst.2017.09.009.

Yoon, Taeyoung, Wooboum Park, Juneseok You, and Sungsoo Na. 2024. "Investigation of Direct Electron Transfer of Glucose Oxidase on a Graphene-CNT Composite Surface: A Molecular Dynamics Study Based on Electrochemical Experiments." *Nanomaterials* 14 (13): 1073.

You, Chaoqun, Like Ning, Zhen Zhang, Hongshuai Wu, Qingli Qu, Fei Wang, Ranhua Xiong, and Chaobo Huang. 2022. "Toxic Reactive Oxygen Species Enhanced Chemodynamic Therapy by Copper Metal-Nanocellulose Based Nanocatalysts." *Carbohydrate Polymers* 289 (August): 119432. https://doi.org/10.1016/j.carbpol.2022.119432.

Yu, Mengting, Na Ji, Yanfei Wang, Lei Dai, Liu Xiong, and Qingjie Sun. 2021. "Starch-based Nanoparticles: Stimuli Responsiveness, Toxicity, and Interactions with Food Components." *Comprehensive Reviews in Food Science and Food Safety* 20 (1): 1075–1100.

Zhang, Bin, Yayuan Yan, Qiujuan Shen, Dong Ma, Langhuan Huang, Xiang Cai, and Shaozao Tan. 2017. "A Colon Targeted Drug Delivery System Based on Alginate Modificated Graphene Oxide for Colorectal Liver Metastasis." *Materials Science and Engineering: C* 79 (October): 185–90. https://doi.org/10.1016/j.msec.2017.05.054.

Zhang, Shan, and Chao Wang. 2023. "Precise Analysis of Nanoparticle Size Distribution in TEM Image." *Methods and Protocols* 6 (4): 63.

Zhao, Dong, Ren-Xi Zhuo, and Si-Xue Cheng. 2012. "Alginate Modified Nanostructured Calcium Carbonate with Enhanced Delivery Efficiency for Gene and Drug Delivery." *Molecular BioSystems* 8 (3): 753–59. https://doi.org/10.1039/C1MB05337J.

Zhao, Yize, Yaqi Pan, Hao Sun, Pengfei Huo, Guangtong Wang, and Shaoqin Liu. 2024. "A Nanoparticle-Coated Cellulose Acetate Membrane for Highly Efficient, Low-Cost Circulating Tumor Cell Detection." *Biosensors* 14 (10): 472. https://doi.org/10.3390/bios14100472.

Zourob, Mohammed. 2010. *Recognition Receptors in Biosensors.* Springer. https://doi.org/10.1007/978-1-4419-0919-0.

7 Bioprospection and Engineering of Lignin- and Chitin-Degrading Enzymes for Industrial Applications

Ashwani Kumar Verma and Arun Goyal

ABBREVIATIONS

AA	Auxiliary activity
AB	Azure B
ABTS	2,2'-Azino-bis(3-ethylbenzothiazoline-6-sulfonic acid)
CAZyme	Carbohydrate-active enzymes
CC	Colloidal chitin
CL	Chemiluminescence
CR	Congo red
CRISPR/cas9	Clustered Regularly Interspaced Short Palindromic Repeats (CRISPR)/CRISPR-associated proteins (cas9)
CrRNA	CRISPR RNA
DA	Degree of acetylation
DGGE	Denaturation gradient gel electrophoresis
EPR	Electron paramagnetic resonance
FISH	Fluorescence *in situ* hybridization
GH	Glycoside hydrolase
GlcNAc	*N*-acetyl glucosamine
HMM	Hidden Markov model
HPLC	High-performance liquid chromatography
HR	Homologous recombination repair
HSQC	Heteronuclear single quantum coherence
IDBA-UD	Iterative De Bruijn Graph De Novo Assembler
Lac	Laccase
LC	Liquid chromatography
LCB	Lignocellulosic biomass
LiP	Lignin peroxidase
MB	Methylene blue
MG	Malachite green

DOI: 10.1201/9781003561682-7

MS	Mass spectrometry
NHEJ	Non-homologous end joining repair
NMR	Nuclear magnetic resonance
PAM	Protospacer adjacent motif
PCR	Polymerase chain reaction
PHAs	Polyhydroxyalkanoates
PR	Phenol red
RAMMCAP	Rapid Analysis of Multiple Metagenomes with a Clustering and Annotation Pipeline
RBBR	Remazol brilliant blue R
sgRNA	Short-guide RNA
ssODNs	Single-stranded oligodeoxynucleotides
TALENs	Transcription activator-like effector editing nucleases
TB	Tryphan blue
TGGE	Temperature gradient gel electrophoresis
TLC	Thin-layer chromatography
tracrRNA	Transactivating CRISPR RNA
T-RFLP	Terminal restriction fragment length polymorphism
VA	Veratryl alcohol
ZFNs	Zinc finger nucleases

7.1 INTRODUCTION

Incessant work inclined toward sustainability from anthropogenic waste remains dedicated toward the development of effective and alternative grounds against the global concern trilemma, i.e., environmental pollution, world hunger, and global warming. Unsustainable consumption of natural resources such as fossil fuels following scrupulous advancements in urbanization is leading toward the brink of environmental destruction, and hence highlighting the relevance of sustainable approaches. An efficient utilization of anthropogenic waste through various biological or thermochemical approaches highlights unprecedented opportunities for future sustainability in the enhanced production of fuel and products from waste. Thermochemical approaches for biomass treatment, such as gasification, hydrogenolysis, pyrolysis (thermolysis), and chemical oxidation methods for waste-derived refineries, are energy-intensive, cost-ineffective, and economically non-viable. Waste arising from industrial and domestic shellfish processing units contributes significantly toward marine pollution, and its limited utilization causes serious resource waste. Marine waste typically contains the exoskeleton of mollusks and crustaceans and is mainly composed of 15–40% chitin, 20–50% calcium carbonate, 0–14% lipid, 20–40%, and 10–40% ash (Venugopal, 2022). Annually, more than 10^{10} tons of insoluble chitin are produced in nature, of which more than 10% is derived from the marine crustaceans. Nearly 6–8 million tons of chitin waste are dumped from various domestic and industrial crustacean processing units into oceans every year due to underutilization, which represents a serious resource waste (Venugopal, 2022).

Lignocellulose (LC)-derived biofuel and bioproducts may provide high-end opportunities for commercial implementation of biorefineries. Plant lignocellulosic biomass (LCB) is comprised of cellulose (35–48 wt%), hemicellulose (22–30 wt%), and lignin (12–25 wt%) (Zhao et al., 2012). In LCB, long cellulose chains remain twisted into external microfibrils, where lignin combines with allosteric hemicellulose with the help of electrostatic interaction. Cellulose and hemicelluloses can be subjected to enzymatic breakdown to generate fermentable sugars after suitable pretreatment approaches; however, aromatic complexity of the chemical structure of lignin imposes reluctance to enzymatic hydrolysis. High carbon/oxygen ratio and natural abundance of lignin in LCB accentuate it as a promising substrate in valorization to bioproducts including vanillin, *cis, cis*-muconic acid (*cis, cis*-MA), polyhydroxyalkanoates (PHAs), lipids, and the other organic acids such as acetic acid, lactic acid, and pyruvic acid (Verma et al., 2024). Unfortunately, <2% of lignin produced from paper/pulp industries (annually 5×10^7 tons of lignin) represents the utilized portion, while the rest is abandoned or burned as low-energy fuels (Haile et al., 2021). Lignin remains underutilized and a challenging target to be subjected to biorefinery applications. The U.S. Energy Security and Independence Act of 2007 estimated production of nearly 6×10^{10} gallons of LC-based biofuel annually by 2030. The data indicate consumption of 7.5×10^8 tons of biomass and generation of approximately 2.25×10^8 tons of unutilized lignin (Congress US. Energy Security and Independence Act of 2007, in 110th Congress, HR 2007).

The exploration of efficient methods for lignin and chitin degradation is essential for the production of waste-derived bioproducts with industrial significance. Nature has evolved diverse catabolic pathways for degradation and depolymerization of complex polysaccharides available in nature, such as lignin and chitin. Many microorganisms, such as bacteria, yeasts, and filamentous fungi, facilitate the depolymerization of lignin and chitin by producing a group of extracellular enzymes referred to as ligninolytic and chitinolytic enzymes, respectively. These enzymes hold immense potential to be commercially exploited at an industrial scale for various biotechnological applications, including food, pharmaceuticals, biofuels, cosmetics, agriculture, and medicine. The viability of such lignin- and chitin-derived biorefineries relies on the exploration of microbes and enzymes capable of withstanding fluctuating stress and diverse parameters such as pH, temperature, and the presence of inhibitors in industrial settings. Various culture-dependent as well as culture-independent approaches allow bioprospecting of novel microorganisms and enzymes displaying ligninolytic and chitinolytic functions from various moderate and extreme environments such as forest soil, hot springs, vermicompost, the guts of insects, animal dung, sea beds, and glaciers (Ji et al., 2024). The enzyme assays, growth-based assays, and instrumental analysis form the basis of culture-dependent techniques, while OMICS technology, including metagenomics, metaproteomics, metatranscriptomics, and metabolomics, breaks rigid boundaries of culture-dependent approaches, allowing intensive research and identification of unculturable microbes and functions. OMICS technology facilitates analysis and probing of extensive data representing the entire function of the genome, RNA, protein, and metabolites in a high-throughput manner (Chettri et al., 2020). The advent of genetic engineering approaches such as heterologous gene expression and a highly versatile gene editing method, Clustered

Regularly Interspaced Short Palindromic Repeats (CRISPR)/CRISPR-associated proteins (Cas9), allows stimulating desired manipulations in the genetic repertoire of microbes, exploiting them more efficiently for industrial requisitions.

This chapter highlights different strategies of bioprospecting novel ligninolytic and chitinolytic enzymes from various environments and the significance of genetic engineering tools for enhancing the efficiency of such enzymes for their applications in biotechnology. The core objective of this chapter is to promote intensive exploration of novel enzymes mediating effective lignin and chitin depolymerization at industrial settings.

7.2 RECALCITRANCE AND COMPLEXITY OF LIGNIN AND CHITIN

7.2.1 Lignin

Lignin is an amorphous, highly branched polyphenol polymer accounting for about 30% of dry biomass weight. Lignin biosynthesis follows enzymatic dehydrogenation of the three essential phenylpropanoid monomers: sinapyl alcohol, coniferyl alcohol, and *p*-coumaryl alcohol (Lin et al., 2012). Random polymerization of phenoxyl radicals generated from such monolignols generates a three-dimensional network of lignin polymer. The deposition of lignin in the cell wall of plants initiates from the middle lamella, which is followed by a primary wall and the S1 layer of the secondary wall, and subsequently gets located in the S2 as well as the S3 layers. As measured by the Klason method (Nicholson et al., 2014), the lignin content is around 25–30% in softwoods, 15–25% in herbaceous plants, and 20–25% in hardwoods (Lin et al., 2012). Three phenylpropane units, viz., *p*-hydroxyphenyl (1H), syringyl (1S), and guaiacyl (1G), form building blocks of lignin structure. Hardwood lignin consists of S- and H-units, while softwood consists only of H-unit lignin. The herbaceous plants contain all three monolignol H-, S-, and G-units, although ferulate and *p*-coumarate also remain incorporated with normal S- and G-units (Zhao et al., 2012). Based on the analysis of lignin degradation products, several side-chain structures such as *p*-coumarate, hydroxycinnamyl alcohol, ferulate, arylglycerol, and hydroxycinnamaldehyde end-units were identified (Zhao et al., 2012). Detection by heteronuclear single quantum coherence (HSQC) nuclear magnetic resonance (NMR) in solution state revealed that cinnamyl alcohol in the end-group (including 5-hydroxyconiferyl alcohol) is present in higher amounts in hardwoods as compared with herbaceous plants and softwoods, while ferulate and *p*-coumarate end-units are abundant in herbaceous plants. Several acylated end-groups, such as hydroxycinnamaldehydes and hydroxycinnamic acids (*p*-hydroxybenzoic acid, sinapic acid, ferulic acid, and p-coumaric acid), are generated from incomplete monolignol biosynthesis and get incorporated in lignin with ester linkages, creating different acylating end-groups such as acetate, *p*-hydroxybenzoate, and γ-acetylate at the γ-OH position in many lignins (Mansfield et al., 2012). Study of degraded low molecular products of monomers, oligomers, and dimers followed by LC-MS, GC-MS, and NMR analyses revealed several lignin unit linkages: β-O-4, β-5, 5-5', and β-β units, with benzodioxane, dibenzodioxocin, and spirodienone as recently discovered units, while other uncommon linkages include 4-O-5 diphenyl, α-O-4, and β-1 linkages (Chakar et al., 2004). The lignin acts as a physical barrier restricting the cellulose accessibility to enzymatic treatment. Lignin can also irreversibly absorb cellulase during enzyme hydrolysis owing to its hydrophobic characteristics, as it contains methoxy groups, polyaromatic rings, and extensive hydrogen bonding. A study conducted by Meng et al. (2017) demonstrated a negative correlation between total lignin content and enzymatic digestibility in natural poplar variants.

7.2.2 Chitin

Chitin represents a linear, high-molecular-weight homopolysaccharide consisting of repetitive units of *N*-acetyl glucosamine (GlcNAc), which remain connected through β-1,4 glycosidic linkage with 100% degree of acetylation (DA) (Iber et al., 2022). Predominantly, chitin is abundantly found in the cell wall of fungi, the exoskeleton of invertebrates, including insects, and a structural component of crustaceans (Venugopal, 2022).

Chitin can be divided into three allomorphic types, i.e., α-chitin, β-chitin, and γ-chitin, based on their arrangement/number of chitin linear chains, degree of hydration, and unit cell size (Hossin et al., 2021). α-Chitin consists of *N, N*'-diacetylchitobiose dimer as its unit cell, which forms two chains in anti-parallel fashion held together by intramolecular hydrogen bonding and intermolecular hydrogen bonding of N-H bonds, resulting in the generation of sheets (Mohan et al., 2021). The co-existence of intra- as well as intermolecular hydrogen bonding in α-chitin makes it highly crystalline and rigid (Mohan et al., 2021). β-Chitin contains the same unit cell as α-chitin, but oriented in a parallel fashion. The absence of intramolecular hydrogen bonding in β-chitin makes it less rigid and less crystalline as compared with α-chitin (Cabrera-barjas et al., 2021). γ-Chitin contains mixed arrangements of the chitin chains, existing in both parallel and antiparallel fashion; thus, it is a combination of both α-chitin and β-chitin (Kaya et al., 2017). Chitin can be efficiently utilized as a sustainable and abundant source of N-containing compounds due to the ubiquity of biologically fixed nitrogen, expanding its paradigm of research in various fields, including food, agriculture, cosmetics, and medicine (Joseph et al., 2021). The preponderance of proficient chitin-degrading microorganisms, associated genes, and enzymes remains the core of several scientific explorations, with chitinolytic enzymes being integral to mediating the chitin degradation. Crustacean shells are major sources of chitin. According to the type of crustacean shell, chitin percentage varies, as shown in Table 7.1.

7.3 LIGNIN- AND CHITIN-DEGRADING ENZYMES

7.3.1 Ligninolytic Enzymes

Ligninolytic or lignin-degrading enzymes play an essential role in the depolymerization of lignin polymer and are produced widely by several microorganisms, plants, and insects. These enzymes enable lignin depolymerization through a sequential action of different enzymes that may involve hydrogen peroxide. These

TABLE 7.1
Proximate Composition of Marine and Freshwater Crustacean Waste

Crustacean shell type	Moisture (%)	Chitin (%)	Lipid (%)	Protein (%)	Ash (%)	Reference
Shrimp shell (*Metapenaeus affinis*)	11.7	21.6	1.9	27.23	33.7	Toliba et al. (2014)
King crab shell (*Paralithodes camtschaticus*)	68.0	5.5	0.9	8.6	17.0	Novikov et al. (2020)
Mud crab shell (*Portunus pelagicus*)	10.9	18.8	4.54	26.0	36.4	Toliba et al. (2014)
Lobster shell (*Thenus orientalis*)	9.5	20.4	3.02	23.2	38.2	Toliba et al. (2014)
Red snow crab shell (*Chionoecetes japonicus*)	11.9	-	0.2	25.0	35.7	Jun et al. (2019)
Snow crab shell (*Chionoecetes opilio*)	53.8	4.6	0.5	25.7	15.1	Novikov et al. (2020)
Freshwater shrimp shell (*Macrobrachium malcolmsonii*)	69.5	0.07	1.0	28.2	2.3	Devi et al. (2015)

enzymes can be divided into two groups: (i) phenol oxidases such as laccases (Lac) and (ii) lignin peroxidases (LiP), which are heme-containing enzymes (Janusz et al., 2017).

7.3.1.1 Laccase

Laccase (E.C. 1.10.3.2) is a multi-copper oxygen: benzenediol oxidoreductase enzyme that possesses the capability to mediate oxidation of a wide range of non-phenolic and phenolic compounds. It is among the oldest enzymes to be studied, being demonstrated first by Yoshida et al. in 1883 and characterized as a metal ion-containing oxidase enzyme by Bertrand et al. (1985). Laccase belongs to auxiliary activity family 1, specifically the AA1_1 subfamily in the CAZy database. Laccase utilizes the molecular O_2 as an electron acceptor and produces a H_2O molecule as a by-product following a four-electron reduction process, which makes it different from other ligninolytic peroxidases (Khatami et al., 2022). The capability of this versatile oxidoreductase to reduce O_2 renders it more relevant than other lignin-degrading peroxidases when inclined toward LC-derived biorefineries (Jones and Solomon, 2015). Laccases are ubiquitously secreted glycoproteins by many bacteria, fungi, plants, and some insects. Fungi such as ascomycetes, basidiomycetes, and deuteromycetes have been shown to produce laccase enzyme, where white-rot basidiomycetes are the most extensively studied (Drula et al., 2022). Some potent laccase-producing basidiomycetes are *Pleurotus ostreatus*, *Trametes versicolor*, *Cerrena unicolor*, *Agaricus bisporus*, *Cyathus bulleri*, *Pycnoporus cinnabarinus*, *Pycnoporus sanguinus*, *Cryptococcus neoformans*, *Schizophyllum commune*, *Fomes fomentarius*, *Panus rudis*, *Rigidoporus microporus*, etc. (Mahuri et al., 2023).

Laccase-producing predominant bacterial genera are *Geobacillus*, *Bacillus*, *Rhodococcus*, *Pseudomonas*, *Staphylococcus*, *Aquisalibacillus*, *Lysinibacillus*, *Delfia*, *Enterobacter*, and *Alteromonas* (Mahuri et al., 2023). Laccase holoenzyme can be a monomer or a dimer, where a single monomer contains four copper (Cu^{2+}) ion sites distributed into three different redox sites designated as Type 1 (T1), Type 2 (T2), and Type 3 (T3) Cu^{2+} ion sites; T1 and T2 contain a single Cu^{2+} ion, while T3 site contains two Cu^{2+} ions. T2 and T3 Cu ion sites combine to form a trinuclear cluster situated between domains 3 and 1, providing residues for coordination of Cu^{2+} ion, while the T1 Cu^{2+} site is located in domain 3, and the three apoprotein domains contain β-barrel architecture. T1 Cu^{2+} ion is paramagnetic and has intense electronic absorption at 610 nm and a well-characterized Electron Paramagnetic Resonance (EPR) signal. The T1 Cu^{2+} ion consists of a trigonal coordination with a cysteine and two histidines as conserved equatorial ligands. The axial ligand can be phenylalanine or leucine in fungal laccases or is methionine in bacterial laccases (CotA). The covalent Cu-Cys bond stimulates intense electronic absorption at T1 site, resulting blue color attributed to the laccase enzyme (Sun et al., 2021).

7.3.1.2 Lignin Peroxidase

Lignin peroxidase (E.C. 1.11.1.14) is a heme-containing (Fe^{3+}), 1, 2-bis (3, 4-dimethoxyphenyl) propane-1, 3-diol: hydrogen-peroxide oxidoreductase enzyme involved in lignin degradation. LiP is secreted by several bacteria and higher fungi, which mediate the breakdown of polymers via an oxidation process; however, some insects, such as the eastern *Subterranean termite* and *Reticulitermes flavipes,* have been studied to secrete LiP to degrade woody debris. Lignin peroxidases belong to auxiliary activity family 2, specifically within class II peroxidase of the CAZy database. The white-rot fungus *Phanerochaete chrysosporium* is the most extensively studied LiP producer due to the presence of multiple genes encoding LiP. These fungi have the capability to produce several LiP isoenzymes, such as H1, H2, H6, H7, H8, and H10 (Shi et al., 2021). Other potent fungal LiP producers are *Phanerochaete sordid*, *Trichoderma versicolor*, *Phlebia radiata*, *Phlebia tremellosa*, *Bjerkandera* sp., *Pycnoporus* sp., and *Trametopsis cervina* (Gu et al., 2024). Bacterial members studied to show LiP activity include *Streptomyces viridosporus*, *Acinetobacter calcoaceticus*, and *Bacillus* (Gu et al., 2024). LiP mediates the oxidation of different inorganic and organic substrates in the presence of H_2O_2 as the electron acceptor. The overall reaction is given below:

$$2S + H_2O_2 + 2e^- \rightarrow 2S_{ox} + 2H_2O$$

where S is a substrate (electron donor) and S_{ox} is the oxidized form of the substrate.

The heme ion of LiP interacts with H_2O_2 to generate intermediate state compounds I and II, proceeding through its catalysis. Overall, the catalytic mechanism of Lip can be summarized as a two-step reaction involving the ferric resting state of the native enzyme: (i) radical cation oxoferryl as unstable intermediate compound I and (ii) impartial oxoferryl intermediate as compound II. Compound I intermediate, i.e., oxoferryl iron porphyrin radical cation [Fe(IV)=O^+] is generated by two-electron oxidation of ferric [Fe (III)]. Lip reduces hydrogen peroxide with a second-order rate constant of 5.4×10^5 M^{-1} S^{-1} in a pH-independent reaction from pH 2–7.5 (Nguyen et al., 2024). After the formation of compound-I, two consecutive single-electron reduction steps of radical cation oxoferryl compound I are initiated using electron donor substrates such as veratryl alcohol (VA), which transfers an electron to reduce [Fe (IV)=O^+] to generate compound II, i.e., [Fe(IV)=O] and VA radical cation [VA^+], thereafter returning to its native oxidation state and completing LiP catalytic cycle (Nguyen et al., 2024).

7.3.2 Chitinolytic Enzymes

Numerous studies have been conducted for bioprospection of chitinolytic enzymes and functions for efficient conversion of chitin waste generated from various crustacean processing units. The CAZy database has grouped chitinolytic enzymes into 15 glycoside hydrolase (GH) families, i.e., GH3, GH5, GH7, GH8, GH18, GH19, GH20, GH23, GH46, GH48, GH75, GH80, GH84, GH109, and GH116. Chitinolytic enzymes have been classified based on two criteria, which include (1) enzyme action site and (2) gene sequence-based classification. Based on the site of enzyme catalysis, these enzymes have been categorized into endochitinases and exochitinases (Poria et al., 2021). The endo-enzymes cleave chitin at random positions along the internal chain and include chitinases (E.C. 3.2.1.14) and chitosanases (E.C. 3.2.1.132), generating *N*-acetyl chitooligosaccharides and deacetylated chitooligosaccharides, respectively, as end products. The exo-acting enzyme includes β-*N*-Acetylhexosaminidases (E.C. 3.2.1.52), which mediate exo-hydrolysis of alternating *β*-1,4 glycosyl linkage, generating *N, N'*-diacetylchitobiose or GlcNAc as an end product. On the other hand, based upon their gene sequence, chitinolytic enzymes have been classified into six types (I-VI), where the classes are determined by parameters including signal peptide (SP), isoelectric pH, N-terminal sequences, enzyme localization, and inducers (Zhang et al., 2021). Type I chitinases are from plants and contain valine- or leucine-rich SP and a cysteine-rich N-terminal; type II chitinases are similar to type I but without the Cys-rich N-terminal and are found in plants, bacteria, and fungi. Type III and type IV chitinases contain different characteristics from types I and II and are smaller in size. Type V and VI chitinases are highly under-characterized and have shown only chitin-binding properties and no catalytic functions (Zhang et al., 2021).

7.3.2.1 Endochitinase

Chitinases or endochitinases (E.C. 3.2.1.14) belong to GH families 18, 19, 23, and 48 and adopt a $(\beta/\alpha)_8$ TIM-barrel fold, where eight β-sheets remain surrounded by 8 α-helices. The strands of β-sheet are arranged into the barrel-like structure, with α-helices forming the ring toward the outer side. Bacterial GH18 chitinases contain a TIM-barrel fold in their catalytic subunit, consisting of three conserved catalytic residues such as Asp-140, Ser-93, and Trp-10 (Jiménez-Ortega et al., 2021). Chitinase is ubiquitous in nature, being produced widely by different bacteria and fungi. Among bacteria, the genera *Serratia*, *Aeromonas*, *Enterobacter*, *Clostridium*, *Streptomyces*, *Bacillus*, *Vibrio*, *Paenibacillus*, *Melghiribacillus*, and *Pseudoalteromonas* are predominant producers of endochitinases. These enzymes mediate random internal cleavage in chitin, resulting in the production of low molecular mass, soluble oligomers of GlcNAc referred to as chitooligosaccharides (COSs), such as chitobiose $(GlcNAc)_2$, chitotriose $(GlcNAc)_3$, chitotetraose $(GlcNAc)_4$, chitopentose $(GlcNAc)_5$, and chitohexose $(GlcNAc)_6$ (Nayak et al., 2021). On the other hand, chitosanases (E.C. 3.2.1.132) are another important type of hydrolytic enzyme belonging to five glycoside hydrolase families: GH5, GH8, GH46, GH75, and GH80. These endolytic enzymes mediate cleavage of partially or fully deacetylated chitin, known as chitosan, resulting in the production of deacetylated COSs. Chitosanase mediates endo-cleavage of *β*-1, 4-linkages of GlcN moieties of chitosan and can be classified into two types, I and II, depending upon their action on the reducing end and non-reducing end, respectively. Chitosanases from different bacteria have been studied, predominantly from the genera *Bacillus*, *Serratia*, *Paenibacillus*, *Streptomyces*, *Acinetobacter*, and *Janthinobacterium* (Nguyen et al., 2025). The gene encoding novel chitosanase (*csnS*) from the marine bacterium, *Serratia* sp. QD07, was overexpressed in *E. coli* BL21 cells (Zheng et al., 2021). This recombinant chitosanase gave maximum activity of 324 U/mL against colloidal chitosan at 60°C and pH 5.8. It showed thermophilic properties by retaining 40% of its activity at 100°C on incubation for 2 h and was also active over a wide range of pH (3.4-8.8). It also indicated psychrophilicity by retaining 38% of enzyme activity at 0°C, and 43% activity at 10°C on incubation for 2 h (Zheng et al., 2021). Its endo-type cleavage action on colloidal chitin was analyzed by TLC, which indicated the presence of $(GlcN)_2$ and $(GlcN)_3$ in the enzyme hydrolysate after 2 h. Similarly, the GH8 family chitosanase was cloned from *Bacillus cereus* TY24 isolated from crab and shrimp shell compost samples and expressed in *Escherichia coli* BL21 DE3 (Zhang et al., 2022). This purified chitosanase (48 kDa) showed specific activity of 1150 U mg^{-1} against colloidal chitin at a pH of 5.5 and 65°C. The relative chitosanase activity was over 55% and 80% of its maximal activity at pH 4.5 and pH 5.0-6.0, respectively, indicating its acidophilic nature. The enzyme also retained 70% and 80% of relative activity at 70°C and 65°C, respectively, whereas it showed no activity above 75°C. The kinetic constants, V_{max} and $K_{m,}$ were determined to be 1401 μmol min^{-1} mg^{-1} and 3 mg mL^{-1}, respectively.

7.3.2.2 Exo-chitinases

Exo-chitinases are *β*-*N*-acetylhexosaminidases (E.C. 3.2.1.52) that include chitobiosidases (E.C. 3.2.1.29) and *β*-1,4-*N*-acetyl glucosaminidase (E.C. 3.2.1.30). Chitobiosidase mediates cleavage of the *β*-1,4 glycosyl linkage in chitin from the non-reducing end, generating *N, N'*-diacetylchitobiose as the end product. *β*-1,4-*N*-acetyl glucosaminidase cleaves the *β*-1,4 glycosyl linkage from the non-reducing end generating a GlcN monomer. *β*-1,4-*N*-Acetyl glucosaminidase enzyme has also been reported to act on oligomeric end products of endochitinase and chitobiosidase, i.e., chitooligosaccharides and *N, N'*-diacetylchitobiose, respectively, producing GlcN as the monomeric product. Both exo-chitinases have been extensively studied from several bacterial genera, such as *Bacillus*, *Streptomyces*, *Ralstonia*, *Chitinolyticbacter*, *Aeromonas*, and *Paenibacillus* and fungi such as *Trichoderma*, *Paracoccidioides*, *Corollospora*, *Penicillium*, and *Conidiobolus coronatus* (Cretoiu et al., 2015).

Bacillus licheniformis strain JS, screened from mushroom cultivation bed left after production of mushroom (*Pleurotus sajor-caju*), was studied for the production of chitobiosidase (Waghmare and Ghosh, 2010). Purified chitobiosidase gave a specific activity of 10227.2 U mg^{-1} against colloidal chitin at a temperature of 60°C and pH 8.0. The enzyme showed activity over a wide range of temperatures from 10 to 80°C and was stable in a wide pH range of 5.0–10.0. High-performance liquid chromatography (HPLC) analysis of the enzyme hydrolysates showed the presence of only $(GlcNAc)_2$ upon degradation of colloidal chitin. Hence, chitinolytic enzymes are important carbohydrate-active enzymes (CAZymes) playing an essential role in chitin waste valorization. Other chitin-degrading enzyme from different microbes are discussed and shown in Table 7.2.

7.4 BIOPROSPECTION OF LIGNINOLYTIC AND CHITINOLYTIC ENZYMES FROM THE ENVIRONMENT

7.4.1 Culture-Dependent Approaches

Culture-dependent methods provide a traditional and stable framework for isolation, screening, and identification of microbes producing ligninolytic and chitinolytic functions. The pre-defined nutritional and functional parameters pave a convenient way to study culturable and industrially relevant microorganisms from various environments. These culturable microorganisms isolated from the environment can be explored for their lignin- or chitin-degrading potentials by various culture-dependent methods. These methods include colorimetric assays, enzymatic assays, liquid chromatography (LC)/mass spectrometry (MS), native gel electrophoresis, thin-layer chromatography (TLC), gas chromatography (GC)/MS, and high-performance liquid chromatography (HPLC) (Verma et al., 2024). Chemiluminescence (CL), along with ^{14}C autoradiography, is among the obsolete conventional techniques for the study of enzyme-mediated transformation of lignin and chitin-based substrates but can be employed in combination with high-throughput approaches. These techniques have multiple advantages and disadvantages, hence highlighting the necessity of proper consideration and selection for satisfactory screening. For instance, native gel electrophoresis and enzymatic assays have shown assessable, safe, and highly convenient results but are less sensitive and time-consuming. On the other hand, GC/MS and LC/MS provide significant tools for biotransformation analysis; however, they are limited due to misleading results generated due to the same *m/z* ratio of different compounds (Nasiri et al., 2021).

Colorimetric analysis provides screening of ligninolytic and chitinolytic enzymes by spectrophotometric detection of enzyme-mediated substrate transformation. It is performed employing various natural substrates such as Malachite Green (MG), colloidal chitin (CC), Congo red (CR), Remazol Brilliant Blue R (RBBR), Methylene Blue (MB), Phenol Red (PR), Tryphan Blue (TB), Azure B (AB), and Indulin AT. Growth medium supplemented with one of these substrates undergoes hydrolytic or oxidative biotransformation by microbial enzymes and undergoes color changes that can be quantified spectrophotometrically. Chitin-degrading microorganisms were isolated and characterized from the gut of the Asian toad, *Duttaphrynnus*

TABLE 7.2
Chitinases Characterized from Different Microorganisms

Microorganism	Substrate	Enzyme Activity	Optimum pH, Tempe-rature	Kinetics Parameters	Chito-oligo Saccharides	Reference
Bacillus thuringiensis serovar konkukian S4	Colloidal chitin, α-chitin, β-chitin, cellulose	65 μmoles min^{-1} against colloidal chitin	5.5, 55°C	$k_{cat} = 28.3\ s^{-1}$, $K_M = 40.3$ mg mL^{-1}	$(GlcNAc)_{2\text{-}4}$	Mehmood et al. (2010)
Pseudoalteromonas sp. DL-6	4MU[a]-(GlcNAc)], 4MU[a]-$(GlcNAc)_2$, 4MU[a]-$(GlcNAc)_3$, colloidal chitin, α-chitin and β-chitin	15.7 U mL^{-1} against colloidal chitin	8.0, 20°C	$k_{cat} = 0.49\ min^{-1}$, $K_M = 233.8$ μM,	$(GlcNAc)_{2\text{-}4}$	Wang et al. (2014)
Coprinopsis cinerea	Chitin powder, chitosan (degree of deacetylation 85%), glycol chitosan, pNGG[#], pNGGG[$]	161 U mg^{-1} against chitosan	5.0, 40°C	$k_{cat} = 1.13\ s^{-1}$ $K_M = 2342$ μM,	$(GlcNAc)_{2\text{-}8}$	Niu et al. (2017)
Trichoderma harzianum GIM 3.442	Colloidal chitin from crab shell	34.5 U mg^{-1} against colloidal chitin	6.0, 45°C	$V_{max} = 36$ μmol min^{-1} mg^{-1} $K_M = 0.7$ mg mL^{-1},	$(GlcNAc)_{2\text{-}4}$	Deng et al. (2019)
Serratia marcescens Bn10	Colloidal chitin, α-chitin, β-chitin	28 U mL^{-1} against colloidal chitin	9.0, 60°C	-	-	Okay and Alshehri (2020)
Clostridium paraputrificum J4	Colloidal chitin, pNG*, pNGG[#], pNGGG[$]	16.6 U mg^{-1} against colloidal chitin	5.5, 60°C	$k_{cat} = 30.0\ s^{-1}$, $K_M = 0.24$ mM	$(GlcNAc)_2$	Dohnálek et al. (2021)
Caldicellulosiruptor acetigenus	Colloidal chitin, pNGG[#], pNGGG[$]	0.2 U mg^{-1} against colloidal chitin	5.0, 70°C	$k_{cat} = 0.28\ s^{-1}$, $K_M = 16.3$ mg.mL^{-1} $V_{max} = 0.25$ μmol min^{-1} mg^{-1}	$(GlcNAc)_{2\text{-}3}$	Qin et al. (2021)
Shewanella khirikhana JW44	Colloidal chitin, glycol chitosan, chitin powder, chitosan	3.9 U mg^{-1} against colloidal chitin	7.0, 40°C	$k_{cat} = 10{,}203\ s^{-1}$, $K_M = 27.2$ μM, $V_{max} = 18$ μmol min^{-1} mg^{-1}	GlcNAc and $(GlcNAc)_2$	Wang et al. (2024)

* 4-Nitrophenyl *N*-acetyl-β-D-glucosaminide; [#]4-nitrophenyl *N, N'*-diacetyl-β-D-chitobiose; [$]4-nitrophenyl *N, N', N''*-triacetyl-β-D-chitotriose; [a]4-Methylumbelliferyl.

melanostictus, following primary degradation of 0.5% (w/v) colloidal chitin supplemented in minimum salt agar medium (Dhole et al., 2022). The bacterial colonies were selected based on the hydrolytic zone of clearance on agar medium. Upon identification, *Bacillus paramycoides* and *Stenotrophomonas maltophila* showed the highest chitinase activity with 1.2 U mL^{-1} and 1.4 U mL^{-1}, respectively, against colloidal chitin. *S. maltophila* also showed the capability to degrade insoluble chitin with chitinase activity of 1.7 U mL^{-1} and 4.2 U mL^{-1} against chitin flakes and prawn shells, respectively. Similarly, the chitosanase-producing capability of a *Bacillus* strain was analyzed using culture-dependent methods (Tang et al., 2024). *Bacillus* sp. HZ20-1 was isolated from mud at the bottom of shrimp and crab shell sewage and was primarily screened by growing in minimal salt agar medium supplemented with 1% (w/v) colloidal chitosan. The bacterium showed maximum chitosanase activity of 2.3 U mL^{-1} against colloidal chitosan, and the enzyme showed maximum activity at a pH range of 8.5-9, indicating that the bacterium produced alkaline chitosanase. Colorimetric assay is an efficient strategy in the qualitative identification of ligninolytic enzymes that belong to the auxiliary activities (AA) family of CAZymes, such as laccase and peroxidases. The ability of bacterial strains *Pseudomonas putida* NX-1, *Ochrobacterium tritici* NX-1, and *Klebsiella pneumoniae* NX-1 screened from leaf molds was investigated for production of laccases and peroxidases employing five different aromatic dyes—AB, RBBR, TB, MG, and MB—supplemented in minimal growth media (Xu et al., 2018). When grown in dye-containing media, the bacteria showed different degrees of dye degradation capabilities. RBBR was only degraded by *K. pneumoniae* with a degradation rate of 17.3% while methylene blue was degraded by *K. pneumoniae* NX-1, *P. putida* NX-1, and *O. tritici* NX-1 at rates of 93.9%, 87.7%, and 93.1%, respectively. Similarly, culture-dependent analysis helped in the identification as well as characterization of lignin peroxidase enzyme from *Bacillus velezensis* CR-502T isolated from plant root soil (Pham et al., 2022). The bacterium was isolated based on its primary capability to degrade methylene blue (1000 mg L^{-1}) up to 99.5% after 48 h of incubation. *Bacillus velezensis* CR-502T showed high lignin peroxidase activity reaching 45 U mL^{-1} and 44 U mL^{-1} against MB under static and shaking conditions, respectively.

The identification of end-products also provides an efficient approach for enzyme identification. For instance, *Bacillus aryabhattai* MCC 3987 was isolated from soil and investigated for its potential to actively produce chitinases using culture-dependent approaches (Subramani et al., 2024). The purified protein-treated colloidal chitin was quantitatively analyzed using TLC, which showed the presence of chitotriose and chitotetraose as chitin hydrolyzed products. LC analysis showed retention time peaks for monomer, which was recorded at 2.3 and 2.4 min; dimer, which was recorded at 2.87 and 3.02 min; trimer recorded at 3.68 and 3.90 min; tetramer at 4.80 and 5.12 min; and pentamer recorded at 6.36 and 6.79 min. Meanwhile, MS analyses showed m/z peaks for monomer to be 243.9, dimer to be 424.8, trimer to be 627.7, tetramer to be 830.6, pentamer to be 517.3 and 1033.5, and hexamer to be 618.7. LC and MS analyses showed efficient production of chitin oligomers.

7.4.2 Culture-Independent Approaches

The nutritional and functional parameters of all microorganisms isolated from the environment remain limited, and such information allows only vivid exploitation of their potential on industrial grounds. Culture-dependent approaches allow identification of only 1% of microbes, while 99% remain overlooked and unstudied. Barriers established by the culture-dependent methods can be overcome by different culture-independent methods, allowing integral study of complex microorganisms and enzymes. Among different culture-independent methods, the role of OMICS technology is often highlighted due to its high potential in exploring novel enzymes from unculturable microorganisms. OMICS technology helps in the collection and study of biological molecules in larger quantities and includes metagenomics, metatranscriptomics, metaproteomics, and metabolomics, as shown in Fig. 7.1. OMICS technology establishes an important framework for process optimization of microbial enzymes relevant in industries, which is essential for their commercialization. The operating process optimization parameters rely on a trial-and-error basis, where the relationship between production performance and cellular physiological requirements is ignored. This eventually leads to low productivity and low economic viability. OMICs technology has been applied in statistics and bioinformatics development of industrial-level tests, which allows data quality control, computational model development, cross-validation, confirmation on an independent dataset, and release of data, code, and fully specified computational procedures (Babar et al., 2018). Hence, OMICS technology monitors cellular performance and metabolism at multiple levels and thus increases microbial production of enzymes and products.

Metagenomics studies the genetic material isolated from the environment, bypassing the limits of traditional genome analysis techniques. The toolbox of metagenomics includes 454 next-generation pyrosequencing, clone library sequencing, fluorescence *in situ* hybridization (FISH), temperature/denaturation gradient gel electrophoresis (TGGE/DGGE), and terminal restriction fragment length polymorphism (T-RFLP) (Zhang et al., 2021). Metagenomics has been widely employed for exploring novel ligninolytic and chitinolytic enzymes from various environments such as soil, hot springs, sewage, insect guts, marine water, and animal dung (Zhang et al., 2021). Insights into LC-degrading microbes and enzymes from a microbial consortium TMC7 enriched from thermophilic compost were developed in a study by employing metagenomics (Wang et al., 2021). The sequencing of the DNA library generated was conducted using the Illumina HiSeq 2500 platform, and Seqpreq was used for the removal of low-quality reads and sequencing adaptors. Iterative De Bruijn Graph De Novo Assembler (IDBA-UD) was employed for *de novo* gene assembly, and gene prediction was conducted using MetaGene. Taxonomic affiliation revealed the predominance of various genera, including *Thermoanaerobacterium* (18.41%), *Ruminiclostridium* (42.85%), *Aeribacillus* (2.65%), *Geobacillus* (10.44%), *Clostridium* (1.26%), and *Caldibacillus* (1.56%) in the sample. Annotation of CAZymes revealed that the microbial consortium encoded a wide array of AA family enzymes such as AA1 laccase, AA2 lignin peroxidase, and AA2 versatile peroxidase, along with several other glycoside hydrolases, including 6 GHs β-glucosidases and 4 GHs exoglucanases involved in cellulose degradation; 3 GHs mannanases and 2 GHs mannosidases acting on mannan; 9 GHs β-xylosidases

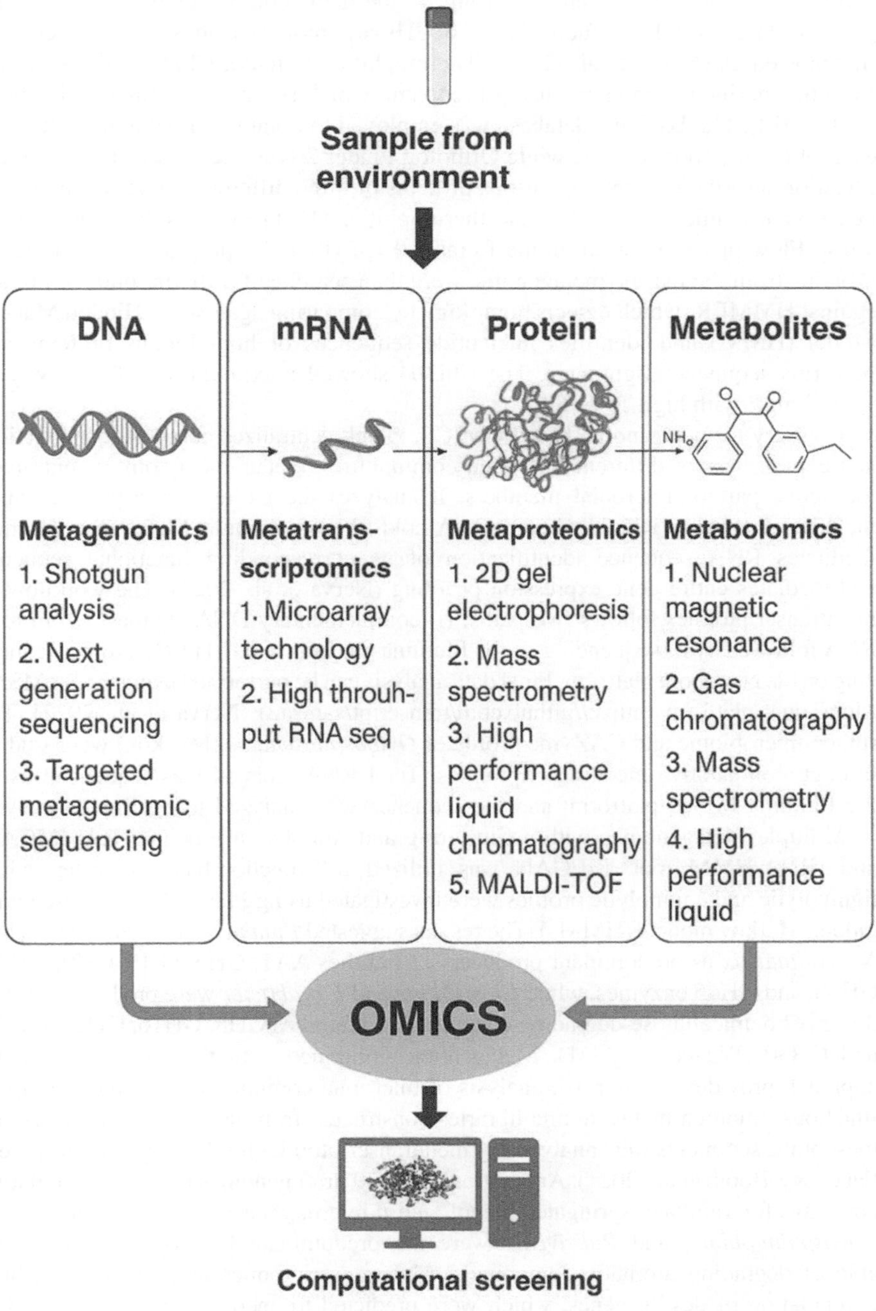

FIGURE 7.1 OMICS for culture-independent analysis of microbes and biological functions from the environment.

and 6 GHs endo-β-1,4-xylanases acting on the hemicellulose main chain; 11 GHs galactosidases; 4 GHs α-fucosidases; 6 GHs arabinofuranosidases; and 3 GHs α-rhamnosidases (Nerva et al., 2022). Thermophilic chitinase, Chi304, was screened from the marine metagenome using Preoptem, which is a deep learning tool (Zhang et al., 2022). The BacDive database was employed to obtain optimal temperatures of each of the microorganisms, while Ortholog Finder 2 (ver. 2.2.7) was used for identification of orthologs among various proteins of three different classes of microbes, i.e., psychrophilic, mesophilic, and thermophilic. The model was generated on the TensorFlow platform based on the Keras 2.0.1, Python 2.7 program. The annotated proteins from the marine metagenome were then searched for thermophilic chitinase against HMMER, which detects homology by comparing against the Hidden Markov Model (HMM) and identifies nucleotide sequences or homologous proteins and performs sequence alignments. The Chi304 showed maximum specific activity of 12.3 U mg^{-1} with high T_m of 89°C.

Contrary to metagenome-based OMICS, which centralizes composition, function, and copiousness of different microbial communities, metatranscriptomics centralizes the active part of microbial members. It analyzes the entire transcriptome, which includes all types of RNAs, where mRNA holds the key module. Metatranscriptomics facilitates RNA sequence identification of its corresponding functional sequence and mediates entire gene expression profiling (Nerva et al., 2022). The workflow of metatranscriptomics follows two paths: (i) complementary DNA synthesis from total RNA followed by its sequencing on the Illumina platform and (ii) mRNA direct sequencing on the Nanopore platform. Final data analysis can be performed using the SAMSA2 MetaTrans platform (https://github.com/transcript/samsa2) (Nerva et al., 2022). The rumen microbiome and CAZyme profile of *Ovibos moschatus* (Muskox) were studied through comparative metatranscriptomics. Total RNA isolated was sequenced using the Illumina MiSeq platform, and the sequence was analyzed using Rapid Analysis of Multiple Metagenomes with a Clustering and Annotation Pipeline (RAMMCAP) and rRNA-HMM tool. MEGAN was utilized for function-based binning, while ligninolytic and chitinolytic profiles were investigated using Pfam HMMer search with hidden Markov models (HMMs). The results suggested *Ruminococcus*, *Piromyces*, and *Neocallimastix* as predominant producers of families AA1, GH6, GH5, GH9, GH18, GH48, and GH45 enzymes, while *Clostridium* and *Fibrobacter* were predominant producers of hemicellulose-degrading enzymes from families GH8, GH16, GH20, GH26, and GH30 (Wu et al., 2021). Metagenome combined with the metatranscriptome approach provides an intensive analysis of microbial communities and their complex functions. Shotgun metagenomic libraries constructed from hot spring sediments and mesophilic sediments were analyzed by metatranscriptomics for their ligninolytic potential (Levy-Booth et al., 2021). Annotation of the 39 draft genomes revealed degradation pathways for vanillate, syringate, phenol, and *p*-hydroxybenzoate. *Altereryhtrobacter, Novosphingobium,* and *Rubrivivax* were the predominant bacterial members capable of degrading aromatic compounds. Metatranscriptomics analysis showed high upregulation of des/lig genes, which were predicted to encode enzymes involved in catabolism of phenolic oligomers, vanillate, and syringate.

Metaproteomics and metabolomics help in the study of microbial proteins and metabolites in an efficient manner. They extend the edge of information provided by metatranscriptomics and metagenomics by developing deeper insights into cellular localization of proteins and enzymatic activity. Metaproteome analysis involves the extraction of protein from the sample, which is followed by peptide fractionation and final analysis by LC/MS (Pattersen et al., 2022). Secretome analysis of the microbial community from the gut of the yellow rice stem borer, *Scirpophaga incertulas,* was conducted by Singh et al. (2019). The microbial community was initially enriched by subjecting it to a rice straw medium followed by 16S rRNA analysis. *Bacteroidetes, Verrucomicrobia, Firmicutes, Actinobacteria,* and *Proteobacteria* were the predominant phyla. Analysis of biomass-bound and soluble protein fractions using LC-MS/MS revealed the presence of different classes of CAZymes, including GHs, AAs, carbohydrate esterases, polysaccharide lyases, and glycosyl transferases. Further analysis of AAs confirmed the presence of laccases (AA1), peroxidases (AA2), xylooligosaccharide oxidases (AA7), and LPMOs (AA10) (Singh et al., 2019).

7.5 GENETIC ENGINEERING FOR ENHANCED EFFICIENCY OF CAZYMES

7.5.1 Heterologous and/or Homologous Gene Expression

Heterologous gene expression employs recombinant DNA technology, where a gene fragment in the form of complementary DNA (cDNA) or complementary RNA (cRNA) encoding a specific protein or function is overexpressed in a suitable prokaryotic or eukaryotic host. The genetic elements encoding the desired functions are taken from native microorganisms and transferred as extra-genomic material in the host, stimulating cellular machinery to mediate their overexpression under suitable conditions. Selection of suitable expression hosts relies on several parameters, including ease of transformation and cultivation, shorter generation time, post-translational modification in yeasts, and high extracellular secretion of recombinant protein. Heterologous gene expression techniques employ several microbes as "expression systems," including bacteria, yeasts, and fungi. The most commonly used are *Escherichia coli, Aspergillus sp., Yarrowia lipolytica, Pichia pastoris*, and *Saccharomyces cerevisiae*, due to their amenability to take external DNA and translate it efficiently for downstream processing (Watts et al., 2021).

Heterologous gene expression is commonly employed for the exploration and investigation of several chitinolytic enzymes from various novel sources. An endochitinase, Chi23, belonging to the GH18 family from a marine bacterium, *Pseudoalteromonas aurantia* DSM6057, was cloned (Wang et al., 2019). The recombinant endochitinase showed activity on both the crystalline and colloidal chitin, with specific activity of 0.1 U mg^{-1} and 1.2 U mg^{-1}, respectively (Verma et al., 2024). The enzyme showed thermostability, with maximum activity at 60°C, retaining over 40% of its maximum enzyme activity after incubating for 1 h at 70°C. TLC analysis revealed $(GlcNAc)_2$ and $(GlcNAc)_3$ as final hydrolyzed products. Similarly, another marine bacterium, *Streptomyces hygroscopicus* R1, was studied to produce a variety of novel chitinolytic enzymes with ideal parameters for industrial settings (Wang et al., 2022). Chitosanase belonging to the GH18 family from *S. hygroscopicus* R1 was heterologously expressed and characterized for its efficient utilization in bioprospection of shrimp waste biomass (Wang et al., 2022). The recombinant enzyme showed specific activity of 1.76 U mg^{-1} against the colloidal chitosan and showed a high protein yield of 3.98 g L^{-1}. The enzyme showed pH and thermal stability with an optimal pH of 5.5 and an optimal temperature of 55°C. The enzyme efficiently hydrolyzed 4% colloidal chitosan, and the hydrolysate showed the presence of chitooligosaccharide with a degree of polymerization from 2 to 6. A novel GH20 family *β-N*-acetyl glucosaminidase (*Cm*NAGase) possessing both hydrolytic and transglycosylating activities from *Chitinolyticbacter meiyuanensis* SYBC-H1 was cloned (Zhang et al., 2020). This recombinant enzyme presented high specific activity of 3157 U mg^{-1} toward the synthetic substrate *p*-nitrophenyl-*N*-acetyl glucosaminide (*p*NP-GlcNAc). Recombinant enzyme showed stability at a pH range of 4.0–7.0, with maximum activity at pH 5.4. The enzyme stability was observed at a temperature range of 25–60°C, with optimal activity at 40°C. It hydrolyzed $(GlcNAc)_{2,}$ generating GlcNAc, while GlcNAc, $(GlcNAc)_2$, and $(GlcNAc)_3$ were released upon degradation of $(GlcNAc)_4$ (Zhang et al., 2020).

Several ligninolytic enzymes have been heterologously expressed and studied, providing a better understanding of the functional parameters of the enzymes. The gene encoding laccase from a fungal strain of *Coriolopsis trogii* strain Mafic-2001 was cloned and overexpressed in *Pichia pastoris* X-33 (Bao et al., 2023). This recombinant laccase showed maximum specific activity of 2.22 U mg^{-1} against 2,2′-azino-bis-(3-ethylbenzothiazoline-6-sulfonic) acid (ABTS). This enzyme showed maximum activity at 40°C and pH 3.0. The recombinant laccase retained 60% of activity in the temperature range of 20–60°C. Similarly, the *lccA* gene encoding acidophilic laccase from *Trichoderma versicolor* was heterologously expressed and characterized (López et al., 2016). Purified laccase with a molecular mass of 58 kDa showed maximum activity of 11,972 U L^{-1} against ABTS. The recombinant enzyme showed maximum activity at 50°C and pH 2.0. The kinetic parameter analysis revealed K_m and V_{max} of 0.43 mM and 82.3 U mg^{-1}, respectively. In another study, a gene encoding lignin-degrading peroxidase enzyme from *Bacillus* sp. strain BL5 was cloned, expressed, and characterized (Khan et al., 2021). The 46 kDa recombinant peroxidase was active over a temperature range 20–50°C with maximum activity at 35°C. It was also stable in a broad pH range of 3.0–8.0, with a maximum at pH 5.0. Kinetic parameter analysis showed V_{max}, $K_{M,}$ and K_{cat} of 519.7 μmol min^{-1} mg^{-1}, 1.06 mM, and 395 s^{-1}, respectively.

Contrary to heterologous gene expression, homologous gene expression involves overexpressing the desired protein or function in its native organism under the influence of a strong promoter. For instance, homologous expression of the lignin-degrading enzyme peroxidase isozyme in the white rot fungus *Pleurotus ostreatus* was conducted in a study (Irie et al., 2001). Recombinant *mnp3* gene constructs were generated under expression signal control of *P. ostreatus sdi1* and co-transformed into wild-type *P. ostreatus* along with carboxin-resistant plasmid pTM1. The monokaryotic transformant TMG9-C1 showed high activity of 1560 U mL^{-1} against guaiacol as a substrate (Irie et al., 2001). Similarly, chitinase (Chi3023) from *Bacillus thuringiensis* serovar morrisoni strain 3023 was homologously expressed using *Escherichia coli*-Bt shuttle vector pHT315 (Okay and Özcengiz, 2011). The overexpressed recombinant enzyme showed 23-fold chitinase activity, 4.3 U mg^{-1}, as against 0.2 U mg^{-1} for the native enzyme.

7.5.2 CRISPR/Cas9-Mediated Genetic Editing

Development of CRISPR/Cas9 represents an orthologous and highly versatile genome manipulation technique, marking a significant breakthrough in the biotechnological ground for desirable exploitation of microbial enzymes and functions. It highlights an innovative strategy in metabolic engineering, synthetic biology, and systems biology targeting specific sites on DNA and programming nucleases for desired attributes. This advanced technique promotes amelioration of cellular tolerance to metabolic stress and helps in dynamic monitoring and controls the fermentation process. Besides, CRISPR/Cas9 also helps in the elimination of competing pathways and allows autonomous control of cell replication and cell size, hence allowing the development of industrially relevant microbial strains, as shown in Fig. 7.2.

This unique gene-editing approach provides opportunities in terms of speed of assembly, speed, targeting efficiency, and multi-site targeting potential compared to traditional DNA-targeting methods such as meganucleases, zinc finger nucleases (ZFNs), and transcription activator-like effector editing nucleases (TALENs) (Verma et al., 2024; Hussen et al., 2023). In the CRISPR/cas9 system, CRISPR RNA (crRNA) forms a complex with transactivating CRISPR RNA (tracrRNA) and CRISPR-associated protein (cas9), which needs to be replaced with synthetically designed short guide RNA (sgRNA) for desired genetic manipulation. The sgRNA is generated by programming a 20 bp region complementary to the NGG protospacer adjacent motif (PAM) loci into the CRISPR array and fused with tracrRNA and crRNA. sgRNA guides cas9 to specific sites on DNA to generate double-stranded DNA breaks (DSBs), which undergo repair either by homologous recombination repair (HR) or non-homologous end joining repair (NHEJ), introducing desired

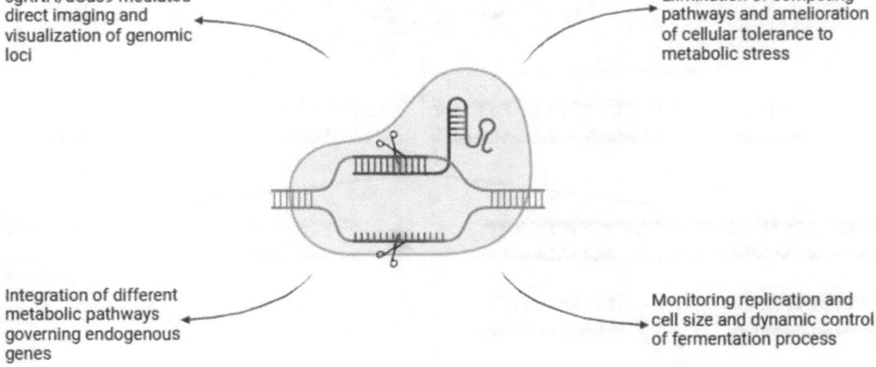

FIGURE 7.2 Application of CRISPR/Cas9 for enhanced efficiency of industrial microorganisms.

insertion/deletion (indel) in the DNA as shown in Fig. 7.3 (Verma et al., 2022). The success of CRISPR/Cas9 hugely relies on the HR mechanism in the case of NHEJ-deficient organisms. Lack of an NHEJ-repair system and introduction of DSBs might be lethal for the host cell; however, the efficiency of HR repair is extremely low. To overcome the low efficiency of HR repair, the newly developed λ-RED recombination is utilized with sgRNA and λ-RED recombinase, which mediates gene editing without promoting DSB-stimulated death (Tahir et al., 2022).

Favorable attributes of the CRISPR/Cas9 system have also been investigated for developing strains producing efficient ligninolytic and chitinolytic enzymes for industrial applications. The CRISPR/Cas9 gene editing platform for *Cerrena unicolor* BBP6 was constructed for its efficient degradation of lignin employing laccase and peroxidases (Zhang et al., 2022). First of all, a *pyrG* gene in *Cerrena unicolor* was disrupted as a screening marker, followed by overexpression of the *lacA* gene and *mnp3*. The enzyme showed 53% and 64% improved laccase and peroxidase activity. Similarly, laccase and peroxidase encoding genes of a ligninolytic fungus, *Dichomitus squalens*, were genetically modified using CRISPR/cas9 (Kowalczyk et al., 2021). HDR-based genetic alteration was introduced by single-stranded oligodeoxynucleotides (ssODNs), which resulted in co-editing of the two loci at the *lcc3* gene. Two knockout mutants were generated for the *lcc3* gene, i.e., lcc3MUT-A and lcc3MUT-B, while for peroxidase, the knockout mutants generated were

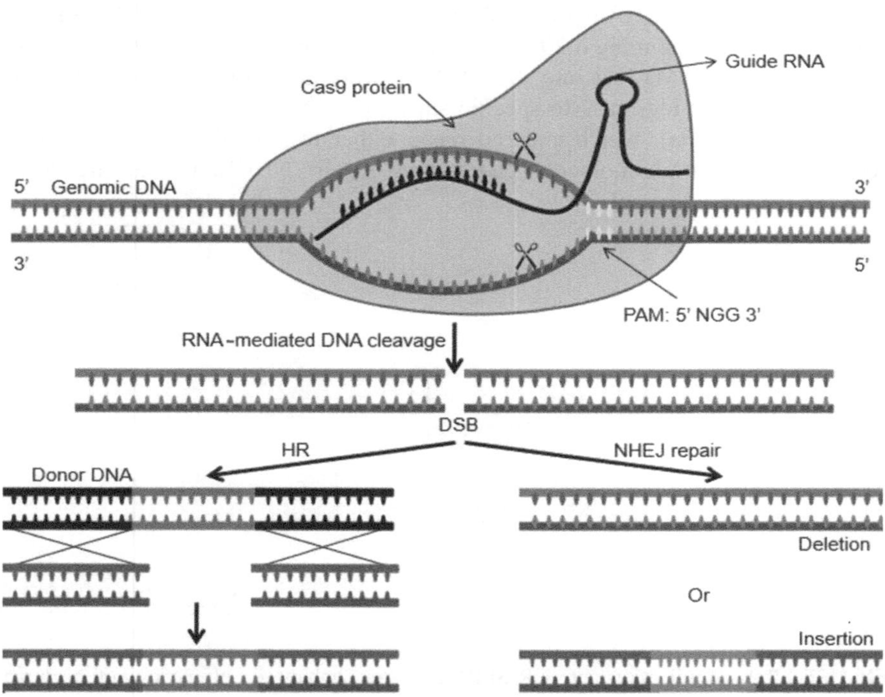

FIGURE 7.3 Mechanism of CRISPR/Cas9-mediated genome editing.

mnp2MUT-A and mnp2MUT-B. The mutant strains showed improved enzyme production as compared with the wild-type strain. On the other hand, overexpression of two endochitinase genes in an endophyte *Phomopsis liquidambaris* was achieved using a CRISPR/Cas9 approach (Zhu et al., 2022). PChi-OE was used as an overexpression plasmid, while the integrated overexpression strain IN-Chi was engineered using the CRISPR/Cas9 system (Zhu et al., 2022). Integration of the Chi gene was confirmed using diagnostic polymerase chain reaction (PCR). The engineered strain showed 38% improved chitinase expression as compared with the wild-type strain.

7.6 CONCLUSION

The efficacy of ligninolytic and chitinolytic enzymes toward valorization of lignin and chitin, respectively, generated as waste due to advancements in industrialization and urbanization, is well studied. The array of enzymes acting on a variety of lignin- and chitin-based substrates, including LCB, marine crustacean biomass, and synthetic chemicals released in the environment, converts them into a group of bioproducts with a broad spectrum of biotechnological applications. Ligninolytic enzymes include laccase and peroxidases, while chitinolytic enzymes include endochitinases, chitosanases, chitobiosidases, and *N*-acetyl-β-D-glucosaminidases. Industrial settings require enzymes that can withstand variations in temperature, pH, salinity, inhibitors, and ions; thus, exploring viable enzymes is essential. Culture-dependent techniques establish a framework for the isolation and screening of culturable microbes that produce lignin and chitin-degrading enzymes. However, the lack of nutritional and functional parameters of unculturable microbes limits the techniques of culturing and paves the way for culture-independent approaches. OMICS technology, which includes metagenomics, metatranscriptomics, metaproteomics, and metabolomics, facilitates intensive investigation of unculturable organisms at cellular and genomic levels. Enzymes bearing favorable attributes can be overexpressed through genetically modified host cells efficiently. Heterologous/homologous gene expression allows expression of genes encoding desired proteins in an organism bearing the potential of high cellular and protein expression. Hence, research oriented around lignin and chitin can promote commercial stability of such enzymes in larger industrial sectors and pave the way for future sustainability platforms.

ACKNOWLEDGEMENTS

The authors thank the Department of Biosciences and Biotechnology, IIT Guwahati, for providing the computational infrastructure and central library facilities for procuring references and plagiarism analysis (Turnitin Plagiarism Detection Software).

AUTHOR CONTRIBUTION STATEMENT

Ashwani Kumar Verma: Writing, original draft preparation, reviewing, and editing.
Arun Goyal: Reviewing and editing.

FUNDING

This research did not receive any specific grant from funding agencies in the public, commercial, or not-for-profit sectors.

CONFLICT OF INTEREST STATEMENT

The authors state that the research was conducted without any commercial or financial relationships that could be construed as potential conflicts of interest.

REFERENCES

Babar, M. M., Afzaal, H., Pothineni, V. R., Najam-us-Sahar, S. Z., Ali, Z., Zahid, M. A., & Gul, A. (2018). Omics approaches in industrial biotechnology and bioprocess engineering. In Debmalya Barh and Vasco Azevedo (Eds.), *Omics Technologies and Bio-Engineering* (pp. 251–269). Academic Press.

Bao, C., Liu, Y., Li, F., Cao, H., Dong, B., & Cao, Y. (2023). Expression and characterization of laccase Lac1 from *Coriolopsis trogii* strain Mafic-2001 in Pichia pastoris and its degradation of lignin. *Applied Biochemistry and Biotechnology*, *195*(10), 6150–6167.

Bertrand, G. (1985). Sur la laccase et sur le pouvoir oxydant de cette diastase. *Comptes Rendus de l'Académie des Sciences (Paris)*, *120*(3), 266–269.

Cabrera-Barjas, G., González, C., Nesic, A., Marrugo, K. P., Gómez, O., Delattre, C., & Cea, J. (2021). Utilization of marine waste to obtain β-chitin nanofibers and films from giant Humboldt squid *Dosidicus gigas*. *Marine Drugs*, *19*(4), 184.

Chakar, F. S., & Ragauskas, A. J. (2004). Review of current and future softwood kraft lignin process chemistry. *Industrial Crops and Products*, *20*(2), 131–141.

Chettri, D., Verma, A. K., & Verma, A. K. (2020). Innovations in CAZyme gene diversity and its modification for biorefinery applications. *Biotechnology Reports*, *28*, e00525.

Cretoiu, M. S., Berini, F., Kielak, A. M., Marinelli, F., & van Elsas, J. D. (2015). A novel salt-tolerant chitobiosidase discovered by genetic screening of a metagenomic library derived from chitin-amended agricultural soil. *Applied Microbiology and Biotechnology*, *99*, 8199–8215.

Deng, J. J., Shi, D., Mao, H. H., Li, Z. W., Liang, S., Ke, Y., & Luo, X. C. (2019). Heterologous expression and characterization of an antifungal chitinase (Chit46) from *Trichoderma harzianum* GIM 3.442 and its application in colloidal chitin conversion. *International Journal of Biological Macromolecules*, *134*, 113–121.

Devi, D. P., Hareesh, K., & Reddy, M. S. (2015). Studies on the proximate composition of tropical freshwater prawn Macrobrachium rosenbergii. *International Journal of Fisheries and Aquatic Studies*, *3*(1), 329–336.

Dhole, N. P., Phuge, S., Dar, M. A., & Pandit, R. S. (2022). Screening of chitin degrading bacteria from the gut of Asian common toad *Duttaphrynus melanostictus*: implication for valorization of chitin rich seafood waste. *Environmental Technology & Innovation*, *28*, 102929.

Dohnálek, J., Dušková, J., Tishchenko, G., Kolenko, P., Skálová, T., Novák, P., & Šimůnek, J. (2021). Chitinase Chit62J4 essential for chitin processing by human microbiome bacterium *Clostridium paraputrificum* J4. *Molecules*, *26*(19), 5978.

Drula, E., Garron, M. L., Dogan, S., Lombard, V., Henrissat, B., & Terrapon, N. (2022). The carbohydrate-active enzyme database: functions and literature. *Nucleic Acids Research*, *50*(D1), D571–D577.

Gu, J., Qiu, Q., Yu, Y., Sun, X., Tian, K., Chang, M., & Huo, H. (2024). Bacterial transformation of lignin: key enzymes and high-value products. *Biotechnology for Biofuels and Bioproducts*, *17*(1), 2.

Haile, A., Gelebo, G. G., Tesfaye, T., Mengie, W., Mebrate, M. A., Abuhay, A., & Limeneh, D. Y. (2021). Pulp and paper mill wastes: utilizations and prospects for high value-added biomaterials. *Bioresources and Bioprocessing*, *8*, 1–22.

Hossin, M. A., Al Shaqsi, N. H. K., Al Touby, S. S. J., & Al Sibani, M. A. (2021). A review of polymeric chitin extraction, characterization, and applications. *Arabian Journal of Geosciences*, *14*(18), 1870.

Hussen, B. M., Rasul, M. F., Abdullah, S. R., Hidayat, H. J., Faraj, G. S. H., Ali, F. A., & Taheri, M. (2023). Targeting miRNA by CRISPR/Cas in cancer: advantages and challenges. *Military Medical Research*, *10*(1), 32.

Iber, B. T., Kasan, N. A., Torsabo, D., & Omuwa, J. W. (2022). A review of various sources of chitin and chitosan in nature. *Journal of Renewable Materials*, *10*(4), 1097.

Irie, T., Honda, Y., Watanabe, T., & Kuwahara, M. (2001). Homologous expression of recombinant manganese peroxidase genes in ligninolytic fungus *Pleurotus ostreatus*. *Applied Microbiology and Biotechnology*, *55*, 566–570.

Janusz, G., Pawlik, A., Sulej, J., Świderska-Burek, U., Jarosz-Wilkołazka, A., & Paszczyński, A. (2017). Lignin degradation: microorganisms, enzymes involved, genomes analysis and evolution. *FEMS Microbiology Reviews*, *41*(6), 941–962.

Ji, C., Ma, Y., Xie, Y., Guo, J., Ba, H., Zhou, Z., & Zheng, W. (2024). Isolation and purification of carbohydrate components in functional food: a review. *RSC Advances*, *14*(32), 23204–23214.

Jiménez-Ortega, E., Kidibule, P. E., Fernández-Lobato, M., & Sanz-Aparicio, J. (2021). Structural inspection and protein motions modelling of a fungal glycoside hydrolase family 18 chitinase by crystallography depicts a dynamic enzymatic mechanism. *Computational and Structural Biotechnology Journal*, *19*, 5466–5478.

Jones, S. M., & Solomon, E. I. (2015). Electron transfer and reaction mechanism of laccases. *Cellular and Molecular Life Sciences*, *72*, 869–883.

Joseph, S. M., Krishnamoorthy, S., Paranthaman, R., Moses, J. A., & Anandharamakrishnan, C. (2021). A review on source-specific chemistry, functionality, and applications of chitin and chitosan. *Carbohydrate Polymer Technologies and Applications*, *2*, 100036.

Jun, J. Y., Jung, M. J., Jeong, I. H., Kim, G. W., Sim, J. M., Nam, S. Y., & Kim, B. M. (2019). Effects of crab shell extract as a coagulant on the textural and sensorial properties of tofu (soybean curd). *Food Science & Nutrition*, *7*(2), 547–553.

Kaya, M., Mujtaba, M., Ehrlich, H., Salaberria, A. M., Baran, T., Amemiya, C. T., & Labidi, J. (2017). On chemistry of γ-chitin. *Carbohydrate Polymers*, *176*, 177–186.

Khan, S. I., Zada, N. S., Sahinkaya, M., Colak, D. N., Ahmed, S., Hasan, F., & Shah, A. A. (2021). Cloning, expression and biochemical characterization of lignin-degrading DyP-type peroxidase from *Bacillus* sp. Strain BL5. *Enzyme and Microbial Technology*, *151*, 109917.

Khatami, S. H., Vakili, O., Movahedpour, A., Ghesmati, Z., Ghasemi, H., & Taheri-Anganeh, M. (2022). Laccase: various types and applications. *Biotechnology and Applied Biochemistry*, *69*(6), 2658–2672.

Kowalczyk, J. E., Saha, S., & Mäkelä, M. R. (2021). Application of CRISPR/Cas9 tools for genome editing in the white-rot fungus *Dichomitus squalens*. *Biomolecules*, *11*(10), 1526.

Levy-Booth, D. J., Hashimi, A., Roccor, R., Liu, L. Y., Renneckar, S., Eltis, L. D., & Mohn, W. W. (2021). Genomics and metatranscriptomics of biogeochemical cycling and degradation of lignin-derived aromatic compounds in thermal swamp sediment. *The ISME Journal*, *15*(3), 879–893.

Lin, L., Zhai, S. R., Xiao, Z. Y., Liu, N., Song, Y., Zhai, B., & An, Q. D. (2012). Cooperative effect of polyethylene glycol and lignin on SiO2 microsphere production from rice husks. *Bioresource Technology*, *125*, 172–174.

iLópez, S. C., Sietiö, O. M., Hildén, K., de Vries, R. P., & Mäkelä, M. R. (2016). Homologous and heterologous expression of basidiomycete genes related to plant biomass degradation. In Monika Schmoll and Christoph Dattenböck (Eds.), *Gene Expression Systems in Fungi: Advancements and Applications* (pp. 119–160). Springer.

Mahuri, M., Paul, M., & Thatoi, H. (2023). A review of microbial laccase production and activity toward different biotechnological applications. *Systems Microbiology and Biomanufacturing*, *3*(4), 533–551.

Mansfield, S. D., Kim, H., Lu, F., & Ralph, J. (2012). Whole plant cell wall characterization using solution-state 2D NMR. *Nature Protocols*, *7*(9), 1579–1589.

Mehmood, M. A., Xiao, X., Hafeez, F. Y., Gai, Y., & Wang, F. (2010). Molecular characterization of an endochitinase from *Bacillus thuringiensis* subsp. *konkukian*. *World Journal of Microbiology and Biotechnology*, *26*, 2171–2178.

Meng, X., Pu, Y., Yoo, C. G., Li, M., Bali, G., Park, D. Y. & Ragauskas, A. J. (2017). An in-depth understanding of biomass recalcitrance using natural poplar variants as the feedstock. *Chemistry Sustainability Energy Materials*, *10*(1), 139–150.

Mohan, K., Muralisankar, T., Jayakumar, R., & Rajeevgandhi, C. (2021). A study on structural comparisons of α-chitin extracted from marine crustacean shell waste. *Carbohydrate Polymer Technologies and Applications*, *2*, 100037.

Nasiri, A., Jahani, R., Mokhtari, S., Yazdanpanah, H., Daraei, B., Faizi, M., & Kobarfard, F. (2021). Overview, consequences, and strategies for overcoming matrix effects in LC-MS analysis: a critical review. *Analyst*, *146*(20), 6049–6063.

Nayak, S. K., Nayak, S., Mohanty, S., Sundaray, J. K., & Mishra, B. B. (2021). Microbial chitinases and their applications: an overview. In Bibhuti Bhusan Mishra, Suraja Kumar Nayak, Swati Mohapatra, and Deviprasad Samantaray (Eds.), *Environmental and Agricultural Microbiology: Applications for Sustainability* (pp. 313–340). Wiley.

Nerva, L., Garcia, J. F., Favaretto, F., Giudice, G., Moffa, L., Sandrini, M., & Chitarra, W. (2022). The hidden world within plants: metatranscriptomics unveils the complexity of wood microbiomes. *Journal of Experimental Botany*, *73*(8), 2682–2697.

Nguyen, O. T. K., Chen, P. T., Nargotra, P., Chen, S. C., Cheng, C. Y., Wang, H. M. D., & Kuo, C. H. (2025). Sustainable bacterial chitosanase production using soybean meal hydrolysate: insights into biochemical and hydrolysis characteristics. *Process Biochemistry*, *150*, 202–212.

Nguyen, T. V. T., Kim, S., Yoo, C. G., Choi, J. W., Leem, G., & Kim, Y. H. (2024). Lignin peroxidase-catalyzed selective cleavage of C–C bonds in lignin at room temperature. *ACS Catalysis*, *14*(15), 11733–11740.

Nicholson, D. J., Leavitt, A. T., & Francis, R. C. (2014). A three-stage klason method for more accurate determinations of hardwood lignin content. *Cellulose Chemistry and Technology*, *48*(1–2), 53–59.

Niu, X., Zhou, J. S., Wang, Y. X., Liu, C. C., Liu, Z. H., & Yuan, S. (2017). Heterologous Expression and characterization of a novel chitinase (ChiEn1) from *Coprinopsis cinerea* and its synergism in the degradation of chitin. *Journal of Agricultural and Food Chemistry*, *65*(32), 6943–6956.

Novikov, V., Derkach, S., & Konovalova, I. (2020). Chitosan technology from crustacean shells of the northern seas. *KnE Life Sciences*, *5*, 65–74.

Okay, S., & Alshehri, W. A. (2020). Overexpression of chitinase a gene from Serratia marcescens in Bacillus subtilis and characterization of enhanced chitinolytic activity. *Brazilian Archives of Biology and Technology*, *63*, e20200061.

Okay, S., & Özcengiz, G. (2011). Molecular cloning, characterization, and homologous expression of an endochitinase gene from *Bacillus thuringiensis* serovar *morrisoni. Turkish Journal of Biology*, *35*(1), 1–7.

Pettersen, V. K., Antunes, L. C. M., Dufour, A., & Arrieta, M. C. (2022). Inferring early-life host and microbiome functions by mass spectrometry-based metaproteomics and metabolomics. *Computational and Structural Biotechnology Journal*, *20*, 274–286.

Pham, V. H. T., Kim, J., Chang, S., & Chung, W. (2022). Biodegradation of methylene blue using a novel lignin peroxidase enzyme producing bacteria, named *Bacillus* sp. React3, as a promising candidate for dye-contaminated wastewater treatment. *Fermentation*, *8*(5), 190.

Poria, V., Rana, A., Kumari, A., Grewal, J., Pranaw, K., & Singh, S. (2021). Current perspectives on chitinolytic enzymes and their agro-industrial applications. *Biology*, *10*(12), 1319.

Qin, X., Xin, Y., Su, X., Wang, X., Zhang, J., Tu, T., & Luo, H. (2021). Heterologous expression and characterization of thermostable chitinase and β-N-acetylhexosaminidase from *Caldicellulosiruptor acetigenus* and their synergistic action on the bioconversion of chitin into N-acetyl-d-glucosamine. *International Journal of Biological Macromolecules*, *192*, 250–257.

Shi, K., Liu, Y., Chen, P., & Li, Y. (2021). Contribution of lignin peroxidase, manganese peroxidase, and laccase in lignite degradation by mixed white-rot fungi. *Waste and Biomass Valorization*, *12*, 3753–3763.

Singh, R., Bennett, J. P., Gupta, M., Sharma, M., Eqbal, D., Alessi, A. M., & Yazdani, S. S. (2019). Mining the biomass deconstructing capabilities of rice yellow stem borer symbionts. *Biotechnology for Biofuels*, *12*, 1–19.

Subramani, A. K., Ramachandra, R., Thote, S., Govindaraj, V., Vanzara, P., Raval, R., & Raval, K. (2024). Engineering a recombinant chitinase from the marine bacterium *Bacillus aryabhattai* with targeted activity on insoluble crystalline chitin for chitin oligomer production. *International Journal of Biological Macromolecules*, *264*, 130499.

Sun, M., Dang, Y., Du, R., Lu, Z., Wang, Y., Wang, X., & Rao, H. (2021). Experiment and theoretical insights into CuNi/CoMoO4 multi-functional catalyst with laccase-like: catalysis mechanism, smartphone biosensing and organic pollutant efficient degradation. *Chemical Engineering Journal*, *425*, 130586.

Tahir, H., Basit, A., Tariq, H., Haider, Z., Ullah, A., Hayat, Z., & Rehman, S. U. (2022). Coupling CRISPR/Cas9 and lambda Red recombineering system for genome editing of *Salmonella gallinarum* and the effect of ssaU knock-out mutant on the virulence of bacteria. *Biomedicines*, *10*(12), 3028.

Tang, Y., Duan, Z., Chen, J., & Zhang, S. (2024). Isolation of a novel *Bacillus* strain with industrial potential of producing alkaline chitosanase. *International Journal of Biological Macromolecules*, *281*, 135725.

Toliba, A. O., Rabie, M. A., & El-Araby, G. M. (2014). Extending the shelf-life of cold stored strawberry by chitosan and carnauba coatings. *Zagazig Journal of Agricultural Research*, *41*(5), 1067–1076.

Venugopal, V. (2022). Green processing of seafood waste biomass towards blue economy. *Current Research in Environmental Sustainability*, *4*, 100164.

Verma, A. K., Chettri, D., & Verma, A. K. (2022). Potential of CRISPR/Cas9-based genome editing in the fields of industrial biotechnology: strategies, challenges, and applications. *Industrial Microbiology and Biotechnology*, 667–690.

Verma, A. K., Chettri, D., & Verma, A. K. (2024). Potential of genetic engineering methods for enhanced biofuel production. In Pradeep Verma (Ed.), *Industrial Microbiology and Biotechnology: A New Horizon of the Microbial World*, 381–403. Springer Nature.

Verma, A. K., Chettri, D., Verma, A. K., Selvaraj, M., & Assiri, M. A. (2024). Bioprospecting lignin for biorefinery: emerging innovations and strategies in microbial technology. *Biomass and Bioenergy*, *181*, 107052.

Waghmare, S. R., & Ghosh, J. S. (2010). Chitobiose production by using a novel thermostable chitinase from *Bacillus licheniformis* strain JS isolated from a mushroom bed. *Carbohydrate Research*, *345*(18), 2630–2635.

Wang, J., Wang, P., Zhu, M., Chen, W., Yu, S., & Zhong, B. (2022). Overexpression and biochemical properties of a GH46 chitosanase from marine *Streptomyces hygroscopicus* R1 suitable for chitosan oligosaccharides preparation. *Frontiers in Microbiology*, *12*, 816845.

Wang, L., Xue, M., Yan, R., Xue, J., Lu, Z., & Wen, C. (2024). Insights into chitin-degradation potential of *Shewanella khirikhana* JW44 with emphasis on characterization and function of a chitinase gene SkChi65. *Microorganisms*, *12*(4), 774.

Wang, X., Zhao, Y., Tan, H., Chi, N., Zhang, Q., Du, Y., & Yin, H. (2014). Characterization of a chitinase from *Pseudoalteromonas* sp. DL-6, a marine psychrophilic bacterium. *International Journal of Biological Macromolecules*, *70*, 455–462.

Wang, Y., Wang, C., Chen, Y., Chen, B., Guo, P., & Cui, Z. (2021). Metagenomic insight into lignocellulose degradation of the thermophilic microbial consortium TMC7. *Journal of Microbiology and Biotechnology*, *31*(8), 1123.

Wang, Y. J., Jiang, W. X., Zhang, Y. S., Cao, H. Y., Zhang, Y., Chen, X. L., & Li, P. Y. (2019). Structural insight into chitin degradation and thermostability of a novel endochitinase from the glycoside hydrolase family 18. *Frontiers in Microbiology*, *10*, 2457.

Watts, A., Sankaranarayanan, S., Watts, A., & Raipuria, R. K. (2021). Optimizing protein expression in heterologous system: strategies and tools. *Meta Gene*, *29*, 100899.

Wu, X., Elekwachi, C. O., Bai, S., Luo, Y., Zhang, K., & Forster, R. J. (2021). Characterizing the alteration in rumen microbiome and carbohydrate-active enzymes profile with forage of muskoxen rumen through comparative metatranscriptomics. *Microorganisms*, *10*(1), 71.

Xu, Z., Qin, L., Cai, M., Hua, W., & Jin, M. (2018). Biodegradation of kraft lignin by newly isolated *Klebsiella pneumoniae*, *Pseudomonas putida*, and *Ochrobactrum tritici* strains. *Environmental Science and Pollution Research*, *25*, 14171–14181.

Yoshida, H. (1883). LXIII.—chemistry of lacquer (Urushi). Part I. communication from the chemical society of Tokio. *Journal of the Chemical Society, Transactions*, *43*, 472–486.

Zhang, A., Mo, X., Zhou, N., Wang, Y., Wei, G., Chen, J., & Ouyang, P. (2020). A novel bacterial β-N-acetyl glucosaminidase from *Chitinolyticbacter meiyuanensis* possessing transglycosylation and reverse hydrolysis activities. *Biotechnology for Biofuels*, *13*, 1–14.

Zhang, L., Chen, F., Zeng, Z., Xu, M., Sun, F., Yang, L., & Xie, Y. (2021). Advances in metagenomics and its application in environmental microorganisms. *Frontiers in Microbiology*, *12*, 766364.

Zhang, R. X., Wu, Z. W., Zhang, S. J., Wei, H. M., Hua, C. W., Li, L., & Yang, T. Y. (2022). Gene cloning and molecular characterization of a thermostable chitosanase from *Bacillus cereus* TY24. *BMC Biotechnology*, *22*(1), 30.

Zhang, X., Yuan, J., Li, F., & Xiang, J. (2021). Chitin synthesis and degradation in crustaceans: a genomic view and application. *Marine Drugs*, *19*(3), 153.

Zhang, Y., Guan, F., Xu, G., Liu, X., Zhang, Y., Sun, J. & Tian, J. (2022). A novel thermophilic chitinase directly mined from the marine metagenome using the deep learning tool Preoptem. *Bioresources and Bioprocessing*, *9*(1), 54.

Zhao, X., Zhang, L., & Liu, D. (2012). Biomass recalcitrance. Part I: the chemical compositions and physical structures affecting the enzymatic hydrolysis of lignocellulose. *Biofuels, Bioproducts and Biorefining*, *6*(4), 465–482.

Zheng, Q., Meng, X., Cheng, M., Li, Y., Liu, Y., & Chen, X. (2021). Cloning and characterization of a new chitosanase from a deep-sea bacterium *Serratia* sp. QD07. *Frontiers in Microbiology*, *12*, 619731.

Zhu, Y. L., Zhang, M. Q., Wang, L. S., Mei, Y. Z., & Dai, C. C. (2022). Overexpression of chitinase in the endophyte *Phomopsis liquidambaris* enhances wheat resistance to *Fusarium graminearum*. *Fungal Genetics and Biology*, *158*, 103650.

8 Scaling Up the Production of Carbohydrate-Active Enzymes for Biotechnological Application

Manswama Boro and Shuchi Singh

ABBREVIATIONS

CAZymes	Carbohydrate-active enzymes
CRISPR	Clustered regularly interspaced short palindromic repeats
GMP	Good manufacturing practice
HTP	High-throughput
ISO	International Organization for Standardization
NGS	Next-generation sequencing
TFs	Transcription factors
TEA	Technoeconomic analysis

8.1 INTRODUCTION

Biocatalysis is a fast-growing technology in the industrial sector that is rapidly replacing the role of chemical catalysis. The global demand for the commercial application of enzymes has increased dramatically in major industrial sectors such as food, agriculture, pharmaceuticals, biofuels, cosmetics, textiles, waste management, etc. These enzymes are mainly of bacterial and fungal origin. Carbohydrate-active enzymes (CAZymes) include all enzymes involved in the modification, degradation, or biosynthesis of carbohydrates and their derivatives. CAZymes play a crucial role in various processes, from breaking down complex sugars for biofuel production to improving food texture and developing new drugs in an efficient and sustainable way. However, scaling up CAZyme production from the lab to a large industrial scale is no easy task.

The process of scaling up CAZyme production and large-scale application is a major challenge. From selecting the right microbial host to fermentation to choosing

 DOI: 10.1201/9781003561682-8

the best downstream processing method, scaling up production to an industrial scale can be difficult in its own way. Each process chosen at each step can impact the efficiency of the enzymes, the cost-effectiveness, and the yield of the final product. Advances in science and technology have greatly improved the stability and efficiency of enzymes. In addition, advances in microbiology research, genetic engineering, synthetic biology, computational methods, and advanced bioreactor design have made it easier to scale up CAZyme production.

This chapter looks at the challenges and breakthroughs in scaling up CAZyme production, highlights the advances that make large-scale production possible, and looks at the future direction of enzyme production in a world increasingly reliant on bio-based technologies.

8.2 CHALLENGES IN SCALING UP CAZYME PRODUCTION

Scaling up CAZyme production to an industrial level is a major challenge. Fig. 8.1 illustrates the complete enzyme production process, highlighting the key stages of upstream, fermentation, and downstream processing. High production costs and low enzyme efficiency are often an obstacle to scaling up enzyme production. Choosing the right microbial host for enzyme production is a challenging task.

Microbial hosts such as *E. coli* and filamentous fungi are commonly used for CAZyme production, but they often require significant genetic modifications to improve productivity and reduce enzyme degradation (Zhang et al., 2020). A major challenge in large-scale production is ensuring the stability and activity of the enzyme under harsh industrial conditions such as extreme pH, high temperatures, and exposure to chemical inhibitors (Ndochinwa et al., 2024). Protein engineering approaches, such as directed evolution and immobilization of enzymes, can help to improve the resilience of enzymes in these environments (Singh et al., 2013).

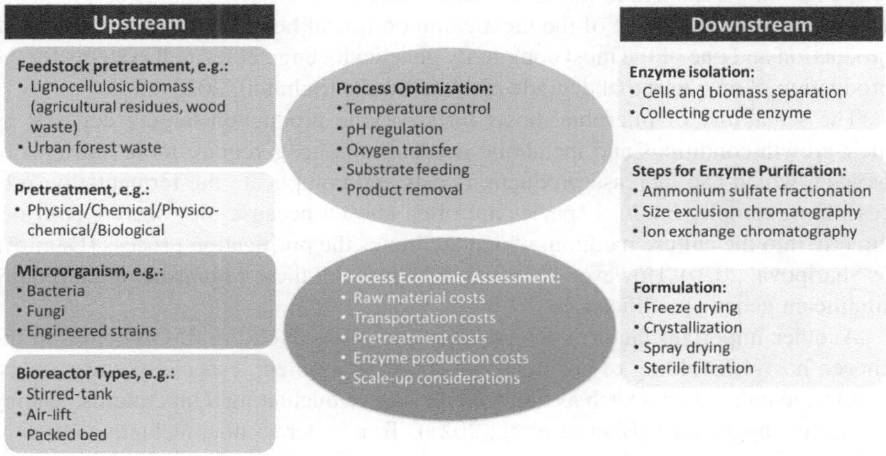

FIGURE 8.1 Overview of CAZymes production from lignocellulosic biomass.

Large-scale fermentation is also challenging because parameters such as aeration, pH, and nutrient supply must be consistently controlled (Anderson, 2009). Optimization of fermentation processes, including fed-batch and continuous methods, along with real-time monitoring, can help address these issues (Shahzadi et al., 2021). In addition, purification and downstream processing are complex and costly. Methods such as membrane filtration, selective precipitation, and affinity chromatography can be adapted to improve efficiency, while advanced labeling techniques streamline these processes.

The cost of large-scale CAZyme production is another obstacle, but the use of renewable feedstocks, recycling of resources, and the use of energy-efficient equipment can make the process more sustainable and cost-effective. Regulatory requirements for enzymes used in food and pharmaceuticals also require compliance with strict quality standards such as GMP and ISO protocols. Intellectual property challenges, including patents on enzymes and production methods, may require strategic partnerships or licensing agreements. By overcoming these challenges with technological innovation and collaborations, the industry can fully utilize CAZymes for various applications.

8.2.1 Microbial Host Selection

Selecting the right microbial host for CAZyme production is a first, but one of the biggest challenges for scaling up enzyme production. The efficiency of the microbial host is directly responsible for enzyme yield, enzyme stability, and production costs.

Bacterial hosts such as *E. coli* are commonly used because of their rapid growth and ease of genetic manipulation (Fakruddin et al., 2013). But *E. coli* lacks the complex post-translational modification mechanisms that some CAZymes require for full and rapid activity, resulting in limited functionality for certain applications. Fungal hosts such as *Pichia pastoris* and other filamentous fungi can perform such modifications, but these organisms have slower growth rates and require precise optimization of fermentation parameters to maximize large-scale productivity (Ahmad et al., 2014). *Trichoderma reesei* is one of the most common fungal hosts for industrial CAZyme production and one of the most commonly genetically engineered cell systems for the production of enzymes that degrade plant cell walls (Schmoll, 2022).

The scalability of microbial hosts for CAZyme production largely depends on their growth conditions and metabolic needs. Some hosts require more nutrients or oxygen, which can increase production costs and complicate the fermentation process. For example, *Bacillus* species are often chosen because they secrete enzymes directly into the culture medium, which facilitates the purification process (Danilova & Sharipova, 2020). However, to ensure a high yield, these organisms often require significant genetic modification.

Another important factor is the stability and correct folding of CAZymes in the chosen host. Misfolding of enzymes is a common problem, especially under industrial fermentation stress such as high cell density or fluctuating temperatures, leading to inactive aggregates (Bhatwa et al., 2021). To counteract this, techniques such as codon optimization and co-expression of chaperones are used to help the enzymes fold correctly. However, they can increase complexity and increase costs (Francis & Page, 2010).

Ultimately, selecting the right microbial host requires a balance between enzyme activity, yield, scalability, and regulatory compliance. Overcoming these challenges through strategic host selection and optimization is key to creating cost-effective and sustainable CAZyme production systems.

8.2.2 Fermentation Optimization

Optimization of fermentation processes is critical for scaling up the production of CAZymes, as it directly affects enzyme yield, activity, and cost. The transition from lab-scale to industrial-scale fermentation brings several challenges, especially maintaining a stable and efficient environment. One of the biggest challenges is managing the dynamic and complex conditions in large fermenters, where pH, temperature, aeration, and nutrient supply are more difficult to control than in smaller systems. These fluctuations can lead to gradients in conditions such as oxygen concentration, which can affect enzyme stability and production.

Oxygen transfer is particularly difficult in large fermenters as its solubility decreases with size (Xia et al., 2021). This can limit cell growth and enzyme production and requires the use of advanced aeration techniques and specialized equipment to ensure even oxygen distribution. Furthermore, optimizing nutrient supply is crucial to avoid toxic byproducts or metabolic imbalances that can affect enzyme yield (Basak et al., 2020). Methods such as fed-batch or continuous fermentation are often used to control nutrient supply but must be carefully tailored to the specific microbial host and CAZyme to be produced.

High cell densities in industrial fermentations can also stress the cells, leading to misfolded enzymes or inactive aggregates. To prevent this, protein folding can be improved by incorporating chaperones or adjusting the fermentation parameters, but this makes the process even more complex (Gasser et al., 2008). Another major challenge is product extraction and downstream processing. Many microbes need to secrete CAZymes into the medium for easy extraction, but not all microbial hosts do this effectively. In addition, waste materials accumulated during fermentation can inhibit enzyme activity, requiring careful waste management and bioreactor design.

For applications such as food or pharmaceuticals, where strict quality control is required, it is critical to maintain consistency between fermentation batches (Siddiqui et al., 2023). Variations in fermentation conditions can lead to differences in enzyme activity or purity. Therefore, real-time monitoring instruments such as sensors for pH, temperature, and oxygen are increasingly being used (Gervais, 2019). However, implementing these systems on a large scale can be expensive. Challenges related to environmental stability, nutrient supply, cell stress, downstream processing, and reproducibility must be overcome when optimizing fermentation for large-scale CAZyme production. Advanced process control and continuous monitoring are essential for efficient and scalable enzyme production (Crater & Lievense, 2018).

8.2.3 Downstream Processing Techniques

Downstream processing is a critical and often difficult phase in scaling up CAZyme production. It involves isolating, purifying, and stabilizing the enzymes after fermentation to meet industrial standards. On a larger scale, this process becomes increasingly complex and costly, impacting both efficiency and cost-effectiveness. A major challenge is to achieve high enzyme purity without sacrificing yield. Fermentation broths usually contain biomass, substrates, and byproducts, which makes it difficult to separate the enzymes without compromising activity. Techniques such as filtration, precipitation, and chromatography, while effective at lab scale, can be expensive and difficult to scale up, so more cost-effective alternatives are needed for larger scale production (Mukherjee, 2019).

Removal of contaminants such as host cell proteins and metabolic byproducts is another challenge, as these can affect enzyme activity and stability. In large-scale systems, inhibitory substances can also accumulate, making effective separation even more difficult. Process optimization, such as improving filtration and centrifugation for large volumes, can help reduce these impurities, but precise control and process development are key for consistent results (Fasim et al., 2021).

Enzyme stability during purification is also an issue, as CAZymes are sensitive to pH, temperature, and ionic strength (Robinson, 2015). Large-scale equipment can cause heat and mechanical stress, potentially denaturing sensitive enzymes. To prevent this, stabilizing agents such as glycerol are used, but these add complexity and cost (Braham et al., 2021). In addition, concentration and formulation processes are required to produce enzymes that are suitable for storage and use. However, these steps can lead to loss of activity, especially when slow techniques such as ultrafiltration are used, which can also lead to protein aggregation.

The economic and environmental impact of downstream processing is also significant, as many purification methods consume large amounts of water, solvents, or chemicals, increasing both costs and environmental impact. Reusing these resources can reduce the impact, but developing an efficient closed-loop system is a challenge. Remaining impurities must meet strict regulatory standards, especially for food, pharmaceutical, or environmental applications.

Consistency between batches is critical to maintaining enzyme quality. Variations in downstream processing conditions can lead to variations in purity or activity. For this reason, real-time monitoring systems are increasingly being used to track purification parameters. Implementing such systems on a large scale requires significant investment. Overcoming these challenges through optimized purification, advanced monitoring, and innovative stabilization techniques is essential for efficient and sustainable CAZyme production on an industrial scale.

8.3 ADVANCEMENTS IN CAZYME PRODUCTION

Researchers around the world are working hard to increase CAZyme production on a larger scale without compromising the efficiency and stability of the enzyme and are looking for a cost-effective method. Researchers are using various techniques, including genetic modification approaches and synthetic biology strategies, to improve enzyme production, efficiency, and stability.

8.3.1 Genetic Engineering Approaches

Microbial genetic studies have revealed a wide variety of genes encoding CAZymes. Many of these genes can be manipulated using genetic engineering methods to increase the production of CAZymes. Advances in the production of CAZymes through genetic engineering have greatly enhanced their industrial application. It is also noteworthy that CAZymes can be easily manipulated for various reactions with high specificity for application in industrial processes such as biofuel production, food processing and pharmaceuticals.

There are several genetic engineering approaches that have been used for advanced CAZyme production. One common practice is gene modification and overexpression, in which microorganisms are genetically engineered to produce CAZymes (Schmid et al., 2015). Genetic engineering is a widely used tool for insertion, deletion, or modification of genes of interest to improve metabolite production, in this case to improve CAZyme production (Kun et al., 2019). An ideal strain is selected on the basis of its specific physiological properties and functional capabilities, providing a high enzyme yield while exhibiting resistance to environmental stress. The improvement of microbial strains by mutations has already been applied to several cellulase-producing strains, but these processes can be quite critical (Kuhad et al., 2016). The cellulase gene has been successfully overexpressed using various genetic engineering techniques. Various microbial strains, such as *Trichoderma reesei*, *Aspergillus niger*, *Saccharomyces cerevisiae*, *Escherichia coli*, and *Bacillus subtilis*, have been genetically engineered to enhance gene expression and CAZyme production. With advances in genomics, the study of the genetic information of key CAZyme-producing microbial strains has opened the door to efficient, targeted modifications. Genetic engineering has enabled the development of highly saccharifiable industrial strains capable of producing a wide range of CAZymes. High cellulolytic activity was shown by a minimal gene set derived from *Caldicellulosiruptor bescii* exoproteome; hence, this primed CAZyme system promised efficient CAZyme production when expressed by engineered microorganisms (Brunecky et al., 2018). Heterologous expression of genes from one species into a host species is a common genetic engineering approach. In heterologous gene expression, codon utilization plays one of the most critical roles, as mutations in the codons can lead to off-frame translations that inhibit protein production (Watts et al., 2021).

By using innovative techniques such as CRISPR and omics-based approaches, researchers have improved the diversity and functionality of CAZymes, leading to improved biofuel production from lignocellulosic biomass. The combined CRISPR and Cas9 system is a simple but precise genetic modification strategy commonly used to improve CAZyme production. CRISPR/Cas9 is a state-of-the-art genome editing tool in the field of genetic engineering that is essentially used to improve enzyme production by using a CRISPR RNA (crRNA) to guide the Cas9 enzyme to make double-stranded cuts in specific target regions of DNA (Singh et al., 2023). The production of cellulase and xylanase enzymes was successfully increased by disrupting the ACE1 gene, which negatively regulates the cellulase and xylanase genes in the thermophilic fungus *Rasamsonia emersonii*, using the CRISPR-Cas9 tool (Singh et al., 2023). The CRISPR/Cas9 system was used to increase lignocellulase production in the thermophilic species *M. thermophila* and *M. heterothallica* by up to five-fold through the simultaneous deletion of four genes in a one-step transformation of the cellulase production pathway (Liu et al., 2017). The functionality of transcriptional regulators was efficiently altered in *A. niger* by generating onsite mutations with CRISPR/Cas9, resulting in increased production of CAZymes (Kun et al., 2020).

Omics-based approaches such as metagenomics, transcriptomics, and proteomics are also widely used for the discovery of new enzymes and the improvement of existing enzymes by understanding their mechanisms and tailoring them for specific applications (Chettri et al., 2020; Ijoma et al., 2021). Tools such as NGS have revolutionized genomic studies and taken CAZyme gene studies and synthetic genomics to a new level (Satam et al., 2023). Beyond nucleic acid sequencing, metaproteomics and metabolomics can analyze the enzymatic activities of microbiomes and their abundance, which may play a key role in the functions of the microbiome (Leggieri et al., 2021).

The industrial CAZyme-producing fungus *T. reesei* has been extensively studied by researchers around the world for improved CAZyme production (especially cellulase and its derivatives). Although considerable progress has been made in understanding the molecular mechanisms, genetic engineering of *T. reesei* to improve cellulase enzyme activity is still being approached comparatively slowly, with the focus tending to be on the stability of the enzyme (Druzhinina & Kubicek, 2017).

Although genetic engineering is a potential technique to obtain efficient CAZyme-producing strains, many of the successes of genetic manipulation of CAZyme are unfortunately limited to laboratory conditions. The chemical changes in the medium are likely to have a strong impact on enzyme production by the engineered strain.

8.3.2 Synthetic Biology Strategies

Synthetic biology is a mature field of biotechnology that focuses on genetic enhancement of desired traits and improved production of biomolecules in organisms using genetic and computational tools (Singh et al., 2019). The main goal of synthetic biology in the context of this chapter is to construct artificial biological systems to produce desired CAZymes using bioengineering approaches. Advances in synthetic biology have greatly improved CAZyme production by increasing microbial efficiency, especially through the optimization of metabolic pathways and the use of non-traditional organisms (Liu et al., 2024). The basic workflow considered in the synthetic biology strategies is shown in Fig. 8.2.

Recent advances in DNA synthesis and sequencing have revolutionized genetic modification, enabling the de novo construction of genes, control elements, and even synthetic genomes (Klein-Marcuschamer et al., 2010; Abdi et al., 2024). Improvements such as error reduction strategies, novel sequencing platforms such as NGS, and the incorporation of synthetic base pairs have increased accuracy and

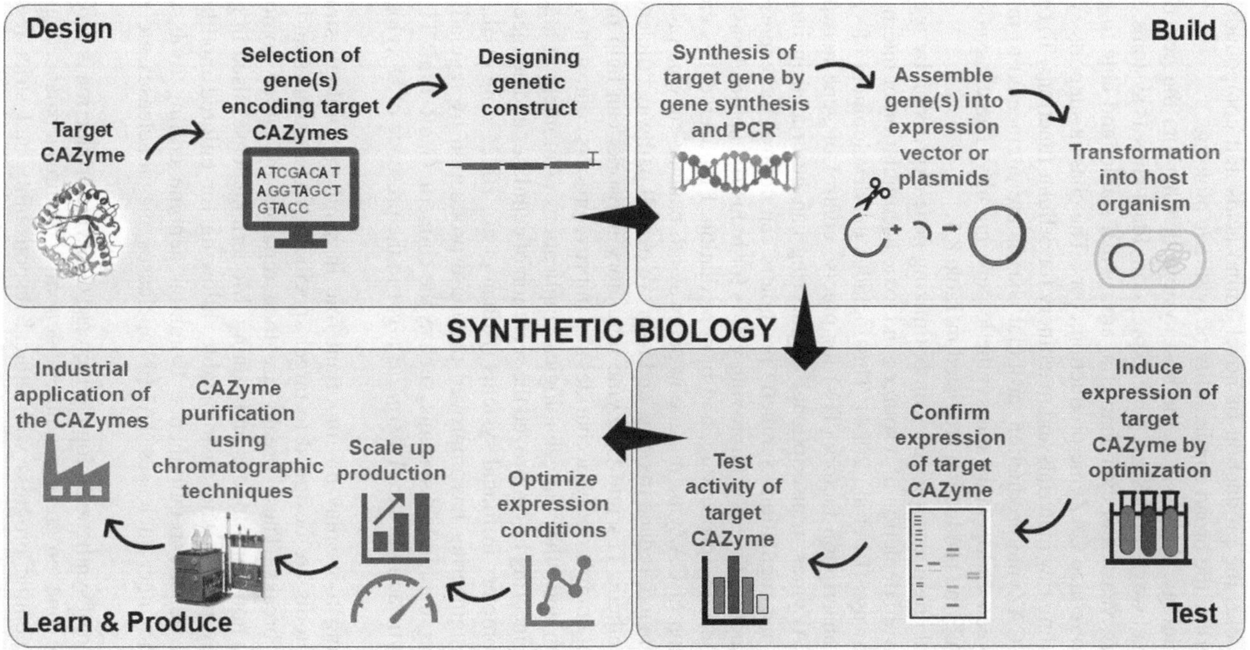

FIGURE. 8.2 A diagram showing the workflow considered in a synthetic biology strategy for the production of CAZymes. The work plan consists of four major categories of several key steps considered in a flow that includes genetic engineering techniques, enzyme assays, microbiology, and fermentations.

efficiency (Satam et al., 2023). Techniques such as green fluorescent protein (GFP)-based error detection and polymerase-based corrections ensure error-free sequences (Kim et al., 2010). The integration of synthetic bases expands the genetic coding potential and enables the production of novel amino acids and compounds. These innovations underline the broad applicability of synthetic biology.

In synthetic biology, a chassis is the host system that harbors the genetic tools and engineered pathways for desired gene expression. Ideal chassis systems, such as *E. coli*, *B. subtilis*, yeast, and cyanobacteria, are robust, stable, and able to grow in minimal media to reduce CAZyme production costs. The chassis supports synthetic systems by providing raw materials and machinery for cellular functions. To optimize the efficiency of CAZyme production, artificial chassis are engineered to minimize competition between endogenous and synthetic gene circuits. The chassis systems developed for CAZyme production are listed in Table 8.1.

Transcriptional engineering focuses on optimizing gene expression by regulating transcription to avoid depletion of resources and toxic accumulation of intermediates in metabolic pathways (Li et al., 2024). This includes the development of synthetic promoters and transcription factors (TFs) for precise control of gene networks. In the context of CAZymes, transcriptional engineering can be crucial in improving their production and regulation. Synthetic promoters can control the expression of CAZyme genes more efficiently, improving yields for industrial applications such as biofuel production, food processing, and bioremediation. Engineered TFs can fine-tune the expression of CAZyme metabolic pathways to balance enzyme production and metabolic demand, reduce toxicity, and improve overall pathway efficiency. The integration of synthetic TFs into CAZyme regulatory networks can help modulate the degradation of complex polysaccharides and improve process control in lignocellulose degradation and other biotechnological applications (Antoniêto et al., 2022). Transcription engineering has improved the secretion of cellulases and hemicellulases by modifying promoters in fungal systems (Adnan et al., 2022; Yang et al., 2023). Transcriptional engineering has improved pectinase production by regulating specific TF promoters in microbial systems, such as the chimeric GaaR-XlnR TF, which induced pectinolytic activities in *Aspergillus niger* in the presence of D-xylose (Kun et al., 2021).

Genome editing tools play an important role in synthetic biology strategies. Genome editing tools are designed to target specific DNA sequences and induce double-stranded breaks. These breaks are then repaired by either homologous recombination or non-homologous end joining. Modern nucleases used in genome editing are highly precise and customizable, allowing for tailored modifications to meet specific genetic requirements. Common genome editing tools include CRISPR/Cas, zinc finger nucleases, TALEN nucleases, meganucleases, etc. (Zhang et al., 2017).

The application of synthetic biology to increase CAZyme production is very promising. At the same time, progress in synthetic genomics requires a balanced approach that combines scientific progress with ethical responsibility and strict regulatory oversight. To harness the benefits of synthetic genomics for society while minimizing the associated risks, the application of responsible research practices is essential.

TABLE 8.1
A List of Some Common Genetically Modified Chassis Developed for Industrial CAZyme Production

Microbial Chassis	Microbial Type	Positive Factors for Consideration	Examples of CAZymes Produced	Applications	References
Escherichia coli	Bacteria	Fast growth, high biomass yield, high protein yield, well-characterized genetics, easy genetic manipulation, strong promoters and expression systems, availability of cloning vectors	Cellulases, Xylanases, Chitinases, Amylases, Laccases	Biofuel production, biorefinery, biomass degradation, starch processing, textile dyeing, bioremediation	Mital et al. (2021), Vojnovic et al. (2024)
Bacillus spp. (e.g., *B. subtilis* and *B. licheniformis*)	Bacteria	GRAS[a] status, High extracellular enzyme secretion capacity, ease in genetic modifications, thermostable enzyme production	Amylases, Cellulases, Xylanases, Laccases	Starch processing, paper industry, biofuel production	Vojnovic et al. (2024)
Corynebacterium glutamicum	Bacteria	GRAS status, low levels of cytoplasmic and extracellular proteases, high protein secretion, robust availability of genetic engineering tools, no endotoxin production, no phage lysis	α-Amylase, endoxylanase	Amino acids production, food industry, starch hydrolysis	Lee et al. (2018), Kiefer et al. (2022)
Streptomyces spp.	Actinobacteria	High secretion efficiency, diverse enzyme production, availability of genetic modification tools,	Cellulases, Hemicellulases, Chitinases	Biodegradation, pharmaceutical, agriculture	Vojnovic et al. (2024), El-Naggar (2021)
Saccharomyces cerevisiae	Yeast	GRAS status, eukaryotic post-translational modifications like glycosylation, tolerance to industrial conditions,	Cellulases, Hemicellulases, β-glucanases, Pectinases, Mannases	Bioethanol production, food industry	Vieira et al. (2020)

(*continued*)

TABLE 8.1 (Continued)
A List of Some Common Genetically Modified Chassis Developed for Industrial CAZyme Production

Microbial Chassis	Microbial Type	Positive Factors for Consideration	Examples of CAZymes Produced	Applications	References
Komagataella phaffii	Yeast	High yield, strong promoters, genetic stability, efficient secretion, post-translational modifications (such as O- and N-linked glycosylation and disulfide bond formation), high cell density fermentation	Xylanases, Mannases, Laccases, Glucanases	Biofuel production, paper industry, textile processing	Unver et al. (2024)
Trichoderma reesei	Filamentous Fungus	High enzyme secretion, cellulase hyper-producer, genetic engineering advancements	Endoglucanases, Exoglucanases, β-Glucosidases	Biomass hydrolysis, biofuel production	Arai et al. (2023)
Aspergillus spp.	Filamentous Fungus	GRAS status, efficient heterologous protein production	Cellulases, Pectinases, Amylases, Xylanases	Food industry, paper pulp processing, bioremediation	Wang et al. (2024)

[a] Generally Recognized as Safe (GRAS) status is given by regulatory agencies, permitting their use in food, pharmaceutical and feed applications.

8.3.3 Improving Enzyme Yields and Stability

Advances in the production of CAZymes have focused on improving their yield, functionality, and stability to meet the needs of industry. Innovations in genetic engineering and synthetic biology have enabled the development of CAZyme variants with improved properties (Chettri et al., 2020). High-throughput screening (HTS) techniques using fluorescent probes and activity-based screening have enabled the rapid identification of CAZyme variants with desirable properties. Fluorescent probes are integral to HTS methodologies, enabling rapid and sensitive detection of enzymatic activity. These probes are designed to emit a fluorescent signal upon interaction with a specific substrate, facilitating real-time monitoring of CAZyme function. The integration of fluorescent probes with advanced technologies, such as microfluidics and synthetic biology, has significantly enhanced the efficiency and effectiveness of CAZyme screening processes. This combination allows for the analysis of extensive enzyme libraries, accelerating the discovery and engineering of CAZymes with desired properties (Wardman et al., 2024). Optimization of microbial strains for enzyme production has further optimized enzyme activity and made processes such as biofuel production more efficient (Boro & Verma, 2024; Chettri & Verma, 2024). Protein engineering with the advances in computational and structural biology has facilitated the redesign of CAZymes for improved performance (Wilken et al., 2019). Together, these approaches improve the applicability and sustainability of CAZyme-based industrial solutions.

8.4 NOVEL TECHNOLOGIES FOR LARGE-SCALE PRODUCTION

The industrial market for CAZymes shows constant growth, both in terms of volume and revenue. This growth is driven by the increasing demand for these enzymes in various industries, including bioenergy, food, pharmaceuticals, and bioremediation. Advances in production technologies coupled with a shift toward sustainable and efficient processes have further fueled this growth, making CAZymes an important component in modern industrial applications.

8.4.1 Expression Systems

Various expression systems are used for the large-scale production of CAZymes, which offer different advantages depending on the specific requirements of the CAZyme and the desired application. The most commonly used systems include bacterial expression systems, fungal expression systems, and yeast expression systems.

Bacterial expression systems are widely used due to their rapid growth, ease of genetic manipulation, and well-established genetic tools that make them cost-effective and efficient for industrial enzyme production. Commonly used bacterial expression systems include *E. coli*, *Pseudomonas*, *Streptomyces*, etc. (Chen, 2012). Fungal expression systems are commonly used for CAZyme production because they naturally secrete large amounts of enzymes, especially cellulases and hemicellulases, making them ideal for industrial scale enzyme production (Singh et al., 2021; Zhang et al., 2020). *T. reesei* is one of the most commonly used fungal systems due to its high secretion capacity, especially for the production of lignocellulolytic enzymes (Sperandio & Filho, 2021). Mixed cultures of *T. reesei* and *A. niger* have also been shown to be effective fungal expression systems for CAZyme production (Fang et al., 2024).

Each microbial expression system has its strengths and limitations, and the choice of system depends on factors such as the type of enzyme, the desired scale of production, cost considerations, and whether post-translational modifications are required.

8.4.2 Bioreactor Designs

Choosing the right bioreactor design is crucial for the efficient and cost-effective production of CAZymes on a large-scale. A well-designed bioreactor ensures effective mixing of the substrates with the host microorganisms, controls shear stress, reduces foaming, ensures good mass transfer, and minimizes energy consumption. Various types of bioreactors are used in industrial CAZyme production, each offering unique advantages. Common designs include stirred-tank reactors, airlift reactors, and fed-batch reactors, with each configuration tailored to the needs of the particular enzyme production process (Tarafdar et al., 2021). Some other bioreactors based on the process requirements are listed in Table 8.2, which provides an overview of various bioreactor types used in CAZyme production, highlighting their unique features.

8.5 ECONOMIC CONSIDERATIONS IN INDUSTRIAL-SCALE ENZYME PRODUCTION

The large-scale production of enzymes, especially CAZymes, has attracted great interest in recent years due to their crucial role in various biotechnological applications, such as biofuel production. The economic feasibility of producing CAZymes on an industrial-scale is influenced by several key factors, including raw-material costs, enzyme production methods, downstream processing, and infrastructure requirements.

8.5.1 Raw Material and Feedstock Costs

The cost of raw materials plays an important role in determining the economic viability of enzyme production. In the case of CAZymes for biofuel production, lignocellulosic biomass, such as agricultural residues and wood waste, is often used as feedstock, and the economics of using lignocellulose depends on the availability, transportation, and pre-treatment costs of these materials. When producing biofuels on a large scale, the selection of biomass feedstocks must be evaluated for cost efficiency. A study by Silva et al. (2022) compared different types of lignocellulosic biomass in terms of their environmental impact and found that urban forest waste is the most efficient option for biofuel production due to its relatively low cost. On the other hand, crops such as sugarcane or palm can increase production costs due to high land and water consumption, leading to food security concerns (Joshi et al., 2017). Using lignocellulosic biomass, which comes from plant waste, is a smart

TABLE 8.2
Some Types of Bioreactors for CAZymes Production

Bioreactor	Key Feature	Application	References
Stirred-Tank Bioreactors	Proper agitation facilitates efficient oxygen transfer and mixing	Cellulase	Ahamed et al. (2008)
Airlift Bioreactors	Low construction cost and the low risk of shearing the mycelium due to the absence of stirring	Cellulase and xylanase	Ritter et al. (2013)
Packed Bed Bioreactors	Cost, space, and labor effective. Often employed in solid state fermentations with the aim of increasing productivity	Glucoamylase production	Melikoglu et al. (2015)
Filamentous fungi bioreactors	Suitable for both native and heterologous protein production. Regulation strategies at transcription and protein levels	Production of various CAZymes	Zhang et al. (2020)
High-throughput (HTP) bioreactors	Rapid screening of enzyme variants. Multiplexing of reactions. Quantitative analysis of enzyme activity	Discovery and optimization of novel CAZymes	Cardoso et al. (2022)
Enrichment bioreactors	Selective pressure for desired microbial communities, Adaptation to specific carbon sources	Discovery of novel CAZymes from environmental samples	Wong et al. (2023)
Metagenomic-based Bioreactors	Utilization of CAZymes discovered through large-scale metagenomic assembly. Exploitation of novel CAZyme diversity	Improved biomass degradation	Auer et al. (2023)
High-rate anaerobic bioreactors	Effective bioconversion of organic compounds. Sequential activity of hydrolytic bacteria, acetogens, and methanogenic archaea. Continuous production	Biogas production, adaptable for CAZyme production	Wong et al. (2023)

and sustainable approach to enzyme production. It fits well with the principles of a circular economy by repurposing agricultural residues, forestry byproducts, and other waste materials that might otherwise go unused. Examples of these feedstocks include corn stover, wheat straw, rice husks, sawdust, wood chips, energy crops like switchgrass and miscanthus, and industrial byproducts such as sugarcane bagasse and paper mill sludge.

However, the cost and availability of these materials can vary widely due to factors like seasonal supply, geographic location, competition with other industries (such as biofuels), transportation costs, and the need for pre-treatment before use. For instance, the price of commonly used feedstocks like poplar wood or glucose can fluctuate based on market conditions and supply chain challenges, adding uncertainty to production costs.

To tackle these challenges and make raw material use more cost-effective, researchers and producers are exploring several strategies. These include diversifying the types of feedstocks used to avoid over-reliance on any single source, improving pre-treatment methods to make cellulose and hemicellulose more accessible for enzyme production, and using genetic engineering to create microbial strains capable of utilizing a wider range of biomass types. Other approaches involve processing feedstocks locally to cut down on transportation expenses and investigating new waste streams as low-cost alternatives.

In addition to raw material innovations, efforts are being made to optimize fermentation processes for better efficiency. This includes fine-tuning the composition of growth media, adopting advanced fermentation techniques like fed-batch or continuous systems, and implementing sophisticated process control technologies. These improvements can help maximize enzyme yields while using fewer resources.

It is important to remember that raw-material costs are just one piece of the puzzle. They need to be weighed against other factors like enzyme yield, activity, and stability. In some cases, spending a bit more on higher quality or specialized substrates might be worthwhile if it leads to significantly better production outcomes.

8.5.2 Enzyme Production Costs

The production of CAZymes on an industrial scale requires the optimization of fermentation conditions, microbial culture techniques, and efficient enzyme extraction and purification methods. The cost of enzyme production is closely linked to microbial cultivation, including the selection of microbial strains, culture media, and fermentation processes. The selection of microorganisms with high enzyme-producing potential, such as those found in the gut microbiomes of herbivores and termites, is critical. Isolating microbes that can thrive in large-scale fermentation systems while producing sufficient quantities of CAZymes is critical to minimizing production costs. In addition, advances in genetic engineering and synthetic biology may provide opportunities to increase the productivity of enzyme-producing microbes, potentially reducing costs (Ben Guerrero et al., 2015).

8.5.3 Downstream Processing and Byproduct Management

Downstream processing and byproduct management are critical aspects of CAZyme production that significantly impact the overall economic viability of the process. These steps involve the recovery, concentration, and purification of enzymes using techniques such as chromatography, phase separation, and cell lysis, depending on the specific enzyme and its intended application. Downstream processing, including enzyme purification, concentration, and formulation, adds to the total cost; thus, technoeconomic analysis (TEA) plays a crucial role in optimizing these processes. For instance, a TEA of enzymatic hydrolysis of municipal solid waste-derived pulp revealed that a two-stage continuous configuration with a total residence time of 54 hours was the most economically favorable option, demonstrating sevenfold higher enzyme productivity and better profit compared to reference systems (Climent et al.,

2022). However, the high cost of enzymes remains a significant challenge, particularly at pilot scales (Climent et al., 2022). To improve economic feasibility, efforts are being made to optimize or reduce enzyme loading, with some studies suggesting targets as low as 5.6 FPU/g for biomass conversion (Brondi et al., 2020; Cheng et al., 2023). Additionally, integrating CAZyme production with other biorefinery processes, such as combining soybean biodiesel and sugarcane bioethanol production, can create valuable byproducts and reduce overall costs (Brondi et al., 2020).

To avoid further financial and environmental costs, the production of byproducts such as microbial biomass, residual nutrients, and wastewater must be effectively managed. The cost of enzymes remains a major challenge, especially in processes that require large quantities for biofuel production. Recent studies have shown the potential of using byproducts and waste streams to offset some of these costs. The potential of peroxidase-producing strains from the gut of termites was investigated in the study by Ali et al. (2021). These strains not only produced CAZyme but also had a coupled pathway for dye degradation, which helped to reduce waste generation. To increase the economic sustainability of enzyme production, research is also being conducted into recycling methods for enzymes and other byproducts. Overall, as the industry continues to evolve, innovative approaches to downstream processing and byproduct utilization will be essential for making large-scale CAZyme production economically viable.

8.5.4 Technological and Infrastructure Considerations

The industrial scale production of CAZymes relies heavily on both technology and infrastructure, with each stage of the process requiring specialized tools and facilities. This includes specialized bioreactors for fermentation, advanced monitoring systems for process control, and infrastructure for the transportation and storage of enzymes and biomass. In the downstream phase, where enzymes are extracted and purified, a variety of technologies come into play, such as chromatography systems for separation and equipment for breaking down cells, depending on the specific enzyme being produced and its intended use. The capital costs associated with scaling up enzyme production can be significant, particularly in terms of equipment, plant development, and human resources. In addition, the transportation of raw materials and enzymes poses a logistical challenge, especially in terms of cost efficiency. The distance between biomass sources and production facilities as well as seasonal fluctuations in biomass yields can lead to cost fluctuations (Balan, 2014; Mungodla et al., 2019). As highlighted by Kesieme et al. (2019), transportation and distribution of biofuels also pose compatibility challenges as biofuels can lead to corrosion in pipelines and incompatibility of engine systems.

In recent years, cutting-edge tools like omics-based approaches (e.g., genomics and proteomics) and HTS have become vital for discovering and designing new CAZymes more efficiently. Beyond the lab, infrastructure becomes even more important when scaling up production. Pilot plants bridge the gap between small-scale experiments and full industrial manufacturing, while dedicated facilities are needed to stabilize and formulate enzymes for commercial use.

As the industry evolves, there's growing interest in integrated biorefineries, which are the facilities designed to process a wide variety of plant-based feedstocks into enzymes and other valuable products. These setups require significant investment in flexible equipment and proper storage solutions but hold great promise for making CAZyme production more efficient and sustainable on a large scale.

8.6 CONCLUSION AND FUTURE PROSPECTS

CAZymes have great promise for use in a variety of industries, such as biofuels, medicines, and environmentally friendly bioprocessing, when produced on an industrial scale. But there are a number of challenges in this industry, ranging from choosing the right microbial hosts to maximizing fermentation and streamlining downstream procedures. Recent advances in synthetic biology, genetic modification, and enzyme stability have significantly increased the efficiency of CAZyme manufacturing. Furthermore, large-scale manufacturing is being made easier by sophisticated bioreactor designs and creative expression systems. Utilizing cutting-edge technologies like precision fermentation, multi-omics, and computational biology to overcome present constraints is key to the future of CAZyme manufacturing. Economically speaking, success will depend on lowering the cost of raw materials, improving enzyme extraction techniques, and efficiently handling byproducts. CAZymes are positioned to revolutionize numerous sectors by providing sustainable substitutes as research advances and interdisciplinary collaboration increases.

ACKNOWLEDGMENT

The authors thank the Department of Microbiology, Sikkim University, for providing thecomputational infrastructure and central library facilities for procuring references and plagiarism analysis (Drillbit PlagiarisDetectionSoftware)

AUTHOR CONTRIBUTION STATEMENT

Manswama Boro and Shuchi Singh contributed equally to writing the original draft and to reviewing and editing the manuscript.

FUNDING

This research did not receive any specific grant from funding agencies in the public, commercial, or not-for-profit sectors.

CONFLICT OF INTEREST

The authors declare that the research was conducted without any commercial or financial relationships that could be construed as potential conflict of interest.

REFERENCES

Abdi, G., Firdous, S., Vasantharekha, R., Singh, G. B., Seetharaman, B., Mudgal, G., Jain, M., & Tarighat, M. A. (2024). Progress on synthetic genomics. In Vijai Singh (Ed.), *Advances in Genomics* (pp. 181–197). Springer Nature Singapore. https://doi.org/10.1007/978-981-97-3169-5_9

Adnan, M., Ma, X., Olsson, S., Wang, J., & Liu, G. (2022). Promoter regulation and genetic engineering strategies for enhanced cellulase expression in Trichoderma reesei. *Microbiological Research*, *259*, 127011. https://doi.org/10.1016/j.micres.2022.127011

Ahamed, A., & Vermette, P. (2008). Enhanced enzyme production from mixed cultures of Trichoderma reesei RUT-C30 and Aspergillus niger LMA grown as fed batch in a stirred tank bioreactor. *Biochemical Engineering Journal*, *42*(1), 41–46. https://doi.org/10.1016/j.bej.2008.05.007

Ahmad, M., Hirz, M., Pichler, H., & Schwab, H. (2014). Protein expression in Pichia pastoris: Recent achievements and perspectives for heterologous protein production. *Applied Microbiology and Biotechnology*, *98*(12), 5301–5317. https://doi.org/10.1007/s00253-014-5732-5

Anderson, T. M. (2009). Industrial fermentation processes. In Moselio Schaechter (Ed.), *Encyclopedia of Microbiology* (pp. 349–361). Elsevier. https://doi.org/10.1016/B978-012373944-5.00369-2

Antoniêto, A. C. C., Maués, D. B., Nogueira, K. M. V., de Paula, R. G., Steindorff, A. S., Kennedy, J. F., Pandey, A., Gupta, V. K., & Silva, R. N. (2022). Engineering of holocellulase in biomass-degrading fungi for sustainable biofuel production Amanda Cristina Campos Antoniêto. *Journal of Cleaner Production*, *371*, 133488. https://doi.org/https://doi.org/10.1007/s00253-021-11428-2

Arai, T., Wada, M., Nishiguchi, H., Takimura, Y., & Ishii, J. (2023). Inducer-free recombinant protein production in Trichoderma reesei: Secretory production of endogenous enzymes and heterologous nanobodies using glucose as the sole carbon source. *Microbial Cell Factories*, *22*(1), 103. https://doi.org/10.1186/s12934-023-02109-y

Auer, E., Lazuka, A., Huguenin-Bizot, B., Jehmlich, N., Déjean, S., Lombard, V., Henrissat, B., O'Donohue, M., & Hernandez-Raquet, G. (2023). Horizontal metaproteomics and CAZymes analysis of lignocellulolytic microbial consortia selectively enriched from cow rumen and termite gut. *ISME Communications*, *3*(1), 134. https://doi.org/10.1038/s43705-023-00339-0

Balan, V. (2014). Current challenges in commercially producing biofuels from lignocellulosic biomass. *International Scholarly Research Notices*, *2014*(1), 463074.

Basak, B., Jeon, B.-H., Kim, T. H., Lee, J.-C., Chatterjee, P. K., & Lim, H. (2020). Dark fermentative hydrogen production from pretreated lignocellulosic biomass: Effects of inhibitory byproducts and recent trends in mitigation strategies. *Renewable and Sustainable Energy Reviews*, *133*, 110338. https://doi.org/10.1016/j.rser.2020.110338

Ben Guerrero, E., Arneodo, J., Bombarda Campanha, R., Abrão de Oliveira, P., Veneziano Labate, M. T., Regiani Cataldi, T., ... & Talia, P. (2015). Prospection and evaluation of (Hemi) cellulolytic enzymes using untreated and pretreated biomasses in two argentinean native termites. *Plos One*, *10*(8), e0136573

Bhatwa, A., Wang, W., Hassan, Y. I., Abraham, N., Li, X.-Z., & Zhou, T. (2021). Challenges associated with the formation of recombinant protein inclusion bodies in Escherichia coli and strategies to address them for industrial applications. *Frontiers in Bioengineering and Biotechnology*, *9*. https://doi.org/10.3389/fbioe.2021.630551

Boro, M., & Verma, A. K. (2024). Optimization of cellulase production by cohnella xylanilytica RU-14 using statistical methods. *Applied Biochemistry and Biotechnology*, *196*(5), 2757–2770. https://doi.org/10.1007/s12010-023-04447-4

Braham, S. A., Siar, E. H., Arana-Peña, S., Bavandi, H., Carballares, D., Morellon-Sterling, R., de Andrades, D., Kornecki, J. F, & Fernandez-Lafuente, R. (2021). Positive effect of glycerol on the stability of immobilized enzymes: Is it a universal fact?. *Process Biochemistry, 102*, 108–121. https://doi.org/10.1016/j.procbio.2020.12.015

Brondi, M. G., Elias, A. M., Furlan, F. F., Giordano, R. C., & Farinas, C. S. (2020). Performance targets defined by retro-techno-economic analysis for the use of soybean protein as saccharification additive in an integrated biorefinery. *Scientific Reports*, *10*(1), 7367. https://doi.org/10.1038/s41598-020-64316-6

Brunecky, R., Chung, D., Sarai, N. S., Hengge, N., Russell, J. F., Young, J., Mittal, A., Pason, P., Vander Wall, T., Michener, W., Shollenberger, T., Westpheling, J., Himmel, M. E., & Bomble, Y. J. (2018). High activity CAZyme cassette for improving biomass degradation in thermophiles. *Biotechnology for Biofuels*, *11*(1), 22. https://doi.org/10.1186/s13068-018-1014-2

Cardoso, V., Brás, J. L., Costa, I. F., Ferreira, L. M., Gama, L. T., Vincentelli, R., Henrissat, B., & Fontes, C. M. (2022). Generation of a library of carbohydrate-active enzymes for plant biomass deconstruction. *International Journal of Molecular Sciences*, *23*(7), 4024. https://doi.org/10.3390/ijms23074024

Chen, R. (2012). Bacterial expression systems for recombinant protein production: E. coli and beyond. *Biotechnology Advances*, *30*(5), 1102–1107. https://doi.org/10.1016/j.biotechadv.2011.09.013

Cheng, M. H., Singh, S., Clennon, A. N. C., Dien, B. S., & Singh, V. (2023). Production of designer xylose-acetic acid enriched hydrolysate from bioenergy Sorghum, oilcane, and energycane bagasses. *Bioresource Technology*, *380*, 129104. https://doi.org/10.1016/j.biortech.2023.129104

Chettri, D., & Verma, A. K. (2024). Statistical optimization of cellulase production from Bacillus sp. YE16 isolated from yak dung of the Sikkim Himalayas for its application in bioethanol production using pretreated sugarcane bagasse. *Microbiological Research*, *281*, 127623. https://doi.org/10.1016/j.micres.2024.127623

Chettri, D., Verma, A. K., & Verma, A. K. (2020). Innovations in CAZyme gene diversity and its modification for biorefinery applications. *Biotechnology Reports*, *28*, e00525. https://doi.org/10.1016/j.btre.2020.e00525

Climent Barba, F., Grasham, O., Puri, D. J., & Blacker, A. J. (2022). A simple techno-economic assessment for scaling-up the enzymatic hydrolysis of MSW pulp. *Frontiers in Energy Research*, *10*, 788534. https://doi.org/10.3389/fenrg.2022.788534

Crater, J. S., & Lievense, J. C. (2018). Scale-up of industrial microbial processes. *FEMS Microbiology Letters*, *365*(13), fny138. https://doi.org/10.1093/femsle/fny138

Danilova, I., & Sharipova, M. (2020). The practical potential of Bacilli and their enzymes for industrial production. *Frontiers in Microbiology*, *11*. https://doi.org/10.3389/fmicb.2020.01782

Druzhinina, I. S., & Kubicek, C. P. (2017). Genetic engineering of Trichoderma reesei cellulases and their production. *Microbial Biotechnology*, *10*(6), 1485–1499. https://doi.org/10.1111/1751-7915.12726

El-Naggar, N. E. A. (2021). Streptomyces-based cell factories for production of biomolecules and bioactive metabolites. In Vijai Singh (Ed.), *Microbial Cell Factories Engineering for Production of Biomolecules* (pp. 183–234). Academic Press. https://doi.org/10.1016/B978-0-12-821477-0.00011-8

Fakruddin, M., Mohammad Mazumdar, R., Bin Mannan, K. S., Chowdhury, A., & Hossain, M. N. (2013). Critical factors affecting the success of cloning, expression, and mass production of enzymes by recombinant E. coli. *ISRN Biotechnology*, *2013*, 1–7. https://doi.org/10.5402/2013/590587

Fang, H., Zhao, C., Li, Y., Song, Y., Wu, Y., Song, X., Wu, J., & Yang, L. (2024). Multi-omics study on the mixed culture of Trichoderma reesei and Aspergillus niger with improved lignocellulase production. *Biomass and Bioenergy*, *190*, 107374. https://doi.org/10.1016/j.biombioe.2024.107374

Fasim, A., More, V. S., & More, S. S. (2021). Large-scale production of enzymes for biotechnology uses. *Current Opinion in Biotechnology*, *69*, 68–76. https://doi.org/10.1016/j.copbio.2020.12.002

Francis, D. M., & Page, R. (2010). Strategies to optimize protein expression in E. coli. *Current Protocols in Protein Science*, *61*(1). https://doi.org/10.1002/0471140864.ps0524s61

Gasser, B., Saloheimo, M., Rinas, U., Dragosits, M., Rodríguez-Carmona, E., Baumann, K., Giuliani, M., Parrilli, E., Branduardi, P., Lang, C., Porro, D., Ferrer, P., Tutino, M. L., Mattanovich, D., & Villaverde, A. (2008). Protein folding and conformational stress in microbial cells producing recombinant proteins: A host comparative overview. *Microbial Cell Factories*, *7*(1), 11. https://doi.org/10.1186/1475-2859-7-11

Gervais, D. (2019). Quality control and downstream processing of therapeutic enzymes. In *Therapeutic Enzymes: Function and Clinical Implications* (pp. 55–80). https://doi.org/10.1007/978-981-13-7709-9_3

Ijoma, G. N., Heri, S. M., Matambo, T. S., & Tekere, M. (2021). Trends and applications of omics technologies to functional characterisation of enzymes and protein metabolites produced by fungi. *Journal of Fungi*, *7*(9), 700. https://doi.org/10.3390/jof7090700

Joshi, G., Pandey, J. K., Rana, S., & Rawat, D. S. (2017). Challenges and opportunities for the application of biofuel. *Renewable and Sustainable Energy Reviews*, *79*, 850–866.

Kiefer, D., Tadele, L. R., Lilge, L., Henkel, M., & Hausmann, R. (2022). High-level recombinant protein production with Corynebacterium glutamicum using acetate as carbon source. *Microbial Biotechnology*, *15*(11), 2744–2757. https://doi.org/10.3389/fmicb.2018.02523

Kim, H., Han, H., Shin, D., & Bang, D. (2010). A fluorescence selection method for accurate large-gene synthesis. *ChemBioChem*, *11*(17), 2448–2452. https://doi.org/10.1002/cbic.201000368

Klein-Marcuschamer, D., Yadav, V. G., Ghaderi, A., & Stephanopoulos, G. N. (2010). De Novo metabolic engineering and the promise of synthetic DNA. In Christoph Wittmann and Rainer Krull (Eds.), *Biosystems Engineering I* (pp. 101–131). Springer Berlin Heidelberg. https://doi.org/10.1007/10_2009_52

Kuhad, R. C., Deswal, D., Sharma, S., Bhattacharya, A., Jain, K. K., Kaur, A., Pletschke, B. I., Singh, A., & Karp, M. (2016). Revisiting cellulase production and redefining current strategies based on major challenges. *Renewable and Sustainable Energy Reviews, 55*, 249–272. https://doi.org/10.1016/j.rser.2015.10.132

Kun, R. S., Garrigues, S., Falco, M. Di, Tsang, A., & De Vries, R. P. (2021). The chimeric GaaR-XlnR transcription factor induces pectinolytic activities in the presence of D-xylose in Aspergillus niger. *Applied Microbiology and Biotechnology*, *105*(13), 5553–5564. https://doi.org/https://doi.org/10.1007/s00253-021-11428-2

Kun, R. S., Gomes, A. C. S., Hildén, K. S., Cerezo, S. S., Mäkelä, M. R., & de Vries, R. P. (2019). Developments and opportunities in fungal strain engineering for the production of novel enzymes and enzyme cocktails for plant biomass degradation. *Biotechnology Advances*, *37*(6), 107361. https://doi.org/10.1016/j.biotechadv.2019.02.017

Kun, R. S., Meng, J., Salazar-Cerezo, S., Mäkelä, M. R., de Vries, R. P., & Garrigues, S. (2020). CRISPR/Cas9 facilitates rapid generation of constitutive forms of transcription factors in Aspergillus niger through specific on-site genomic mutations resulting in increased saccharification of plant biomass. *Enzyme and Microbial Technology, 136*, 109508. https://doi.org/10.1016/j.enzmictec.2020.109508

Lee, M. J., & Kim, P. (2018). Recombinant protein expression system in Corynebacterium glutamicum and its application. *Frontiers in Microbiology*, *9*, 2523. https://doi.org/10.3389/fmicb.2018.02523

Leggieri, P. A., Liu, Y., Hayes, M., Connors, B., Seppälä, S., O'Malley, M. A., & Venturelli, O. S. (2021). Integrating systems and synthetic biology to understand and engineer microbiomes. *Annual Review of Biomedical Engineering*, *23*(1), 169–201. https://doi.org/10.1146/annurev-bioeng-082120-022836

Li, Y., Liu, M., Yang, C., Fu, H., & Wang, J. (2024). Critical reviews in biotechnology engineering microbial metabolic homeostasis for chemicals production. *Critical Reviews in Biotechnology*, 1–20. https://doi.org/10.1080/07388551.2024.2371465

Liu, Q., Gao, R., Li, J., Lin, L., Zhao, J., Sun, W., & Tian, C. (2017). Development of a genome-editing CRISPR/Cas9 system in thermophilic fungal Myceliophthora species and its application to hyper-cellulase production strain engineering. *Biotechnology for Biofuels*, *10*(1), 1. https://doi.org/10.1186/s13068-016-0693-9

Liu, W., Wang, P., Zhuang, X., Ling, Y., Liu, H., Wang, S., Yu, H., Ma, L., Jiang, Y., Zhao, G., Yan, X., Zhou, Z., & Zhang, G. (2024). RDBSB: A database for catalytic bioparts with experimental evidence. *Nucleic Acids Research*, *53*, https://doi.org/10.1093/nar/gkae844

Melikoglu, M., Lin, C. S. K., & Webb, C. (2015). Solid state fermentation of waste bread pieces by Aspergillus awamori: Analysing the effects of airflow rate on enzyme production in packed bed bioreactors. *Food and Bioproducts Processing*, *95*, 63–75. https://doi.org/10.1016/j.fbp.2015.03.011

Mital, S., Christie, G., & Dikicioglu, D. (2021). Recombinant expression of insoluble enzymes in Escherichia coli: A systematic review of experimental design and its manufacturing implications. *Microbial Cell Factories*, *20*, 1–20. https://microbialcellfactories.biomedcentral.com/articles/10.1186/s12934-021-01698-w

Mukherjee, S. (2019). Isolation and purification of industrial enzymes: Advances in enzyme technology. In RamSarup Singh ,Reeta Rani Singhania, Ashok Pandey and Christian Larroche (Eds.), *Advances in Enzyme Technology* (pp. 41–70). Elsevier. https://doi.org/10.1016/B978-0-444-64114-4.00002-9

Mungodla, S. G., Linganiso, L. Z., Mlambo, S., & Motaung, T. (2019). Economic and technical feasibility studies: technologies for second generation biofuels. *Journal of Engineering, Design and Technology*, *17*(4), 670–704.

Ndochinwa, O. G., Wang, Q.-Y., Amadi, O. C., Nwagu, T. N., Nnamchi, C. I., Okeke, E. S., & Moneke, A. N. (2024). Current status and emerging frontiers in enzyme engineering: An industrial perspective. *Heliyon*, *10*(11), e32673. https://doi.org/10.1016/j.heliyon.2024.e32673

Ritter, C. E. T., Fontana, R. C., Camassola, M., da Silveira, M. M., & Dillon, A. J. P. (2013). The influence of sorbitol on the production of cellulases and xylanases in an airlift bioreactor. *Bioresource Technology*, *148*, 86–90. https://doi.org/10.1016/j.biortech.2013.08.125

Robinson, P. K. (2015). Enzymes: Principles and biotechnological applications. *Essays in Biochemistry*, *59*, 1. https://doi.org/10.1042/bse0590001

Satam, H., Joshi, K., Mangrolia, U., Waghoo, S., Zaidi, G., Rawool, S., Thakare, R. P., Banday, S., Mishra, A. K., Das, G., & Malonia, S. K. (2023). Next-generation sequencing technology: Current trends and advancements. *Biology*, *12*(7), 997. https://doi.org/10.3390/biology12070997

Schmid, J., Sieber, V., & Rehm, B. (2015). Bacterial exopolysaccharides: Biosynthesis pathways and engineering strategies. *Frontiers in Microbiology*, *6*, 496. https://doi.org/10.3389/fmicb.2015.00496

Schmoll, M. (2022). Trichoderma reesei. *Trends in Microbiology*, *30*(4), 403–404. https://doi.org/10.1016/j.tim.2021.12.008

Shahzadi, I., Al-Ghamdi, M. A., Nadeem, M. S., Sajjad, M., Ali, A., Khan, J. A., & Kazmi, I. (2021). Scale-up fermentation of Escherichia coli for the production of recombinant endoglucanase from Clostridium thermocellum. *Scientific Reports*, *11*(1), 7145. https://doi.org/10.1038/s41598-021-86000-z

Siddiqui, S. A., Erol, Z., Rugji, J., Taşçı, F., Kahraman, H. A., Toppi, V., Musa, L., Di Giacinto, G., Bahmid, N. A., Mehdizadeh, M., & Castro-Muñoz, R. (2023). An overview of fermentation in the food industry-looking back from a new perspective. *Bioresources and Bioprocessing*, *10*(1), 85. https://doi.org/10.1186/s40643-023-00702-y

Singh, A., Bajar, S., Devi, A., & Pant, D. (2021). An overview on the recent developments in fungal cellulase production and their industrial applications. *Bioresource Technology Reports*, *14*, 100652. https://doi.org/10.1016/j.biteb.2021.100652

Singh, R., Tiwari, M., Singh, R., & Lee, J.-K. (2013). From protein engineering to immobilization: Promising strategies for the upgrade of industrial enzymes. *International Journal of Molecular Sciences*, *14*(1), 1232–1277. https://doi.org/10.3390/ijms14011232

Singh, S. P., Bansal, S., & Pandey, A. (2019). Basics and roots of synthetic biology. In Sudhir P. Singh ,Ashok Pandey,Goucheng Du, and Sudesh Kumar (Eds.), *Current Developments in Biotechnology and Bioengineering* (pp. 3–22). Elsevier. https://doi.org/10.1016/B978-0-444-64085-7.00001-0

Singh, V., Raheja, Y., Basotra, N., Sharma, G., Tsang, A., & Chadha, B. S. (2023). CRISPR/Cas9 mediated gene editing of transcription factor ACE1 for enhanced cellulase production in thermophilic fungus Rasamsonia emersonii. *Fungal Biology and Biotechnology*, *10*(1), 18. https://doi.org/10.1186/s40694-023-00165-y

Sperandio, G. B., & Filho, E. X. F. (2021). An overview of Trichoderma reesei co-cultures for the production of lignocellulolytic enzymes. *Applied Microbiology and Biotechnology*, *105*(8), 3019–3025. https://doi.org/10.1007/s00253-021-11261-7

Tarafdar, A., Sirohi, R., Gaur, V. K., Kumar, S., Sharma, P., Varjani, S., Pandey, H. O., Sindhu, R., Madhavan, A., Rajasekharan, R., & Sim, S. J. (2021). Engineering interventions in enzyme production: Lab to industrial scale. *Bioresource Technology*, *326*, 124771. https://doi.org/10.1016/j.biortech.2021.124771

Unver, Y., & Dagci, I. (2024). Komagataella phaffii (Pichia pastoris) as a powerful yeast expression system for biologics production. *Frontiers in Bioscience-Elite*, *16*(2), 19. https://doi.org/10.31083/j.fbe1602019

Vieira, E. F., & Delerue-Matos, C. (2020). Exploitation of Saccharomyces cerevisiae enzymes in food processing and preparation of nutraceuticals and pharmaceuticals. In Naveen Kumar Arora, Jitendra Mishra, and Vaibhav Mishra (Eds.), *Microbial Enzymes: Roles and Applications in Industries* (pp. 41–62). Springer Nature. https://doi.org/10.1007/978-981-15-1710-5_2

Vojnovic, S., Aleksic, I., Ilic-Tomic, T., Stevanovic, M., & Nikodinovic-Runic, J. (2024). Bacillus and Streptomyces spp. as hosts for production of industrially relevant enzymes. *Applied Microbiology and Biotechnology*, *108*(1), 185. https://doi.org/10.1007/s00253-023-12900-x

Wang, Z. D., Wang, B. T., Jin, L., Ruan, H. H., & Jin, F. J. (2024). Implications of carbon catabolite repression for Aspergillus-based cell factories: A review. *Biotechnology Journal*, *19*(2), 2300551. https://doi.org/10.1002/biot.202300551

Wardman, J., & Withers, S. G. (2024). Carbohydrate-active enzyme (CAZyme) discovery and engineering via (Ultra) high-throughput screening. *RSC Chemical Biology*, *5*. https://doi.org/10.1039/D4CB00024B

Watts, A., Sankaranarayanan, S., Watts, A., & Raipuria, R. K. (2021). Optimizing protein expression in heterologous system: Strategies and tools. *Meta Gene*, *29*, 100899. https://doi.org/10.1016/j.mgene.2021.100899

Wilken, S. E., Swift, C. L., Podolsky, I. A., Lankiewicz, T. S., Seppälä, S., & O'Malley, M. A. (2019). Linking 'omics' to function unlocks the biotech potential of non-model fungi. *Current Opinion in Systems Biology*, *14*, 9–17. https://doi.org/10.1016/j.coisb.2019.02.001

Wong, M. T., Nesbø, C. L., Wang, W., Couturier, M., Lombard, V., Lapebie, P., Terrapon, N., Henrissat, B., Edwards, E. A., & Master E. R. (2023). Taxonomic composition and carbohydrate-active enzyme content in microbial enrichments from pulp mill anaerobic granules after cultivation on lignocellulosic substrates. *Frontiers in Microbiomes*, *2*, 1094865. https://doi.org/10.3389/frmbi.2023.1094865

Xia, J., Wang, G., Fan, M., Chen, M., Wang, Z., & Zhuang, Y. (2021). Understanding the scale-up of fermentation processes from the viewpoint of the flow field in bioreactors and the physiological response of strains. *Chinese Journal of Chemical Engineering*, *30*, 178–184. https://doi.org/10.1016/j.cjche.2020.12.004

Yang, J., Yue, H.-R., Pan, L.-Y., Feng, J.-X., Zhao, S., Suwannarangsee, S., Champreda, V., Liu, C.-G., & Zhao, X.-Q. (2023). Fungal strain improvement for efficient cellulase production and lignocellulosic biorefinery: Current status and future prospects. *Bioresource Technology*, *385*, 129449. https://doi.org/10.1016/j.biortech.2023.129449

Zhang, H., Zhang, J., Lang, Z., Botella, J. R., & Zhu, J.-K. (2017). Genome editing—principles and applications for functional genomics research and crop improvement. *Critical Reviews in Plant Sciences*, *36*(4), 291–309. https://doi.org/10.1080/07352689.2017.1402989

Zhang, T., Liu, H., Lv, B., & Li, C. (2020). Regulating strategies for producing carbohydrate active enzymes by filamentous fungal cell factories. *Frontiers in Bioengineering and Biotechnology*, *8*, 691. https://doi.org/10.3389/fbioe.2020.00691

9 Structure, Classes, Databases, and Bioinformatics Tools Associated with Carbohydrate-Active Enzymes

Sandip Das, Abhijit Reang, and Arabinda Ghosh

ABBREVIATIONS

AA	Auxiliary activities
BLAST	Basic Local Alignment Search Tool
CAZymes	Carbohydrate-active enzymes
CBM	Carbohydrate-binding module
CE	Carbohydrate esterases
cryo-EM	Cryo-electron microscopy
EC	Enzyme Commission
GH	Glycoside hydrolases
GT	Glycosyltransferases
HMM	Hidden Markov Model
KEGG	Kyoto Encyclopedia of Genes and Genomes
K_m	Michaelis Constant (indicator of enzyme affinity)
LPMO	Lytic polysaccharide monooxygenases
NMR	Nuclear magnetic resonance
PDB	Protein Data Bank
PH	Potential of hydrogen
PL	Polysaccharide lyases
Swiss-Prot	SWISS-PROT Protein Sequence Data Bank
TrEMBL	Translation of the EMBL nucleotide sequence database
UniParc	UniProt Archive
UniProt	Universal Protein Resource
UniProtKB	UniProt Knowledgebase
UniRef	UniProt Reference Clusters
V_{max}	Maximum reaction velocity (indicator of enzyme efficiency)

DOI: 10.1201/9781003561682-9

9.1 INTRODUCTION

9.1.1 Overview of Carbohydrate-Active Enzymes (CAZymes)

Complex carbs are synthesized, modified, and broken down by a varied collection of enzymes known as carbohydrate-active enzymes (CAZymes). They play a role in a number of biological functions, including host-pathogen interactions, cell wall remodeling, and energy metabolism. These enzymes are essential for breaking down polysaccharides into simpler sugars, which organisms can use for structural and energy-producing processes (Henrissat & Davies, 1997; Lombard et al., 2014). The several functional classes of CAZymes include GHs, GTs, PLs, CEs, and AAs. Different substrate specificities and enzymatic processes are displayed by each of these classes. For example, GH breaks down glycosidic bonds in polysaccharides, whereas GTs are in charge of oligosaccharide and polysaccharide formation (Bornscheuer & Kazlauskas, 2004; Cantarel et al., 2009). Numerous creatures, including bacteria, fungi, plants, and mammals, contain these enzymes. Because of their potential for use in industrial processes including food processing, pharmaceutical development, and biofuel generation, microbial CAZymes in particular have attracted a lot of attention (Lynd et al., 2002). Numerous bacteria and fungi are important contributors to the carbon cycle and renewable energy sources because they generate CAZymes, which break down plant biomass. Developments in bioinformatics and computational biology have made it easier to identify and categorize CAZymes, which has resulted in the creation of specialized databases and research instruments. In addition to improving our knowledge of how carbohydrates are metabolized, ongoing research into CAZymes advances biotechnology, medicine, and sustainable business.

9.1.2 Importance of CAZymes in Biological and Industrial Processes

Because of their capacity to effectively digest carbohydrates, CAZymes are essential in both biological and industrial situations. They support vital metabolic processes in biological systems, which make it possible for complex carbohydrates to be broken down and synthesized. The synthesis of cellular energy, structural integrity, and intercellular communication all depend on these enzymatic processes (Cantarel et al., 2009; Henrissat & Davies, 1997). CAZymes aid in the breakdown and rebuilding of cell walls in microbial and plant systems, impacting defense, growth, and symbiosis (Bornscheuer & Kazlauskas, 2004; Lombard et al., 2014). CAZymes serve a crucial role in digestion in both human and animal physiology, helping to hydrolyze dietary carbohydrates like cellulose and starch into sugars that may be absorbed. Furthermore, by dissolving or altering glycan structures, CAZymes contribute to immune responses by regulating interactions with microbial pathogens (Lynd et al., 2002). CAZymes have numerous industrial uses in a variety of industries, including biofuels, food processing, medicines, textiles, and agriculture. In the biofuel business, lignocellulosic biomass is broken down into fermentable sugars by cellulases and hemicellulases, which makes it easier to produce bioethanol and other renewable fuels (Bornscheuer & Kazlauskas, 2004; Henrissat & Davies, 1997). Amylases and pectinases are examples of CAZymes that are used in the food industry to improve food texture, increase processing efficiency, and create useful ingredients (Cantarel et al., 2009; Lombard et al., 2014). Antibiotics and enzyme replacement treatments for metabolic diseases are among the bioactive glycan-based medications that CAZymes help to synthesize in the pharmaceutical industry (Lynd et al., 2002). Waste management and bioremediation are two important industrial applications for CAZymes. By turning waste materials into useful products, enzymes that can break down complex polysaccharides are employed to treat industrial and agricultural waste, hence boosting sustainability (Bornscheuer & Kazlauskas, 2004; Cantarel et al., 2009). Furthermore, CAZymes are used in the paper and textile sectors to improve processing methods and alter fibers without the use of harsh chemicals.

9.1.3 Applications of CAZymes: Biofuels, Food, and Pharmaceuticals

CAZymes are widely used in a variety of industries, including food, pharmaceuticals, and biofuels. Because of their capacity to hydrolyze, alter, and synthesize carbohydrates, they are vital instruments for enhancing product quality, sustainability, and industrial efficiency. CAZymes convert plant biomass into fermentable sugars, which is essential for the synthesis of biofuels. These enzymes are a crucial part of renewable energy systems because they help turn lignocellulosic materials, like forestry residues and agricultural waste, into bioethanol and other biofuels (Himmel et al., 2007; Lynd et al., 2002). Cellulases, hemicellulases, and lignin-modifying enzymes are important CAZyme types in the synthesis of biofuel. Hemicellulases break down hemicellulose into xylose and arabinose, whereas cellulases hydrolyze cellulose into glucose monomers. By dissolving lignin structures, lignin-modifying enzymes such as laccases and peroxidases increase biomass accessibility and boost fermentation and saccharification efficiency (Kuhad et al., 2011; Y. H. P. Zhang et al., 2007). Recent developments in synthetic biology and enzyme engineering have produced extremely effective CAZymes with enhanced stability and catalytic activity. By improving substrate selectivity and lowering the need for enzyme loading, these modified enzymes make it possible to produce biofuel at a reasonable cost (Lynd et al., 2002; Magrey et al., 2018). Consolidated bioprocessing (CBP), a method that combines fermentation, biomass hydrolysis, and enzyme synthesis into a single phase, also makes use of CAZymes. This strategy improves the economic feasibility of biofuel technologies and drastically reduces production costs (Himmel et al., 2007; Y. H. P. Zhang et al., 2007). Researchers continue to increase the sustainability and scalability of biofuel production by refining CAZyme formulations and utilizing biotechnology advancements, which aids in the global shift to renewable energy sources (Kuhad et al., 2011; Magrey et al., 2018). The food industry makes extensive use of CAZymes to improve food processing, texture, and nutritional content. These enzymes are essential for altering polysaccharides, dissolving complex carbohydrates, and enhancing food products' digestibility (de Souza & de Oliveira Magalhães, 2010; Rastall & Gibson, 2015). Amylases, cellulases, pectinases, and xylanases are important CAZymes used in the food sector. Amylases are vital for use in baking, brewing, and confections because they hydrolyze starch into simple sugars. Pectinases improve the extraction of fruit-based goods and help clarify fruit juice. The fiber content and texture of baked foods and nutritional supplements are enhanced by cellulases and xylanases (Gänzle, 2014; Jayani et al., 2005). The creation of thermostable and pH-stable CAZymes, which increase processing effectiveness and product stability, is the result of recent developments in enzyme engineering. Low-calorie sweeteners, gluten-free goods, and functional meals with improved health benefits can all be produced thanks to these designed enzymes (de Souza & de Oliveira Magalhães, 2010). In order to enhance lactose hydrolysis and increase the digestibility of dairy products for those who are lactose intolerant, CAZymes are also used in dairy processing. They also contribute to the tenderization of meat and the creation of plant-based protein substitutes, meeting the rising demand for sustainable food sources (Jayani et al., 2005; Rastall & Gibson, 2015). Researchers are constantly coming up with new ways to improve food quality, increase shelf life, and satisfy consumer demands for healthier and more environmentally friendly food products by maximizing CAZyme uses in food manufacturing (Gänzle, 2014). CAZymes have significant applications in the pharmaceutical industry, where they are utilized for drug development, therapeutic enzyme production, and vaccine formulation. These enzymes contribute to the synthesis of bioactive glycoconjugates, the degradation of pathogenic biofilms, and the modification of glycoproteins for targeted therapies (Astronomo & Burton, 2010; Varki, 1993). Particularly crucial to drug manufacturing are GH and GTs, which make it possible to produce antiviral and anticancer medications. Glycosylated antibiotics have better stability and effectiveness, while oligosaccharides produced by enzymes are essential ingredients in immunomodulatory medications (Khatiwada & Pokharel, 2009; Moremen et al., 2012). CAZymes aid in the alteration of polysaccharide antigens during vaccine production, improving immunological response and boosting vaccine efficacy. Pneumococcal and meningococcal illnesses are among the bacterial pathogens for which conjugate vaccines are made using these enzymes (Astronomo & Burton, 2010; Khatiwada & Pokharel, 2009). Furthermore, CAZymes are involved in enzyme replacement treatments for lysosomal storage illnesses such as Fabry and Gaucher diseases. Patients with severe genetic disorders can get life-saving therapy with recombinant enzymes, which are designed for increased stability and activity (Khatiwada & Pokharel, 2009; Varki, 1993). CAZymes continue to spur innovation in the creation of new therapies by utilizing developments in enzyme engineering and biopharmaceutical research, enhancing drug efficacy and broadening the range of disorders that can be treated (Jennings & Lugowski, 1981; Moremen et al., 2012).

9.2 STRUCTURE OF CARBOHYDRATE-ACTIVE ENZYMES

9.2.1 General Structural Features

The catalytic processes and substrate selectivity of CAZymes are significantly influenced by their structural features. Catalytic domains and carbohydrate-binding modules (CBMs) make up the modular structures of these enzymes, which help with substrate recognition and enzymatic activity (Henrissat & Davies, 1997; Lombard et al., 2014). In order to stabilize substrate contacts and promote hydrolysis, GH frequently have catalytic domains with α/β-barrel folds (Bornscheuer & Kazlauskas, 2004; Cantarel et al., 2009). The transfer of sugar moieties to target molecules is facilitated by the frequent adoption of GT-A or GT-B folds by glycosyltransferases (Lynd et al., 2002). Large polysaccharide chains can be accommodated by the β-helix architectures of polysaccharide lyases, which facilitate the effective cleavage of glycosidic linkages (Cantarel et al., 2009; Lombard et al., 2014). CAZymes are frequently linked to carbohydrate-binding modules (CBMs), which increase their catalytic effectiveness and substrate affinity. Enzymes can bind selectively to polysaccharide substrates thanks to their non-catalytic regions, which raise local substrate concentrations and speed up reactions (Bornscheuer & Kazlauskas, 2004; Lynd et al., 2002). CAZymes' structural diversity adds to their functional adaptability, making them essential instruments for industrial processes like food processing, pharmaceutical research, and biofuel manufacturing (Cantarel et al., 2009; Henrissat & Davies, 1997). The logical design of modified enzymes with improved stability, specificity, and catalytic efficiency is facilitated by an understanding of these structural characteristics.

9.2.2 Active Sites and Substrate Specificity

Usually made up of conserved amino acid residues involved in substrate binding, catalysis, and transition state stabilization, the active sites of CAZymes are responsible for determining their catalytic activity and selectivity for various carbohydrate substrates (Cantarel et al., 2009; Henrissat & Davies, 1997). The stereochemistry of the hydrolysis reaction is determined by the two main catalytic mechanisms that GH displays: retaining and inverting mechanisms. Aspartate and glutamate, two active site residues, are essential for nucleophilic assault and proton donation (Bornscheuer & Kazlauskas, 2004; Lombard et al., 2014). Depending on the shape of their active sites, GTs catalyze the creation of glycosidic bonds by either maintaining or inverting mechanisms. Precise glycan synthesis is ensured by interactions between the enzyme and the sugar moieties, which dictate the donor and acceptor substrate specificities (Lynd et al., 2002). The β-elimination process used by PLs allows uronic acid-containing polysaccharides to be cleaved by active site residues, which frequently include arginine and lysine (Bornscheuer & Kazlauskas, 2004; Cantarel et al., 2009). Many carbohydrate-active enzymes (CAZymes) contain non-catalytic domains called carbohydrate-binding modules (CBMs). Their primary function is to bind selectively to certain polysaccharides, which facilitates the enzyme's better localization and substrate adhesion. CBMs improve the efficiency and selectivity of the enzyme by establishing a close and stable contact between the catalytic domain and the polysaccharide. CBMs achieve this through non-covalent interactions, primarily CH–π stacking between aromatic residues (e.g., tryptophan, tyrosine) and sugar rings, as well as hydrogen bonding between polar residues and hydroxyl groups of the carbohydrate. Based on their binding site topology, CBMs are classified as Type A (binding to flat crystalline surfaces like cellulose), Type B (interacting with internal glycan chains in clefts), or Type C (recognizing chain termini or small sugars via pocket-shaped sites). These interactions are enthalpically favorable, By raising the concentration of substrate at the enzyme surface and orienting the enzyme in the best possible catalytic orientation, this enhances reaction speeds and influences substrate specificity (Boraston et al., 2004). Enzyme engineering requires an understanding of the active site architecture and substrate selectivity of CAZymes in order to create optimal biocatalysts for commercial uses such as pharmaceutical synthesis, food processing, and biofuel generation (Bornscheuer & Kazlauskas, 2004; Henrissat & Davies, 1997).

9.2.3 Structural Adaptations for Function

CAZyme structural modifications are essential for maximizing enzyme activity in a range of physiological and commercial settings. By improving substrate binding, catalytic efficiency, and enzyme stability, these modifications enable CAZymes to function in a variety of settings (Henrissat & Davies, 1997; Lombard et al., 2014). The existence of carbohydrate-binding modules (CBMs), which tie the enzyme to the surfaces of polysaccharides and raise substrate affinity, is one important adaptation. By keeping the enzyme close to its substrate and lowering diffusion restrictions, CBMs promote efficient catalysis (Bornscheuer & Kazlauskas, 2004; Cantarel et al., 2009). Furthermore, under particular circumstances, CAZymes exhibit changes in their catalytic domains that increase activity. Extremophilic thermophilic CAZymes, for instance, have more disulfide bonds, salt bridges, and hydrophobic interactions, all of which support the stability of the enzyme at high temperatures (Lombard et al., 2014; Lynd et al., 2002). pH tolerance, which allows CAZymes to continue to function in both acidic and alkaline environments, is another important adaptation. These enzymes can work well in industrial processes like biomass conversion and food production because of structural changes including flexible loop regions and changed surface charge distributions (Cantarel et al., 2009; Lynd et al., 2002). One of the primary mechanisms involves modifications in the active site, where key catalytic residues such as glutamate or aspartate have their pKa values modulated through the local microenvironment, allowing them to maintain appropriate protonation states (Turner et al., 2007). These residues often reside within conserved motifs that are protected by flexible loop regions, which dynamically adjust to pH-induced conformational changes without compromising substrate binding or catalysis (Saratale et al., 2020). Also, altered surface charge distributions on CAZymes help repel excess H^+ or OH^- ions, reducing the risk of denaturation or loss of structural integrity (Zhou et al., 2018). This is achieved through substitution of surface amino acids with more stable residues, reducing the number of pH-sensitive side chains exposed to the solvent. Furthermore, increased formation of salt bridges, hydrogen bonds, and disulfide bonds helps maintain the overall tertiary structure under extreme pH conditions (Zhang et al., 2010). Some CAZymes also exhibit hydrophobic core packing and domain separation strategies, which shield the catalytic core from environmental pH fluctuations and prevent unfolding or aggregation. These structural modifications are often naturally evolved in extremophilic microorganisms or are engineered in industrial strains for enhanced enzyme stability and function. Additionally, some CAZymes have structural flexibility, which enables them to adapt to a variety of substrates via induced-fit processes. In GTs, where precise sugar transfer processes are made possible by dynamic conformational changes, this adaptability is especially crucial (Bornscheuer & Kazlauskas, 2004; Lynd et al., 2002). Since changes might improve CAZyme performance in biotechnological applications such as biofuel generation, food processing, and pharmaceutical development, an understanding of these structural adaptations is crucial for enzyme engineering (Cantarel et al., 2009; Henrissat & Davies, 1997).

9.3 CLASSES OF CARBOHYDRATE-ACTIVE ENZYMES

The functional classification of carbohydrate-active enzymes is shown in Figure 9.1.

9.3.1 Glycoside Hydrolases

The hydrolysis of glycosidic bonds in carbohydrates is carried out by a broad class of enzymes known as glycoside hydrolases. In many different biological systems, these enzymes are essential for the breakdown of biomass, energy metabolism, and glycan processing (Henrissat & Davies, 1997; Lombard et al., 2014). In the CAZy database, GHs are categorized into many families according to sequence similarities. Every family has structural folds and conserved catalytic processes, including the α/β-barrel fold that is frequently present in several GHs (Bornscheuer & Kazlauskas, 2004; Cantarel et al., 2009). The stereochemistry of the hydrolysis product is determined by the two main paths that the hydrolytic mechanism of GHs follows: the retaining mechanism and the inverting mechanism (Lombard et al., 2014; Lynd et al., 2002). Important catalytic residues found in GHs' active sites, such as glutamate and aspartate, play a role in proton donation and nucleophilic attack during the breakdown of glycosidic bonds (Bornscheuer & Kazlauskas, 2004; Henrissat & Davies, 1997). The shape of the enzyme's binding pocket, which holds particular sugar residues and guarantees accurate hydrolysis, determines the substrate specificity in GHs (Cantarel

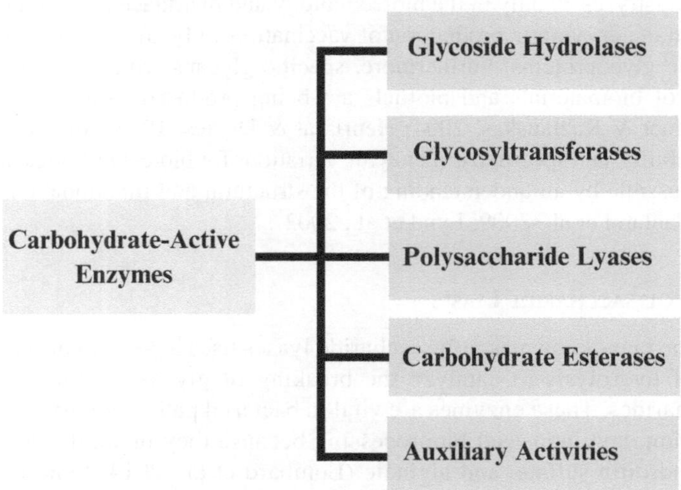

FIGURE 9.1 Functional classification of carbohydrate-active enzymes: glycoside hydrolases, glycosyltransferases, polysaccharide lyases, carbohydrate esterases, and auxiliary activities.

et al., 2009; Lynd et al., 2002). GHs have industrial uses in the food processing, medicinal, and biofuel manufacturing sectors. Amylases are commonly used in the processing of starch, whilst cellulases and hemicellulases from the GH families are utilized in the breakdown of lignocellulosic biomass for the manufacture of bioethanol (Bornscheuer & Kazlauskas, 2004; Lombard et al., 2014). GHs aid in the enzymatic alteration of glycolipids and glycoproteins for medicinal purposes in the pharmaceutical industry (Henrissat & Davies, 1997; Lynd et al., 2002). Enzyme engineering initiatives to improve catalytic efficiency, stability, and substrate specificity for industrial applications depend on an understanding of the structure and function of GHs (Bornscheuer & Kazlauskas, 2004; Cantarel et al., 2009).

9.3.2 Glycosyltransferases

A class of enzymes known as glycosyltransferases catalyzes the formation of glycosidic linkages by moving sugar moieties from active donor molecules to acceptor substrates. These enzymes are crucial for the manufacture of polysaccharides, oligosaccharides, glycoproteins, and glycolipids, which support a variety of physiological functions including pathogen recognition, immunological response, and cell signaling (Bornscheuer & Kazlauskas, 2004; Cantarel et al., 2009). In the CAZy database, GTs are categorized into various families according to their catalytic processes and sequence similarities. Their manner of sugar transfer is defined by the GT-A and GT-B folds, which are their two primary structural features. Whereas GT-B enzymes have two Rossmann-like domains, GT-A enzymes usually only use one (Lynd et al., 2002). The spatial arrangement of important active-site residues determines whether the catalytic mechanism of GTs is retaining or inverting. The ability of GTs to identify donor and acceptor substrates determines their specificity, which guarantees accurate glycan synthesis (Lombard et al., 2014; Lynd et al., 2002). GTs are widely used in industry, especially in the biotechnology and pharmaceutical industries. They are used in the enzymatic production of vaccinations, glycosylated medications, and therapeutic glycoproteins. Furthermore, specific glycans with potential uses in the synthesis of biomaterials and biofuels are being produced using customized GTs (Bornscheuer & Kazlauskas, 2004; Henrissat & Davies, 1997). Enhancing the efficiency, stability, and specificity of enzyme variations for biotechnological applications is made possible by an understanding of the structural and functional characteristics of GTs (Cantarel et al., 2009; Lynd et al., 2002).

9.3.3 Polysaccharide Lyases

An enzyme class known as polysaccharide lyases uses a β-elimination mechanism instead of hydrolysis to catalyze the breaking of glycosidic linkages in acidic polysaccharides. These enzymes are vital in bacterial pathogenicity, plant cell wall modification, and industrial bioprocessing because they break down pectin, heparin, chondroitin sulfate, and alginate (Lombard et al., 2014; Lynd et al., 2002). In the CAZy database, PLs are categorized into various families according to their catalytic processes and sequence similarities. Conserved lysine, arginine, and histidine residues found in their active sites aid in proton abstraction and β-elimination, which results in the production of unsaturated sugar compounds (Cantarel et al., 2009; Henrissat & Davies, 1997). In terms of structure, PLs frequently take on right-handed β-helix folds, which enable them to hold big polysaccharide chains and promote effective cleavage. The active site design, which identifies uronic acid-containing polysaccharides like pectin and alginate, determines their substrate specificity (Bornscheuer & Kazlauskas, 2004; Lynd et al., 2002). Applications for PLs are numerous in the food, medicinal, and biofuel sectors. Alginate lyases may find value in wound healing and medication delivery, whereas pectinases, a subclass of PLs, are extensively employed in fruit juice clarity. By converting plant-derived polysaccharides into fermentable sugars, PLs aid in the breakdown of biomass during the creation of biofuel (Bornscheuer & Kazlauskas, 2004; Henrissat & Davies, 1997). Enzyme engineering initiatives to maximize PLs' catalytic performance and broaden their commercial applications depend on an understanding of their structural and functional characteristics (Cantarel et al., 2009; Lynd et al., 2002).

9.3.4 Carbohydrate Esterases

A class of enzymes known as CEs hydrolyzes the ester linkages found in carbohydrates, changing the structures of polysaccharides and aiding in their breakdown. In order to convert biomass, modify cell walls, and increase pathogenic virulence, these enzymes are essential for the deacetylation of plant and microbial polysaccharides (Bornscheuer & Kazlauskas, 2004; Cantarel et al., 2009). The CAZy database groups CEs into families according to their catalytic mechanism and sequence similarity. Through a traditional serine hydrolase mechanism, their active sites frequently contain conserved serine, histidine, and aspartate residues that aid in the hydrolysis of ester bonds (Henrissat & Davies, 1997; Lynd et al., 2002). The catalytic domains of CEs vary structurally, and many of them adopt an α/β-hydrolase fold that offers stability and adaptability for interactions with substrates. These enzymes affect the physicochemical characteristics of polysaccharides by targeting a variety of substrates, such as pectins, acetylated xylans, and chitin derivatives (Cantarel et al., 2009; Lombard et al., 2014). There are numerous uses for carbohydrate esterases in both industry and biotechnology. They are used in food processing to alter the characteristics of dietary fiber, in pharmaceuticals for drug formulation and prodrug activation, and in biofuel production to increase the digestibility of lignocellulosic biomass. Furthermore, the potential of CEs in sustainable chemical synthesis and bioremediation is being investigated (Bornscheuer & Kazlauskas, 2004; Lynd et al., 2002). To maximize CEs' catalytic efficiency and increase its industrial biotechnology applications, it is crucial to comprehend their structural and functional characteristics (Cantarel et al., 2009; Henrissat & Davies, 1997).

9.3.5 Auxiliary Activities

A class of redox enzymes known as "auxiliary activities" helps break down complex carbohydrates by producing reactive oxygen species or altering the structure of lignocellulose to make it easier for other CAZymes to break it down. These enzymes include flavin-dependent oxidases, lignin-modifying peroxidases, and lytic polysaccharide monooxygenases (LPMOs) (Henrissat & Davies, 1997; Lynd et al., 2002). By causing oxidative cleavage in polysaccharide chains, LPMOs contribute significantly to the breakdown of lignocellulose and increase the effectiveness of GH. By using metal ions like copper to propel oxidation processes, these enzymes make plant material more accessible for enzymatic digestion (Bornscheuer & Kazlauskas, 2004; Lombard et al., 2014). One of the main structural elements of plant cell walls, lignin, is broken down by lignin-modifying peroxidases and oxidases. These enzymes increase cellulose and hemicellulose's accessibility for enzymatic hydrolysis by altering lignin structures, which makes them useful for biofuel generation and bioremediation (Cantarel et al., 2009; Lynd et al., 2002). Paper bleaching, textile processing, and biofuel generation are just a few of the industrial uses for AAs. They are crucial parts of enzyme cocktails used for effective biomass conversion because of their capacity to accelerate the breakdown of lignocellulose (Bornscheuer & Kazlauskas, 2004; Henrissat & Davies, 1997). Optimizing AAs' activity and incorporating them into industrial processes for sustainable biotechnological applications requires an understanding of their structural and functional characteristics (Lombard et al., 2014; Lynd et al., 2002).

9.4 DATABASES FOR CARBOHYDRATE-ACTIVE ENZYMES

9.4.1 CAZy Database: Overview and Classification

A thorough resource, the Carbohydrate-Active Enzymes (CAZy) database categorizes and lists CAZymes according to their enzymatic function, structural characteristics, and sequence similarity. For researchers looking into the evolution of enzymes and the metabolism of carbohydrates, it is an essential resource (Henrissat & Davies, 1997; Lombard et al., 2014). GH, GTs, PLs, CEs, AAs, and carbohydrate-binding modules (CBMs) are the six primary categories into which the CAZy database divides CAZymes. Based on catalytic processes and sequence homology, each category is further subdivided into families (Bornscheuer & Kazlauskas, 2004; Cantarel et al., 2009). Through bioinformatics techniques, the CAZy classification system is updated on a regular basis to include newly found sequences from genome metagenome research. Additionally, the database offers structural and functional annotations to facilitate biotechnological applications and enzyme engineering (Lombard et al., 2014; Lynd et al., 2002). The CAZy database is frequently used by researchers to investigate the diversity of CAZymes, find new enzymes for use in industry, and examine the processes by which carbohydrates are processed in various organisms. A deeper understanding of enzyme processes and substrate selectivity is made possible by the integration of CAZy with structural databases like the Protein Data Bank (PDB) (Bornscheuer & Kazlauskas, 2004; Henrissat & Davies, 1997). CAZy contributes significantly to the advancement of glycomics research and the growth of bio-based enterprises by providing a carefully curated and ordered repository of carbohydrate-active enzymes (Figure 9.2) (Cantarel et al., 2009; Lynd et al., 2002).

Databases for Carbohydrate-Active Enzymes (CAZymes)	
CAZy Database	The CAZy database systematically classifies and annotates carbohydrate-active enzymes to support research in glycomics, enzyme evolution, and biotechnological applications.
UniProt	UniProt provides comprehensive functional and sequence annotations for proteins, integrating curated and computational data to support enzyme function, structure, and evolutionary studies.
PDB	The Protein Data Bank (PDB) provides atomic-level 3D structural data on biomolecules, supporting studies on enzyme-substrate interactions, molecular docking, and enzyme engineering for biotechnology and medicine.
BRENDA	BRENDA provides enzymic functional and kinetic data for research in characterization, drug development, and biotechnology.
KEGG	KEGG maps metabolic pathways, linking CAZymes to enzyme function and biotechnology.

FIGURE 9.2 Five databases are: CAZy Database, UniProt, PDB, BRENDA, and KEGG for carbohydrate-active enzymes (CAZymes).

9.4.2 UniProt: Functional and Sequence Annotations

A popular database that offers comprehensive functional and sequence annotations for proteins, including CAZymes, is the Universal Protein Resource (UniProt). It is a useful tool for researching enzyme function, structure, and evolution because it incorporates carefully chosen data from several sources, such as computational predictions and experimental data (Bateman et al., 2023). UniProtKB (Knowledgebase), UniRef (Reference Clusters), and UniParc (Archive) are the three primary parts of UniProt. Swiss-Prot (manually curated entries) and TrEMBL (automatically annotated sequences) are two further divisions of UniProtKB. TrEMBL offers high-throughput sequence data that is pending manual curation, whereas Swiss-Prot entries have comprehensive functional descriptions, sequence characteristics, and experimental annotations (Lombard et al., 2014). Enzymatic activity, substrate selectivity, catalytic processes, and post-translational modifications are all covered by functional annotation in UniProt. Furthermore, UniProt facilitates integrated analyses of CAZymes in metabolic pathways and structural investigations by offering cross-references to various bioinformatics databases, including CAZy, KEGG, and PDB (Bateman et al., 2023; Henrissat & Davies, 1997). UniProt is used by researchers for evolutionary studies of CAZymes, homology modeling, and sequence-based comparisons. UniProt improves our knowledge of enzyme activity and its function in industrial applications and carbohydrate metabolism by fusing computational predictions with experimental data (Bateman et al., 2023).

9.4.3 PDB (Protein Data Bank): Structural Data

The Protein Data Bank (PDB), which offers atomic-level information on biomolecular structures, including CAZymes, is a crucial tool for structural biology. It is a publicly available resource that houses 3D structural data gathered using methods like cryo-electron microscopy (cryo-EM), nuclear magnetic resonance (NMR) spectroscopy, and X-ray crystallography (Burley et al., 2017; Jumper et al., 2021). Understanding CAZyme processes and substrate specificity requires knowledge of enzyme-substrate interactions, structural coordinates, and functional annotations found in PDB entries. PDB data is used by researchers for protein-ligand interaction investigations, molecular docking, and logical enzyme design (Andreeva et al., 2020; Lombard et al., 2014). Clarifying active site topologies and conformational changes upon substrate binding is one of the main uses of PDB in CAZyme research. Enzyme engineering attempts to increase catalytic stability, efficiency, and industrial application are supported by these insights (Cantarel et al., 2009; Holm, 2020). Comprehensive structural investigations of CAZymes are made possible by the regular integration of PDB with other bioinformatics tools, such as PyMOL for molecular visualization and Swiss-Model for homology modeling. Our comprehension of enzyme function and adaptability is improved by the capacity to compare homologous structures and examine evolutionary links (Andreeva et al., 2020; Berman et al., 2000). PDB is essential to the advancement of CAZyme research because it offers comprehensive structural information, which helps with discoveries in biotechnology, medicine, and renewable energy applications (Henrissat & Davies, 1997; Jumper et al., 2021).

9.4.4 BRENDA: Enzyme Functional and Kinetic Data

One of the most extensive databases of enzymes is BRENDA, which offers thorough functional and kinetic information on CAZymes and other enzymes. It gathers information on enzyme characteristics from source literature and experimental research, including reaction kinetics, substrate specificity, inhibitor profiles, and ideal reaction conditions (Jeske et al., 2019; Schomburg et al., 2013). The Enzyme Commission (EC) system, which links enzymes to their structural characteristics, cofactors, and metabolic pathways, is the basis for the information on enzyme classification included in BRENDA. In order to provide a comprehensive understanding of enzyme function and application, it also incorporates cross-references to other bioinformatics databases, including KEGG, UniProt, and PDB. BRENDA's large dataset on enzyme kinetics, which includes turnover numbers (kcat), inhibition constants (Ki), and Michaelis-Menten parameters (Km, Vmax), is one of its main advantages. Applications such as metabolic modeling, medication development, and enzyme engineering require this knowledge (Schomburg et al., 2013). BRENDA is used by researchers to characterize enzymes, compare homologous enzymes, and optimize enzymatic reactions for use in medicinal and industrial settings. Reliability and usefulness in biochemical research are guaranteed by the database's vast collection of manually selected experimental data (Jeske et al., 2019). BRENDA is a vital tool for the research and use of CAZymes in biotechnology and medicine since it provides thorough insights into enzyme kinetics and functional characteristics (Schomburg et al., 2013).

9.4.5 KEGG: Pathway Mapping and CAZyme Roles in Metabolism

A comprehensive database called the Kyoto Encyclopedia of Genes and Genomes (KEGG) maps metabolic pathways, including those involving CAZymes, by combining genomic, biochemical, and functional data. Researchers can examine how CAZymes contribute to carbohydrate metabolism and associated biological processes by using KEGG, which makes it easier to comprehend the roles of enzymes in metabolic networks (Kanehisa et al., 2014, 2021). Curated maps of metabolic pathways, including polysaccharide breakdown, the pentose phosphate pathway, and glycolysis, are available through KEGG PATHWAY. These pathways demonstrate the role of CAZyme in the biosynthesis and degradation of complex carbohydrates, highlighting their importance in microbial metabolism, energy production, and host-pathogen interactions (Kanehisa, 2019; Kanehisa & Goto, 2000). CAZymes are further divided into functional units by the KEGG MODULE according to their functions in particular carbohydrate-processing systems. Enzyme synergies, such as the collaboration between GH and CEs in the breakdown of plant biomass, can be found due to this modular structure (Kanehisa et al., 2014; Kanehisa & Sato, 2020). The research of CAZyme specificity and catalytic processes is improved by the comprehensive annotations of carbohydrate structures and enzyme-substrate interactions provided by KEGG COMPOUND and KEGG GLYCAN. KEGG facilitates logical enzyme engineering efforts to optimize biocatalytic processes for industrial applications by combining route knowledge with enzyme kinetics data (Kanehisa, 2019; Kanehisa et al., 2021). KEGG is frequently used by researchers in synthetic biology, metabolic modeling, and biotechnological applications. KEGG facilitates a systems-level understanding of carbohydrate metabolism by connecting CAZymes to larger metabolic networks. This information also guides the development of new enzymatic pathways for the manufacturing of biofuel, medications, and food processing (Kanehisa & Goto, 2000; Kanehisa & Sato, 2020).

9.5 BIOINFORMATICS TOOLS FOR CAZYME ANALYSIS

Figure 9.3 shows the bioinformatics tools for CAZymes analysis.

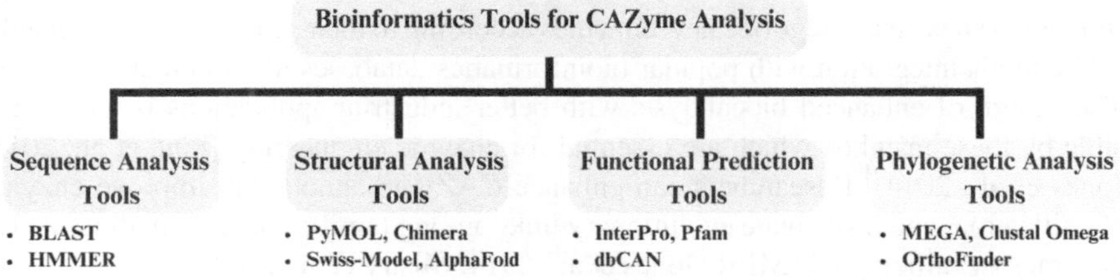

FIGURE 9.3 Bioinformatics tools for CAZymes analysis are divided into four categories. For sequence analysis there is BLAST and HMMER, structural analysis tools include PyMOL, Chimera, Swiss-Model, AlphaFold, tools for functional prediction are InterPro, Pfam, dbCAN, and for phylogenetic analysis there is MEGA, Clustal Omega, OrthoFinder.

9.5.1 Sequence Analysis Tools

9.5.1.1 BLAST and FASTA for Homology Search

Two popular bioinformatics tools for sequence alignment and homology search are BLAST (Basic Local Alignment Search Tool) and FASTA. By comparing query sequences to extensive protein and nucleotide databases, these methods make it easier to find homologous sequences, which supports functional annotation and evolutionary research on CAZymes (Altschul et al., 1997; Pearson & Lipman, 1988). BLAST is a favored tool for large-scale sequence analysis because it quickly finds local sequence alignments using a heuristic technique. It has several variations, including BLASTx (converting nucleotide sequences into proteins), BLASTn (comparing nucleotides), and BLASTp (comparing proteins) (Camacho et al., 2009; Mount, 2007). Another alignment tool called FASTA uses efficient scoring matrices and a hash-based search approach to find sequence similarities. FASTA is helpful for more sensitively identifying distant homologs and sequence motifs, even though it is typically slower than BLAST (Lipman & Pearson, 1985; Pearson, 1990). In order to identify conserved functional domains, evolutionary links, and possible biotechnological uses, BLAST and FASTA are essential tools for studying CAZymes. To make comparative genomics and enzyme characterisation easier, these techniques are commonly connected with databases like UniProt, PDB, and KEGG (Pearson, 1990; Pearson & Lipman, 1988). Researchers can use BLAST and FASTA to guide protein engineering efforts, anticipate the function of novel CAZymes, and investigate carbohydrate metabolism in a variety of organisms (Camacho et al., 2009; Lipman & Pearson, 1985).

9.5.1.2 HMMER for Domain and Family Detection

A potent bioinformatics tool for identifying protein domains and categorizing protein families according to sequence homology is HMMER. It is especially useful for examining CAZyme families and their evolutionary links because it uses profile hidden Markov models (HMMs) to find conserved motifs (Eddy, 1998; Finn et al., 2011). Because it has a higher sensitivity than conventional sequence alignment tools like BLAST, HMMER is frequently utilized in the structural categorization and functional annotation of CAZymes. It is particularly helpful for finding conserved domains that define CAZyme activity and for recognizing distant homologs (El-Gebali et al., 2019; Mistry et al., 2021). Researchers may investigate CAZyme domain structures, forecast functional sites, and categorize new enzymes according to their sequence profiles thanks to the tool's integration with popular bioinformatics databases like Pfam and InterPro. The design of enhanced biocatalysts with better industrial applications is made possible by these insights, which are essential for enzyme engineering (Finn et al., 2011; Jones et al., 2014). Researchers can enhance CAZyme annotation, improve enzyme classification, and investigate evolutionary links among families of carbohydrate-active enzymes by utilizing HMMER (Jones et al., 2014; Mistry et al., 2021).

9.5.2 Structural Analysis Tools

9.5.2.1 PyMOL and Chimera for Visualization

Researchers can examine and work with the three-dimensional structures of CAZymes using popular molecular visualization programs like PyMOL and Chimera. Insights into enzyme function and catalytic mechanisms are provided by these methods, which make structural modeling, molecular docking investigations, and protein-ligand interaction analysis easier (Delano, 2002; Pettersen et al., 2004). PyMOL is a robust, easy-to-use program that facilitates scripting for sophisticated visualization and excellent molecular graphics. It enables users to estimate the separations between active site residues, investigate protein-ligand interactions, and highlight structural elements that are important for enzyme stability and activity (Seeliger & De Groot, 2010). The University of California, San Francisco (UCSF) created Chimera, which has sophisticated features like surface representations, electrostatic potential mapping, and comparison modeling. It is especially helpful for integrating structural data from many sources and displaying massive macromolecular complexes (Meng et al., 2006; Pettersen et al., 2004). For CAZyme research, PyMOL and Chimera are both necessary tools that support protein engineering, structural comparisons, and inhibitor design docking experiments. Their capacity to produce animations and graphics of publishing quality improves the comprehension of structural data and facilitates logical enzyme design approaches (Meng et al., 2006). Researchers can improve enzyme applications in biotechnology and medicine by employing these visualization methods to better understand CAZyme structures, substrate interactions, and functional adaptations (Delano, 2002; Seeliger & De Groot, 2010).

9.5.2.2 Swiss-Model and AlphaFold for Modeling

Swiss-Model and AlphaFold are sophisticated tools for protein structure prediction and homology modeling. By using these computational methods, scientists can forecast the structures of CAZymes, which makes enzyme engineering and functional analysis easier (Jumper et al., 2021; Waterhouse et al., 2018). A popular homology modeling server called Swiss-Model creates three-dimensional protein structures by aligning sequences with pre-existing templates. It offers top-notch models that facilitate comprehension of substrate interactions, active site layouts, and CAZyme activity (Bertoni et al., 2017). DeepMind's AlphaFold uses deep learning to make remarkably accurate protein structure predictions. By offering extremely dependable models even in the absence of homologous templates, it transforms structural biology and is especially helpful for CAZymes with little structural information (Jumper et al., 2021; Senior et al., 2020). By facilitating structural comparisons, directing site-directed mutagenesis, and assisting with logical enzyme design for commercial applications, Swiss-Model and AlphaFold both enhance CAZyme research. To improve structural and functional annotations, these tools easily interface with other bioinformatics databases like PDB and UniProt (Senior et al., 2020; Waterhouse et al., 2018). Researchers can improve enzyme stability, investigate new biotechnological applications, and obtain a deeper understanding of CAZyme mechanisms by utilizing homology modeling (Bertoni et al., 2017).

9.5.3 Functional Prediction Tools

9.5.3.1 InterPro and Pfam for Domain Annotation

Two popular bioinformatics tools for functional categorization and protein domain annotation are InterPro and Pfam. By using these approaches, scientists can find conserved domains in CAZymes, which makes it easier to classify enzymes and conduct evolutionary study (El-Gebali et al., 2019; Finn et al., 2016). InterPro offers a thorough understanding of protein domain topologies by integrating several protein signature databases, such as Pfam, SMART, and CATH-Gene3D. InterPro supports the annotation and structural analysis of CAZymes by reliably predicting their functional domains through the use of hidden Markov models (HMMs) and pattern recognition techniques (Hunter et al., 2009; Mitchell et al., 2019). Proteins are categorized using conserved sequence motifs and domain designs by the protein family database Pfam. In order to identify new enzymes, it uses HMM-based searches to find evolutionary links between CAZymes and other proteins that are active on carbohydrates (El-Gebali et al., 2019). Protein function, structure, and phylogeny may be easily cross-referenced thanks to InterPro and Pfam's integration with popular bioinformatics systems like UniProt and Ensembl. By offering important insights into enzyme processes, active site layouts, and substrate-binding characteristics, these databases improve CAZyme research (Finn et al., 2016). Researchers can enhance CAZyme functional annotation, improve enzyme categorization, and investigate novel enzymatic activities for industrial and biotechnological applications by utilizing InterPro and Pfam (Hunter et al., 2009; Mitchell et al., 2019).

9.5.3.2 dbCAN for Automated CAZyme Annotation

One popular computational resource for the automated annotation of carbohydrate-active enzymes (CAZymes) is the dbCAN database. To find CAZyme groups based on domain architecture, functional annotations, and sequence homology, it combines several bioinformatics techniques (Yin et al., 2012; H. Zhang et al., 2018). The dbCAN pipeline includes hidden Markov model (HMM)-based searches against the CAZy database, sequence-based annotations using DIAMOND, and conserved domain discovery through Pfam and TIGRFAMs. Researchers can forecast enzyme function and evolutionary links thanks to these complementary approaches, which guarantee accurate CAZyme classification (Drula et al., 2022). Rapid identification of carbohydrate-active enzyme repertoires in a variety of biological materials is made possible by the widespread usage of dbCAN in metagenomics and microbial genome research. Exploring novel CAZymes with potential uses in biotechnology, biofuel production, and human microbiome research requires this skill (Lombard et al., 2014; H. Zhang et al., 2018). dbCAN speeds up functional genomics research, facilitates comparative enzyme analysis, and improves our comprehension of the metabolism of carbohydrates in many organisms by automating CAZyme annotation (Yin et al., 2012).

9.5.4 Phylogenetic Analysis Tools

9.5.4.1 MEGA and Clustal Omega for Phylogenetic Studies

Two popular bioinformatics tools for phylogenetic research that shed light on the evolutionary links among CAZymes are MEGA (Molecular Evolutionary Genetics Analysis) and Clustal Omega. These techniques facilitate comparative genomics and enzyme categorization by facilitating evolutionary distance estimation, phylogenetic tree construction, and sequence alignment (Kumar et al., 2018; Sievers et al., 2011). Maximum likelihood, neighbor-joining, and Bayesian techniques are among the many phylogenetic inference procedures that MEGA provides. Researchers can examine the divergence and functional evolution of the CAZyme family thanks to its statistical models for evolutionary analysis (Hall, 2013; Tamura et al., 2021). Clustal Omega is an effective multiple sequence alignment program that produces high-quality alignments by utilizing hidden Markov models (HMMs) and progressive alignment algorithms. Large CAZyme datasets may be aligned with its help, which makes precise phylogenetic comparisons and domain conservation research possible (Hall, 2013; Sievers et al., 2011). Clustal Omega and MEGA are both included into larger bioinformatics workflows, facilitating evolutionary research, functional annotation, and enzyme discovery. These resources help with logical enzyme engineering for commercial applications by improving our knowledge of CAZyme diversification and adaption across many organisms (Kumar et al., 2018; Madeira et al., 2019). Researchers can improve protein family classifications, recreate CAZyme evolutionary pathways, and investigate new enzymatic functions in diverse biological situations by using MEGA and Clustal Omega (Hall, 2013; Tamura et al., 2021).

9.5.4.2 OrthoFinder for Comparative Genomics

A popular bioinformatics tool for comparative genomics, OrthoFinder makes it possible to identify orthologous and paralogous genes in a variety of species. By making it easier to categorize the evolutionary links and functional conservation among enzymes that are active in carbohydrates, it is essential to the study of CAZymes (Emms & Kelly, 2015, 2019). OrthoFinder builds orthogroups, infers gene trees, and detects duplication events using sophisticated phylogenetic techniques. Researchers can learn more about horizontal gene transfer, gene expansion, and species-specific adaptations in glucose metabolism by examining CAZyme families (Altenhoff et al., 2018; Eddy, 2011). OrthoFinder's scalability, which enables genome-wide comparisons across hundreds of species, is one of its main advantages. It offers a thorough framework for researching CAZyme evolution and functional divergence and interfaces well with other bioinformatics tools like KEGG and PANTHER (Emms & Kelly, 2015; Huerta-Cepas et al., 2017). Researchers can improve CAZyme annotation, investigate evolutionary trends, and discover new enzyme activities important to biotechnology, agriculture, and medicine by utilizing OrthoFinder (Altenhoff et al., 2018; Emms & Kelly, 2019).

9.6 CONCLUSION

CAZymes, or carbohydrate-active enzymes, are vital in industrial and biological contexts. Innovations in food processing, medicinal development, and biofuels have been made possible by their capacity to break down, alter, and synthesize complex polysaccharides. The applications of CAZymes will keep growing into new biotechnology fields as our understanding of them is deepened by sophisticated computational methods, bioinformatics resources, and structural biology. The stability, specificity, and catalytic power of enzymes will be improved by the creation of high-efficiency

CAZymes by protein engineering, synthetic biology, and machine learning. Industrial scalability and sustainability will be further enhanced by developments in metabolic engineering, multi-enzyme systems, and enzyme immobilization. The potential for new uses in biomedicine, environmental sustainability, and synthetic carbohydrate chemistry is enormous when CAZymes are combined with cutting-edge technologies like nanotechnology and artificial intelligence. To fully utilize the potential of CAZymes, however, issues like high production costs, enzyme stability in industrial settings, and regulatory barriers need to be resolved. Overcoming these challenges and realizing the full potential of CAZyme-based technologies will require sustained investment in research, interdisciplinary cooperation, and industrial partnerships. CAZymes will continue to lead biotechnology innovation in the upcoming years, supporting innovative treatment approaches, effective bioprocessing, and sustainable bio-based companies. In order to guarantee that these extraordinary enzymes continue to propel advancements in science and industry, their increasing significance highlights the necessity for additional research, development, and optimization. Despite the extensive industrial use of CAZymes, several challenges persist, including limited substrate specificity, high production costs, and reduced enzyme stability under extreme conditions. Addressing these issues requires continued progress in biotechnology and enzyme engineering. Recent advances in artificial intelligence (AI) and machine learning (ML) are revolutionizing the discovery and design of CAZymes by enabling the prediction of enzyme structures, substrate interactions, and catalytic efficiencies with high accuracy. These tools facilitate rational enzyme design, directed evolution, and high-throughput screening, thereby accelerating the development of robust, tailored CAZymes for industrial needs. In combination with protein engineering and computational modeling, AI and ML are expected to significantly enhance CAZyme performance and adaptability. As a result, CAZymes are poised to play an increasingly central role in sustainable and high-efficiency production processes across the food, biofuel, and pharmaceutical industries through ongoing innovation and research.

ACKNOWLEDGMENT

The authors thank the Department of Molecular Biology and Bioinformatics, Tripura University, for providing the computational infrastructure and central library facilities for procuring references and plagiarism analysis (Original Plagiarism Detection Software).

AUTHOR CONTRIBUTION STATEMENT

Sandip Das: Writing, original draft preparation, Abhijit Reang: Writing, original draft preparation, reviewing and editing. Dr. Arabinda Ghosh: Writing, supervision, reviewing and editing.

FUNDING

This research did not receive any specific grant from funding agencies in the public, commercial, or not-for-profit sectors.

CONFLICT OF INTEREST STATEMENT

The author declares that the research was conducted without any commercial or financial relationships that could be construed as potential conflicts of interest.

REFERENCES

Altenhoff, A. M., Glover, N. M., Train, C. M., Kaleb, K., Warwick Vesztrocy, A., Dylus, D., De Farias, T. M., Zile, K., Stevenson, C., Long, J., Redestig, H., Gonnet, G. H., & Dessimoz, C. (2018). The OMA orthology database in 2018: Retrieving evolutionary relationships among all domains of life through richer web and programmatic interfaces. *Nucleic Acids Research, 46*(D1). https://doi.org/10.1093/nar/gkx1019

Altschul, S. F., Madden, T. L., Schäffer, A. A., Zhang, J., Zhang, Z., Miller, W., & Lipman, D. J. (1997). Gapped BLAST and PSI-BLAST: A new generation of protein database search programs. *Nucleic Acids Research, 25*(17). https://doi.org/10.1093/nar/25.17.3389

Andreeva, A., Kulesha, E., Gough, J., & Murzin, A. G. (2020). The SCOP database in 2020: Expanded classification of representative family and superfamily domains of known protein structures. *Nucleic Acids Research, 48*(D1). https://doi.org/10.1093/nar/gkz1064

Astronomo, R. D., & Burton, D. R. (2010). Carbohydrate vaccines: Developing sweet solutions to sticky situations? *Nature Reviews Drug Discovery, 9*(4). https://doi.org/10.1038/nrd3012

Bateman, A., Martin, M. J., Orchard, S., Magrane, M., Ahmad, S., Alpi, E., Bowler-Barnett, E. H., Britto, R., Bye-A-Jee, H., Cukura, A., Denny, P., Dogan, T., Ebenezer, T. G., Fan, J., Garmiri, P., da Costa Gonzales, L. J., Hatton-Ellis, E., Hussein, A., Ignatchenko, A., … Zhang, J. (2023). UniProt: The universal protein knowledgebase in 2023. *Nucleic Acids Research, 51*(D1). https://doi.org/10.1093/nar/gkac1052

Berman, H. M., Westbrook, J., Feng, Z., Gilliland, G., Bhat, T. N., Weissig, H., Shindyalov, I. N., & Bourne, P. E. (2000). The protein data bank. *Nucleic Acids Research, 28*(1). https://doi.org/10.1093/nar/28.1.235

Bertoni, M., Kiefer, F., Biasini, M., Bordoli, L., & Schwede, T. (2017). Modeling protein quaternary structure of homo- and hetero-oligomers beyond binary interactions by homology. *Scientific Reports, 7*(1). https://doi.org/10.1038/s41598-017-09654-8

Boraston, A. B., Bolam, D. N., Gilbert, H. J., & Davies, G. J. (2004). Carbohydrate-binding modules: Fine-tuning polysaccharide recognition. *Biochemical Journal, 382*(3), 769–781. https://doi.org/10.1042/BJ20040892

Bornscheuer, U. T., & Kazlauskas, R. J. (2004). Catalytic promiscuity in biocatalysis: Using old enzymes to form new bonds and follow new pathways. *Angewandte Chemie – International Edition, 43*(45). https://doi.org/10.1002/anie.200460416

Burley, S. K., Berman, H. M., Kleywegt, G. J., Markley, J. L., Nakamura, H., & Velankar, S. (2017). Protein Data Bank (PDB): The single global macromolecular structure archive. *Methods in Molecular Biology, 1607*. https://doi.org/10.1007/978-1-4939-7000-1_26

Camacho, C., Coulouris, G., Avagyan, V., Ma, N., Papadopoulos, J., Bealer, K., & Madden, T. L. (2009). BLAST+: Architecture and applications. *BMC Bioinformatics, 10*. https://doi.org/10.1186/1471-2105-10-421

Cantarel, B. I., Coutinho, P. M., Rancurel, C., Bernard, T., Lombard, V., & Henrissat, B. (2009). The Carbohydrate-Active EnZymes database (CAZy): An expert resource for glycogenomics. *Nucleic Acids Research, 37*(SUPPL. 1). https://doi.org/10.1093/nar/gkn663

de Souza, P. M., & de Oliveira Magalhães, P. (2010). Application of microbial α-amylase in industry – a review. *Brazilian Journal of Microbiology, 41*(4). https://doi.org/10.1590/s1517-83822010000400004

Delano, W. L. (2002). The PyMOL molecular graphics system. *CCP4 Newsletter on Protein Crystallography*, *40*(1).

Drula, E., Garron, M. L., Dogan, S., Lombard, V., Henrissat, B., & Terrapon, N. (2022). The carbohydrate-active enzyme database: Functions and literature. *Nucleic Acids Research*, *50*(D1). https://doi.org/10.1093/nar/gkab1045

Eddy, S. R. (1998). Profile hidden Markov models. *Bioinformatics*, *14*(9). https://doi.org/10.1093/bioinformatics/14.9.755

Eddy, S. R. (2011). Accelerated profile HMM searches. *PLoS Computational Biology*, *7*(10). https://doi.org/10.1371/journal.pcbi.1002195

El-Gebali, S., Mistry, J., Bateman, A., Eddy, S. R., Luciani, A., Potter, S. C., Qureshi, M., Richardson, L. J., Salazar, G. A., Smart, A., Sonnhammer, E. L. L., Hirsh, L., Paladin, L., Piovesan, D., Tosatto, S. C. E., & Finn, R. D. (2019). The Pfam protein families database in 2019. *Nucleic Acids Research*, *47*(D1). https://doi.org/10.1093/nar/gky995

Emms, D. M., & Kelly, S. (2015). OrthoFinder: Solving fundamental biases in whole genome comparisons dramatically improves orthogroup inference accuracy. *Genome Biology*, *16*(1). https://doi.org/10.1186/s13059-015-0721-2

Emms, D. M., & Kelly, S. (2019). OrthoFinder: Phylogenetic orthology inference for comparative genomics. *Genome Biology*, *20*(1). https://doi.org/10.1186/s13059-019-1832-y

Finn, R. D., Clements, J., & Eddy, S. R. (2011). HMMER web server: Interactive sequence similarity searching. *Nucleic Acids Research*, *39*(SUPPL. 2). https://doi.org/10.1093/nar/gkr367

Finn, R. D., Coggill, P., Eberhardt, R. Y., Eddy, S. R., Mistry, J., Mitchell, A. L., Potter, S. C., Punta, M., Qureshi, M., Sangrador-Vegas, A., Salazar, G. A., Tate, J., & Bateman, A. (2016). The Pfam protein families database: Towards a more sustainable future. *Nucleic Acids Research*, *44*(D1). https://doi.org/10.1093/nar/gkv1344

Gänzle, M. G. (2014). Enzymatic and bacterial conversions during sourdough fermentation. *Food Microbiology*, *37*. https://doi.org/10.1016/j.fm.2013.04.007

Hall, B. G. (2013). Building phylogenetic trees from molecular data with MEGA. *Molecular Biology and Evolution*, *30*(5). https://doi.org/10.1093/molbev/mst012

Henrissat, B., & Davies, G. (1997). Structural and sequence-based classification of GH. *Current Opinion in Structural Biology*, *7*(5). https://doi.org/10.1016/S0959-440X(97)80072-3

Himmel, M. E., Ding, S. Y., Johnson, D. K., Adney, W. S., Nimlos, M. R., Brady, J. W., & Foust, T. D. (2007). Biomass recalcitrance: Engineering plants and enzymes for biofuels production. *Science*, *315*(5813). https://doi.org/10.1126/science.1137016

Holm, L. (2020). Using dali for protein structure comparison. *Methods in Molecular Biology*, *2112*. https://doi.org/10.1007/978-1-0716-0270-6_3

Huerta-Cepas, J., Forslund, K., Coelho, L. P., Szklarczyk, D., Jensen, L. J., Von Mering, C., & Bork, P. (2017). Fast genome-wide functional annotation through orthology assignment by eggNOG-mapper. *Molecular Biology and Evolution*, *34*(8). https://doi.org/10.1093/molbev/msx148

Hunter, S., Apweiler, R., Attwood, T. K., Bairoch, A., Bateman, A., Binns, D., Bork, P., Das, U., Daugherty, L., Duquenne, L., Finn, R. D., Gough, J., Haft, D., Hulo, N., Kahn, D., Kelly, E., Laugraud, A., Letunic, I., Lonsdale, D., … Yeats, C. (2009). InterPro: The integrative protein signature database. *Nucleic Acids Research*, *37*(SUPPL. 1). https://doi.org/10.1093/nar/gkn785

Jayani, R. S., Saxena, S., & Gupta, R. (2005). Microbial pectinolytic enzymes: A review. In *Process Biochemistry* (Vol. 40, Issue 9). https://doi.org/10.1016/j.procbio.2005.03.026

Jennings, H. J., & Lugowski, C. (1981). Immunochemistry of groups A, B, and C meningococcal polysaccharide-tetanus toxoid conjugates. *The Journal of Immunology*, *127*(3). https://doi.org/10.4049/jimmunol.127.3.1011

Jeske, L., Placzek, S., Schomburg, I., Chang, A., & Schomburg, D. (2019). BRENDA in 2019: A European ELIXIR core data resource. *Nucleic Acids Research*, *47*(D1). https://doi.org/10.1093/nar/gky1048

Jones, P., Binns, D., Chang, H. Y., Fraser, M., Li, W., McAnulla, C., McWilliam, H., Maslen, J., Mitchell, A., Nuka, G., Pesseat, S., Quinn, A. F., Sangrador-Vegas, A., Scheremetjew, M., Yong, S. Y., Lopez, R., & Hunter, S. (2014). InterProScan 5: Genome-scale protein function classification. *Bioinformatics*, *30*(9). https://doi.org/10.1093/bioinformatics/btu031

Jumper, J., Evans, R., Pritzel, A., Green, T., Figurnov, M., Ronneberger, O., Tunyasuvunakool, K., Bates, R., Žídek, A., Potapenko, A., Bridgland, A., Meyer, C., Kohl, S. A. A., Ballard, A. J., Cowie, A., Romera-Paredes, B., Nikolov, S., Jain, R., Adler, J., … Hassabis, D. (2021). Highly accurate protein structure prediction with AlphaFold. *Nature*, *596*(7873). https://doi.org/10.1038/s41586-021-03819-2

Kanehisa, M. (2019). Toward understanding the origin and evolution of cellular organisms. *Protein Science*, *28*(11). https://doi.org/10.1002/pro.3715

Kanehisa, M., & Goto, S. (2000). KEGG: Kyoto encyclopedia of genes and genomes. Nucleic Acids Research, *28*(1). https://doi.org/10.1093/nar/28.1.27

Kanehisa, M., & Sato, Y. (2020). KEGG Mapper for inferring cellular functions from protein sequences. *Protein Science*, *29*(1). https://doi.org/10.1002/pro.3711

Kanehisa, M., Furumichi, M., Sato, Y., Ishiguro-Watanabe, M., & Tanabe, M. (2021). KEGG: Integrating viruses and cellular organisms. *Nucleic Acids Research*, *49*(D1). https://doi.org/10.1093/nar/gkaa970

Kanehisa, M., Goto, S., Sato, Y., Kawashima, M., Furumichi, M., & Tanabe, M. (2014). Data, information, knowledge and principle: Back to metabolism in KEGG. *Nucleic Acids Research*, *42*(D1). https://doi.org/10.1093/nar/gkt1076

Khatiwada, B., & Pokharel, A. (2009). Lysosomal storage disease. *Journal of the Nepal Medical Association*, *48*(175). https://doi.org/10.31729/jnma.193

Kuhad, R. C., Gupta, R., & Singh, A. (2011). Microbial cellulases and their industrial applications. *Enzyme Research*, *2011*(1). https://doi.org/10.4061/2011/280696

Kumar, S., Stecher, G., Li, M., Knyaz, C., & Tamura, K. (2018). MEGA X: Molecular evolutionary genetics analysis across computing platforms. *Molecular Biology and Evolution*, *35*(6). https://doi.org/10.1093/molbev/msy096

Lipman, D. J., & Pearson, W. R. (1985). Rapid and sensitive protein similarity searches. *Science*, *227*(4693). https://doi.org/10.1126/science.2983426

Lombard, V., Golaconda Ramulu, H., Drula, E., Coutinho, P. M., & Henrissat, B. (2014). The carbohydrate-active enzymes database (CAZy) in 2013. *Nucleic Acids Research*, *42*(D1). https://doi.org/10.1093/nar/gkt1178

Lynd, L. R., Weimer, P. J., van Zyl, W. H., & Pretorius, I. S. (2002). Microbial cellulose utilization: Fundamentals and biotechnology. *Microbiology and Molecular Biology Reviews*, *66*(4). https://doi.org/10.1128/mmbr.66.4.739.2002

Madeira, F., Park, Y. M., Lee, J., Buso, N., Gur, T., Madhusoodanan, N., Basutkar, P., Tivey, A. R. N., Potter, S. C., Finn, R. D., & Lopez, R. (2019). The EMBL-EBI search and sequence analysis tools APIs in 2019. *Nucleic Acids Research*, *47*(W1). https://doi.org/10.1093/nar/gkz268

Magrey, A., Sahay, S., & Gothalwal, R. (2018). Cellulases for biofuel: A review. *International Journal of Recent Trends in Science And Technology, P-Special Issue*. *ACAEE*, 17–25.

Meng, E. C., Pettersen, E. F., Couch, G. S., Huang, C. C., & Ferrin, T. E. (2006). Tools for integrated sequence-structure analysis with UCSF Chimera. *BMC Bioinformatics*, *7*. https://doi.org/10.1186/1471-2105-7-339

Mistry, J., Chuguransky, S., Williams, L., Qureshi, M., Salazar, G. A., Sonnhammer, E. L. L., Tosatto, S. C. E., Paladin, L., Raj, S., Richardson, L. J., Finn, R. D., & Bateman, A. (2021). Pfam: The protein families database in 2021. *Nucleic Acids Research*, *49*(D1). https://doi.org/10.1093/nar/gkaa913

Mitchell, A. L., Attwood, T. K., Babbitt, P. C., Blum, M., Bork, P., Bridge, A., Brown, S. D., Chang, H. Y., El-Gebali, S., Fraser, M. I., Gough, J., Haft, D. R., Huang, H., Letunic, I., Lopez, R., Luciani, A., Madeira, F., Marchler-Bauer, A., Mi, H., … Finn, R. D. (2019). InterPro in 2019: Improving coverage, classification and access to protein sequence annotations. *Nucleic Acids Research*, *47*(D1). https://doi.org/10.1093/nar/gky1100

Moremen, K. W., Tiemeyer, M., & Nairn, A. V. (2012). Vertebrate protein glycosylation: Diversity, synthesis and function. *Nature Reviews Molecular Cell Biology*, *13*(7). https://doi.org/10.1038/nrm3383

Mount, D. W. (2007). Using the basic local alignment search tool (BLAST). *Cold Spring Harbor Protocols*, *2007*(7). https://doi.org/10.1101/pdb.top17

Pearson, W. R. (1990). Rapid and sensitive sequence comparison with FASTP and FASTA. *Methods in Enzymology*, *183*(C). https://doi.org/10.1016/0076-6879(90)83007-V

Pearson, W. R., & Lipman, D. J. (1988). Improved tools for biological sequence comparison. *Proceedings of the National Academy of Sciences of the United States of America*, *85*(8). https://doi.org/10.1073/pnas.85.8.2444

Pettersen, E. F., Goddard, T. D., Huang, C. C., Couch, G. S., Greenblatt, D. M., Meng, E. C., & Ferrin, T. E. (2004). UCSF Chimera – A visualization system for exploratory research and analysis. *Journal of Computational Chemistry*, *25*(13). https://doi.org/10.1002/jcc.20084

Rastall, R. A., & Gibson, G. R. (2015). Recent developments in prebiotics to selectively impact beneficial microbes and promote intestinal health. *Current Opinion in Biotechnology*, *32*. https://doi.org/10.1016/j.copbio.2014.11.002

Saratale, R. G., Oh, S. E., Ghodake, G., & Saratale, G. D. (2020). Exploiting structural adaptations in CAZymes for sustainable biomass degradation: Recent advances and challenges. *Bioresource Technology*, *299*, 122574. https://doi.org/10.1016/j.biortech.2019.122574

Schomburg, I., Chang, A., Placzek, S., Söhngen, C., Rother, M., Lang, M., Munaretto, C., Ulas, S., Stelzer, M., Grote, A., Scheer, M., & Schomburg, D. (2013). BRENDA in 2013: Integrated reactions, kinetic data, enzyme function data, improved disease classification: New options and contents in BRENDA. *Nucleic Acids Research*, *41*(D1). https://doi.org/10.1093/nar/gks1049

Seeliger, D., & De Groot, B. L. (2010). Ligand docking and binding site analysis with PyMOL and Autodock/Vina. *Journal of Computer-Aided Molecular Design*, *24*(5). https://doi.org/10.1007/s10822-010-9352-6

Senior, A. W., Evans, R., Jumper, J., Kirkpatrick, J., Sifre, L., Green, T., Qin, C., Žídek, A., Nelson, A. W. R., Bridgland, A., Penedones, H., Petersen, S., Simonyan, K., Crossan, S., Kohli, P., Jones, D. T., Silver, D., Kavukcuoglu, K., & Hassabis, D. (2020). Improved protein structure prediction using potentials from deep learning. *Nature*, *577*(7792). https://doi.org/10.1038/s41586-019-1923-7

Sievers, F., Wilm, A., Dineen, D., Gibson, T. J., Karplus, K., Li, W., Lopez, R., McWilliam, H., Remmert, M., Söding, J., Thompson, J. D., & Higgins, D. G. (2011). Fast, scalable generation of high-quality protein multiple sequence alignments using Clustal Omega. *Molecular Systems Biology*, *7*. https://doi.org/10.1038/msb.2011.75

Tamura, K., Stecher, G., & Kumar, S. (2021). MEGA11: Molecular evolutionary genetics analysis version 11. *Molecular Biology and Evolution*, *38*(7). https://doi.org/10.1093/molbev/msab120

Turner, P., Mamo, G., & Karlsson, E. N. (2007). Potential and utilization of thermophiles and thermostable enzymes in biorefining. *Microbial Cell Factories*, *6*, 9. https://doi.org/10.1186/1475-2859-6-9

Varki, A. (1993). Biological roles of oligosaccharides: All of the theories are correct. Glycobiology, *3*(2). https://doi.org/10.1093/glycob/3.2.97

Waterhouse, A., Bertoni, M., Bienert, S., Studer, G., Tauriello, G., Gumienny, R., Heer, F. T., De Beer, T. A. P., Rempfer, C., Bordoli, L., Lepore, R., & Schwede, T. (2018). SWISS-MODEL: Homology modelling of protein structures and complexes. *Nucleic Acids Research*, *46*(W1). https://doi.org/10.1093/nar/gky427

Yin, Y., Mao, X., Yang, J., Chen, X., Mao, F., & Xu, Y. (2012). DbCAN: A web resource for automated carbohydrate-active enzyme annotation. *Nucleic Acids Research*, *40*(W1). https://doi.org/10.1093/nar/gks479

Zhang, H., Yohe, T., Huang, L., Entwistle, S., Wu, P., Yang, Z., Busk, P. K., Xu, Y., & Yin, Y. (2018). DbCAN2: A meta server for automated carbohydrate-active enzyme annotation. *Nucleic Acids Research*, *46*(W1). https://doi.org/10.1093/nar/gky418

Zhang, Y. H. P., Ding, S. Y., Mielenz, J. R., Cui, J. B., Elander, R. T., Laser, M., Himmel, M. E., McMillan, J. R., & Lynd, L. R. (2007). Fractionating recalcitrant lignocellulose at modest reaction conditions. *Biotechnology and Bioengineering*, *97*(2). https://doi.org/10.1002/bit.21386

Zhang, Y. H. P., Himmel, M. E., & Mielenz, J. R. (2010). Outlook for cellulase improvement: Screening and selection strategies. *Biotechnology Advances*, *28*(6), 562–572. https://doi.org/10.1016/j.biotechadv.2010.04.001

Zhou, Y., Wang, Y., Chu, X., & Ren, H. (2018). Engineering pH stability of enzymes for industrial applications. *Journal of Molecular Catalysis B: Enzymatic*, *150*, 206–213. https://doi.org/10.1007/s11274-025-04568-4

10 Carbohydrate-Active Enzymes Inhibitors

Mechanisms and Mitigation

Di Cai, Yankun Wang, Honggang Zhang, Hongjia Zhang, Yunxing Gao, Zicheng Liao, and Changjing Chen

ABBREVIATIONS

LDA	Lignin deposited Avicel
CBM	Cellulose-binding domain
EHLR	Enzyme hydrolysis lignin residues
DA	Dilute acid
β-GLs	β-Glucosidases
pI	Isoelectric point
FA	Ferulic acid
CBH	Cellobiohydrolase
SF	Severity factor
CAOSA	Cooking with active oxygen and solid alkali
MA	Maleic acid
WIS	Water-insoluble solids
LHW	Liquid hot water
HOBt	1-Hydroxybenzotriazole
BSA	Bovine serum albumin
PEG	Polyethylene glycol
CnTAB	Cetyltrimethylammonium bromide
AM	Acrylamide
DMC	[2-(Methacryloyloxy)ethyl] trimethylammonium chloride
CCR	Corn cob residue
QAP	Quaternary ammonium phosphate
EHL	Enzymatic hydrolysis lignin
LQAP	Lignin-grafted quaternary ammonium phosphate
UCST	Upper critical solution temperature

10.1 INTRODUCTION

In the face of the energy crisis, climate change, and adverse effects on the environment caused by petrochemical industries, the development of sustainable chemicals,

 DOI: 10.1201/9781003561682-10

alternative fuels, and recyclable composites from renewable biomass materials has attracted significant attention in recent years (Ma et al., 2019). Due to abundant reserves on earth, the development and implementation of the second-generation biorefineries from lignocelluloses is considered as one of the most feasible ways to realize the circular economy and replace the fossil-based routes (Ayodele et al., 2020).

Lignocelluloses are primarily contained in the plant cell wall, which consists of three polymeric components: lignin, cellulose, and hemicellulose (Wang et al., 2024b). The mass ratios of these components in lignocellulose are diverse from different species, tissues, age, and even the harvest places of the plants (Cai et al., 2016; Luo et al., 2017). Among the chemical components, cellulose and hemicellulose are the main carbohydrate fractions that can be hydrolyzed into pentose, hexose, and other monosugars. In biorefinery processes, these sugars are further valorized into various renewable biochemicals via either the biological or the chemical catalysis routes (Dharmapriya and Huang, 2024). Lignin, however, is a non-carbohydrate polymer that is mainly composed of cross-linked phenylpropane units, and is the largest natural aromatic source on earth (Pei et al., 2024).

Lignocelluloses exhibit a tightly bonded structure, where the hemicellulose serves as a binding agent and acts as 'wire mesh' to support the 'reinforced concrete' like matrix (Qaseem et al., 2021). Cellulose act as the 'rebars' in lignocellulose and connected with the hemicellulose through hydrogen bonds and van der Waals interactions (Zhang et al., 2011a). Lignin, acting as the 'concrete', bonds with the hemicellulose by covalent ferulate-ether linkage and provides hydrophobicity and structural rigidity to lignocellulose (Albornoz-Palma et al., 2020). This intricate combination of the three important components protected the plant cell wall from the microbial attack, and thereby realized a strong resistance to the degradation of lignocelluloses (Mohanram et al., 2013).

Therefore, in the process of monosugars production from lignocelluloses, an essential step is to destroy the original stubborn cell wall structure of the plant before the enzymatic saccharification of carbohydrate fractions is carried out; otherwise, the access of the carbohydrates' active enzymes, namely cellulase, to the cellulose and hemicellulose fractions would be largely hindered (Leu and Zhu, 2013). This lignocellulose depolymerization process is called pretreatment (Roy et al., 2020). To date, various methods, including physical, chemical, biological, and their combinations have been suggested to pretreat the lignocellulose, which significantly promoted the subsequent enzymatic saccharification process (Zhou and Tian, 2022).

Although most of these pretreatment techniques allowed partial delignification and dissolved the lignin debris into the liquidous phase, the lignin fraction can hardly be completely removed from the carbohydrate-rich pulps due to the dense structure of lignocellulose, unless conducted under strict conditions (Yuan et al., 2021). Meanwhile, the lignin debris in liquidous phase, mainly the low molecular weight phenolic compounds, is also difficult to be completely separated from the pulp after typical solid-liquid separation, and involved in the following enzymatic saccharification (Qin et al., 2016). However, the residual lignin always exhibits adverse influences to the enzymatic saccharification of the carbohydrates, including steric hindrance as well as non-productive adsorption to cellulase through hydrogen bonding, electrostatic, and hydrophobic interactions (Cai et al., 2023). Therefore, the inhibition of lignin to cellulase requires proper mitigation; otherwise, the inactivation of cellulase would severely reduce the lignocellulosic sugar productivity (Zhai et al., 2022).

The primary lignin inhibition mitigation strategies can be divided into three categories. The first is the proactive measures that aim to reduce the inhibitors' production or toxicity in the pretreatment stage. The second is the passive measures that reduce the negative influence of the lignin to cellulase in saccharification process. The last is improving the tolerance of cellulase to achieve high carbohydrate hydrolysis activity of the enzymes. Hence, this chapter will concentrate on the inhibition of the residual lignin from both the insoluble large polymers in pulps and the soluble low-molecule debris to the carbohydrate-active enzymes. The mechanisms for the inhibition will be discussed, and the lignin inhibition mitigation strategies will also be included. We hope this chapter can bring inspiration to the relevant investigations on reducing the inhibition of lignin to the carbohydrate-active enzymes, so that enabling lower cost lignocellulosic sugars production in the near future.

10.2 MECHANISMS FOR LIGNIN INHIBITION TO CARBOHYDRATE-ACTIVE ENZYMES

Enzymatic hydrolysis of lignocellulose is confronted by a number of obstacles that reduce the efficiency of hydrolysis. Interaction of phenolic derivatives of lignin with cellulase is considered to be one of the major factors that reduce the enzymatic hydrolysis efficiency. There are three main inhibitions of cellulase by lignin derivatives: (1) steric hindrance, (2) non-productive adsorption, and (3) inhibitory effect of soluble phenols on enzyme activity, as shown in Fig. 10.1.

10.2.1 Steric Hindrance

Pretreatment is a crucial step in lignocellulosic monosugars production chain, whereas most of the existing pretreatment techniques cannot completely remove lignin fraction from the carbohydrate-rich pulp obtained in the solid phase (Lin et al., 2024). Even in the lignin-first biorefining schemes that could effectively passivate lignin and eliminate the re-condensation of the active intermediates, normally, there was still less than 10 wt% of residual lignin in polymer form remaining in the fractionated pulps (Wu et al., 2024; Wu et al., 2023). In typical pretreatment processes, the polymerized lignin is hydrolyzed by cleavage of the ether bonds and migrates from the interior structure to the surface, and redistributes the entire matrix structure (Li et al., 2014).

The steric hindrance of the residual lignin always occurs in the pretreatment processes under acidic or high-temperature conditions, e.g., the hydrothermal, the organosolv, and the acid fractionation processes (Lin et al., 2024). In these pretreatment processes, the lignin intermediates are more easily re-condensed into stable polymers and adhered on the surface of pulps (Li et al., 2014). For instance, Selig et al. (2007) observed the spherical structures remaining on the surface of pulp after dilute acid pretreatment. Further experiments indicated the 'lignin droplet' was

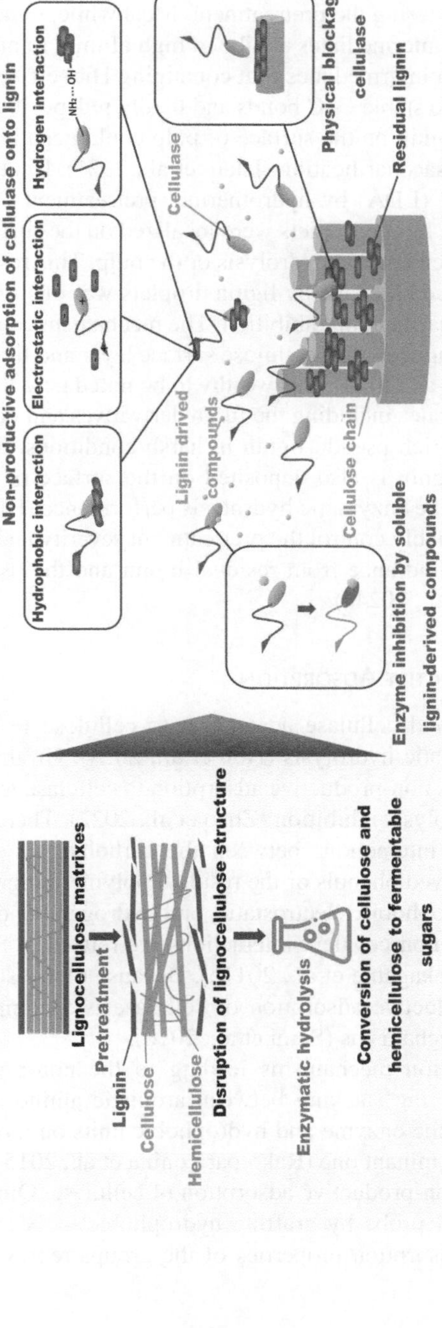

FIGURE 10.1 Schematic illustration of the carbohydrate-active enzymes-lignin interaction in lignocellulosic monosugars production.

formed in high temperatures (~130°C). Under more severe conditions (>150°C), the droplets could be redeposited on the surface of pulp. Similar phenomenon was also observed by other researches for hydrothermal or dilute acid pretreatment (Selig et al., 2007). This is partly attributed to the morphological change of lignin from glassy to rubbery state during the pretreatment. Meanwhile, under acidic conditions, carbon cation in lignin intermediates exhibited high affinity to nucleophiles (Trajano et al., 2013). The lignin intermediates that containing Hibbert ketone structures tend to be repolymerized into stable C–C bonds and finally redeposit (Haldar et al., 2022).

The re-condensed lignin on the surface of pulp could negatively influence on the subsequent enzymatic saccharification (Islam et al., 2021). Li et al. (2014) prepared lignin-deposited Avicel (LDA) by hydrothermal pretreatment of Avicel and poplar wood, and observed the lignin droplets were localized on the surface of LDA and significantly hindered the enzymatic hydrolysis of the pulp. This research also indicated that the cellulose surface blockage by lignin droplets was one of the critical reasons for the enzymatic saccharification inhibition. The mechanism could be the prevention of enzymes from moving along the cellulose surface layer and from entering the inner layer of cellulose (Li et al., 2014). It is worthy to be noted here that except for lignin, the degraded carbohydrates including the furan derivatives and organic acids can also form insoluble carbon-rich pseudo-lignin in harsh conditions. Similar to the lignin droplets, the pseudo-lignin is also deposited on the surface of pulp and possesses negative influences on the enzymatic hydrolysis performance (Hu et al., 2013). Thus, it is necessary to reasonably control the pretreatment severity and conditions in order to alleviate the steric hindrance from residual lignin and the pseudo-lignin (Li and Zheng, 2017; Wang et al., 2024a).

10.2.2 Non-productive Adsorption

As aforementioned, good cellulase adsorption to cellulose is the prerequisite for highly effective enzymatic hydrolysis (Hou et al., 2024). On this basis, it has been pointed out that lignin's non-productive adsorption to cellulase is an additional factor for the enzymatic hydrolysis inhibition (Zhao et al., 2022). There are primarily three types of non-covalent interactions between the carbohydrate-active enzymes and lignin (either the dissolved phenols or the residual polymeric lignin on the surface of the pulp), namely hydrophobic, electrostatic, and hydrogen bonding adsorption (Liu et al., 2016). All these non-covalent interactions are related to the physicochemical properties of lignin (Nakagame et al., 2011b). Because of the structural complexity of lignin, the non-productive adsorption of cellulase is a comprehensive result of different interaction mechanisms (Saini et al., 2016).

Among the interaction mechanisms leading to the non-productive adsorption of cellulase by lignin, ring stacking between aromatic amino acid residues in the molecular structure of the enzyme and hydrophobic units on the surface of lignin is recognized as the predominant one (Rakotoarivonina et al., 2015). To get an in-depth understanding of the non-productive adsorption of cellulase, Qin et al. (2014) modified the tip of an AFM probe by grafting hydrophobic –OH and –COOH groups, so that compare the adsorption properties of the groups representing hydrophobic, hydrogen bonding and electrostatic interactions in lignin, respectively. Among the active groups, the probes with hydrophobic tips showed the strongest adhesion to cellulase, which highlighted the influence of hydrophobic action on the non-productive binding of cellulase and lignin (Qin et al., 2014). Besides, the hydrophobic interaction between lignin and carbohydrate-active enzymes is proved to be mediated by the cellulose-binding domain (CBM) of cellulase (Palonen et al., 2004). The CBM in cellulase usually contains hydrophobic residues such as tryptophan, phenylalanine, and tyrosine. The binding of CBM to cellulose is driven by the stacking (hydrophobic) interaction when hydrophobic residual align with the pyranose ring of cellulose (Rahikainen et al., 2013). Whereas, this hydrophobic interaction also occurs in the non-productive adsorption of cellulase on lignin (aromatic units from lignin may also participate in stacking with the hydrophobic residual of the CBM).

Early studies believed that lignin has higher hydrophobicity than cellulose, which resulted in higher adsorption probability of lignin to cellulase (Hodgson and Berg, 1988). However, the current literature indicated that the physicochemical properties of lignin dependent on the pretreatment process. Sun et al. (2016) investigated the relationship between poplar lignin and cellulase binding toward various pretreatment methods, and observed that the condensed aromatic rings in lignin strengthened the hydrophobic interaction to cellulase. On this basis, Chandra et al. (2015) successfully prevented the cellulase adsorption onto lignin by sulfonation, whilst Nakagame et al. (2011a) avoided the non-productive adsorption by carboxyl modification of lignin. Therefore, the decrease of lignin hydrophobicity could effectively eliminate the non-productive adsorption of cellulase, which can be realized by proactive measures in pretreatment stage. This will be discussed in Section 10.3.1.

Hydrogen bonding is an additional driver for lignin adsorption by cellulase. It has shown the hydroxyl groups in lignin possessed a crucial role in cellulase-lignin interactions, especially the phenolic hydroxyl groups (Pan, 2008). For instance, the organosolv lignin is more efficiently adsorbed on cellulases than steam-exploded lignin, owing to the higher phenolic hydroxyl content (Saini et al., 2016). To have deeper insight into the mechanisms, Zheng et al. (2024) investigated the adsorption of cellulases by enzyme hydrolysis lignin residues (EHLR) of sulfite-pretreated (SPORL) and dilute acid pretreated (DA)-sugarcane bagasse. FTIR results indicated that the SPORL-EHLR-containing more hydroxyl groups (including phenolic hydroxyl groups and aliphatic hydroxyl groups) and possessed stronger hydrogen bonding to β-glucosidases (β-GLs) (Zheng et al., 2024). In another study, Lan et al. (2020) found that after the addition of lignin isolated from p-TsOH pretreatment process in the following carbohydrate hydrolysis, the complex formed by substrate and lignin had more O–H bonds than the substrate alone, inferring that the hydrogen bond between cellulase and the complex was significantly improved (Lan et al., 2020). Therefore, to eliminate the hydrogen bonding between lignin and cellulase, one of the potential strategies is the blocking phenolic hydroxyl groups in lignin. For instance, the phenolic hydroxyl group in lignin will be converted to aliphatic hydroxyl group after hydroxypropylization (Yang and Pan, 2016). Meanwhile, the induced formation of hydrogen bonds between additives with the phenolic groups in lignin could also positively affect the reduction of the non-productive adsorption. This can be realized by adding surfactant as an assistant agent (Zhang et al., 2011b).

Both lignin and cellulases contain charged chemical groups (Rahikainen et al., 2013). As for the electrostatic interaction, the dissociation or association of the active groups (such as carbonyl, carboxyl, hydroxyl, amino) is the primary reason for the charge formation on the surface (Liu et al., 2016). Isoelectric point (pI) is a unique character of specific cellulase. Therefore, different cellulases in the same solution exhibited different charging characteristics (Yao et al., 2022). For instance, at pH 4.8 (the optimized pH for cellulose hydrolysis), Cel6A and Cel5A are positively charged, while Cel7A and Cel7B are negatively charged (Saini et al., 2016). When the pH of buffer is higher than pI, the overall charge of the enzyme is negative. In contrast, the overall charge of the enzymes transformed into positive when pH of the buffer was lower than pI. Therefore, cellulases and lignin bind together by electrostatic interactions when they have the opposite charges (Lan et al., 2013). In general, lignin fractions typically exhibit negative charges due to the ionization of carboxyl and phenolic hydroxyl groups (Lan et al., 2020); the electrostatic interaction between enzyme and lignin could be different with the pI of enzyme (Yu et al., 2015). For instance, Zhang et al. (2020) analyzed the adsorption behavior of glucanases (TvEG and TrCel7A) on lignins isolated from raw poplar and dilute acid pretreatments. Since the pI of TrCel7A was 4.58, its adsorption on lignin in a pH 4.5 buffer is affected by both electrostatic and hydrophobic interactions, while TvEG (pI=4.4) undergoes electrostatic repulsion toward lignin (Zhang et al., 2020). The previous researches also showed that the increased pH of buffer could reduce the non-productive binding by increasing electrostatic repulsion between negatively charged enzymes and lignin (Saini et al., 2016). For example, Lou et al. (2013) showed that the significant enhanced enzymatic hydrolysis of cellulose in pretreated pulp at pH 5.5–6.0 was due to increased surface charge of lignin. Consequently, the increase of lignin's hydrophilicity resulted in the weakening of nonspecific lignin-cellulase binding. At the same time, the increased negative charge also strengthened the electrostatic repulsion between lignin and cellulases (Lou et al., 2013). Therefore, modifying the charged chemical groups in enzymes and lignin is a passive measure to reduce unproductive enzyme binding (Saini et al., 2016).

10.2.3 Inhibitory Effect of Soluble Phenols on Enzyme Activity

In addition to the steric hindrance and non-productive adsorption caused by residual lignin and pseudo-lignin on the surface of pulp, as aforementioned, the low molecular weight phenolic compounds that are degraded from lignin and dissolved in the pulping liquor also exhibit inhibition to carbohydrate-active enzymes' activity in saccharification process (Ximenes et al., 2013). Particularly, it has been indicated that monophenols, such as vanillin, syringic acid, and 4-hydroxybenzyl alcohol, exhibited an obviously higher inhibitory effect on enzyme activity, in contrast to other by-products degraded from carbohydrates, such as furan derivatives and organic acids. The activity of cellulase was sharply reduced even in conditions with extremely low concentrations of monophenols in buffer (Kim, 2018). For instance, Ximenes et al. (2010) investigated the effects of four commonly detected phenolic compounds in pulping liquor, including vanillin, syringaldehyde, trans-cinnamic acid, and hydroxybenzoic acid, on the enzymatic hydrolysis of cellulose. Results indicated that the cellulose hydrolysis rate was reduced by ~50% after adding 4 mg of vanillin into a buffer that contained 1 mg of cellulase.

The relationship between phenol compounds and the enzymes involved in the saccharification of cellulose is currently poorly understood. Most of the current research speculated that the inhibition of phenolic compounds could be the covalently or non-covalently binding toward one or more sites on enzyme, and the covalent binding leads to the precipitation of cellulases (Jiang et al., 2022). For the direct inhibition of the dissolvable lignins, the decreased enzyme activity can be manifested by reversible and irreversible inhibition (Qin et al., 2016). The reversible inhibition is caused by rapid binding equilibrium between the inhibitor and the enzymes, while the irreversible inhibition (inactivation) is ascribed to an exponential decrease in enzyme activity as the contact time between the inhibitor and the enzyme increases (Kim et al., 2011).

Actually, enzymatic hydrolysis of cellulose is actually a complex process that requires the coordination of multi-carbohydrate-active enzymes (including endo-β-1,4-glucanase, exo-β-1,4-glucanase, β-GLs, and the assistant enzymes such as the β-xylosidase) (Jeoh et al., 2006), while the phenolic compounds as the inhibitors are also a complex mixture in buffer (Table 10.1). Therefore, it is challenging to accurately identify the individual enzyme's inhibition by the specific phenolic compounds (Jiang et al., 2022). Indeed, model phenolic compounds are always suggested to be used in the exploration of phenol inhibition on single enzyme (Vohra et al., 1980).

TABLE 10.1
Effect of Different Enzymes in the Hydrolysis of Lignocellulose (Ilić et al., 2023)

Enzyme Type	Substrate	Product	Mechanism
Endoglucanase (Endo-1,4-β-glucanase)	Amorphous cellulose	Long-chain oligosaccharides	Random cleaves internal β-1,4-glycosidic bonds in cellulose chain, producing shorter cellulose fragments.
Exoglucanase (Exo-1,4-β-glucanase)	Crystalline cellulose	Cellobiose	Sequentially cleaves β-1,4-glycosidic bonds from the ends of the cellulose chain, releasing cellobiose.
β-Glucosidase (β-Gls)	Cellobiose and cellooligosaccharides	Glucose	Hydrolysis cellobiose and other oligosaccharides into glucose.
β-Xylosidase	Xylobiose	Xylose	Hydrolysis the degradation products of xylan into xylose.

Most of the literature showed that the β-GLs are more susceptible to the phenolic inhibitors in buffer (Ximenes et al., 2011). For instance, Michelin et al. (2016) suggested the acetone-extracted phenols from the hydrothermal pretreatment liquor could strongly inactivate β-GLs and β-xylosidase activities (Michelin et al., 2016). However, there are also some studies indicating the phenols severely inhibited other types of enzymes, instead of the β-GLs. Liu et al. (2019) suggested p-coumaric acid (p-CA) and ferulic acid (FA) exhibited little inhibition to the activity of exoglucanase and β-GLs, but possessed severer toxicity to the endoglucanase and xylanase activities. The quantitative analysis result indicated the activities of endoglucanase and xylanase were reduced by 16% and 18–19%, after adding 0.5 g/L FA and p-CA to the buffer, respectively (Liu et al., 2019). In another research, vanillin and 4-hydroxybenzaldehyde were suggested to have stronger inhibition to the cellobiohydrolase (CBH) activity than other enzymes, and the activity inhibition rates by the two phenols were 61.4% and 20.8%, respectively (Jiang et al., 2022). This controversy regarding the specific inhibition of phenolic compounds on enzymes could be ascribed to the differences of the spatial construction of the enzymes that are secreted from different species of microorganisms, although these enzymes have the similar functions (Liu et al., 2019).

It is also suggested that the pretreatment strategies and conditions could not only influence the constitution of the soluble phenolic compounds but also affect the activity of carbohydrate-active enzymes involved in saccharification process (Cai et al., 2023). Research suggested that differences in functional group characteristics in phenolic molecules could be the primary reason for the different inhibitions to enzymes (Qin et al., 2016). For instance, Chen et al. (2020) investigated the effects of soluble phenol compounds from four different pretreatment processes, including the dilute acid, hydrothermal, ammonia fiber expansion, and alkaline pretreatment, on enzymatic hydrolysis. Results suggested the phenolic compounds produced by AFEX had a higher inhibitory effect on enzymatic hydrolysis of pulp. The authors speculated that the introduction of amide groups into molecular structure of phenols. For instance, the formation of -C=O or $CONH_2$, the strong dipoles, could be the reason for the strong interaction and thereby inhibiting the cellulase activity (Chen et al., 2020). In another work, Tian et al. (2013) found that phenolic acid could promote the activity of cellulase more than other phenols. This enhancement is ascribed to the binding of phenolic acid molecule to cellulase, which resulted in the increase of hydrophilicity of cellulose-binding domain (to maintain the proximity of the enzyme domain to cellulose on the water-cellulose surface). In addition, the covalent binding of phenolic acid to cellulase changes the protein structure of cellulase (the content of α-helix is reduced) and the affinity of cellulase to cellulose substrate is increased (Tian et al., 2013). As for the inhibition mechanisms of the phenolic aldehydes to carbohydrate-active enzymes, because of the higher hydrophobicity than other types of phenols, the phenolic aldehydes decrease the catalytic activity of enzymes by competing with the binding sites and destroying the interaction between CBH and substrates (Jiang et al., 2022).

It worthy to note here that most of the existing researches focused on revealing the inhibition of phenolic compounds to cellulase has emphasized the interaction between phenolic compounds and enzymes and has primarily demonstrated that the inhibitory effect is generally attributed to the interaction between phenolic compounds and enzymes via activity changes. However, in fact, there are three component (substrate, cellulase, and phenolic compounds) involved in the enzymatic hydrolysis of the pretreated pulp. Once simplifying this process into a two-component participated system (cellulase and phenolic compounds), the interactions between substrate and phenolic compounds are always ignored (Stamogiannou et al., 2021). Therefore, it is recommended to further clarify the mechanism of interaction between phenolic compounds and carbohydrate-rich pulps.

10.3 MITIGATION STRATEGIES OF THE PHENOL INHIBITORS

A number of mitigation strategies have been developed to counteract the inhibitory effects of lignin-derived phenolics on cellulases. These strategies generally include interventions in pretreatment processes and hydrolysis processes, as well as modifications of cellulases.

10.3.1 Proactive Measures: The Intervention in Pretreatment Stage

The primary purpose for the proactive measures for eliminating the negative influences of lignin on the carbohydrate-active enzymes is the reduction of lignin toxicity in the pretreatment stage. Although in most cases, a simple post-washing process could effectively remove the dissolved phenolic compounds in the pretreated liquor, the large wastewater or solvent discharge and the loss of polysaccharides led to uneconomic and environmentally unfriendly processes (Zhai et al., 2022). Therefore, various researches have been done aimed at intervening in the lignocellulose pretreatment to realize the improved pulp's enzymatic saccharification (Cai et al., 2023).

The most classic strategy for mitigating the lignin inhibition of cellulase is the regulation of pretreatment type and severity. For instance, the decrease of acid pretreatment temperature and period could reduce the dissolution of phenols and other by-products, whilst the warmer conditions avoided the generation of pseudo-lignin that hindered the access of cellulase to cellulose (Duret et al., 2013). For instance, it was suggested after increasing the severity factor (SF) of wheat straw pretreatment from 2.1 to 3.3, the concentration of phenolic compounds in the aqueous phase was increased to 2.22 g/L from 1.08 g/L (Jiang et al., 2022). In another work, the phenolic compounds dissolution was also increased when the pretreatment temperature in the acidic process increased from 160°C to 180°C (Moxley et al., 2012). The increase in phenol dissolution was attributed to the promotion of bond cleavage under higher severities.

Nonetheless, the excessive decrease of pretreatment severity led to incomplete depolymerization of the initial recalcitrant lignocellulose structure, which is also unfavorable for monosugar yield. To deal with the gaming, a potential strategy is to decrease the phenols generation in relatively higher severity conditions by modifying the pretreatment protocols. For instance, after taking freeze grinding of lignocellulose before the hydrothermal pretreatment, the phenolic inhibitors of cellulase were significantly decreased in the received liquor (Maitra and Singh, 2021). Ilanidis et al. suggested that the addition of O_2 might raise the severity of the pretreatment, resulting in more widespread lignin degradation into phenolics, while the pretreatments without gas addition or protected by inert gases (e.g., N_2) decrease the lignin release (Ilanidis, 2021).

Another strategy is the chemical modification of lignin structure during the pretreatment process, so that decrease the toxicity of lignin to cellulase activity. For instance, Zhu et al. (2009) developed the sulfite pretreatment to overcome recalcitrance of lignocellulose that utilizes sulfite or bisulfite solutions as pulping liquor. In contrast to the dilute acid and alkali pretreatments, the enzymatic hydrolysis of the SPORL pulp of switchgrass that pretreated under acidic conditions was significantly improved because of the almost full removal of hemicellulose, partial dissolution of lignin, and catalytic sulfonation of lignin into hydrophobic molecules (Li et al., 2012b). In addition, the introduction of carboxyl groups and carbonyl groups into lignin molecule is another method to improve enzymatic hydrolysis, aiming to increase the hydrophilicity of lignin. (Cai et al., 2023; Ying et al., 2018). A representative pretreatment method is namely cooking with active oxygen and solid alkali (CAOSA) pretreatment. Yang et al. (2012) and Ding et al. (2021) investigated the mechanism of lignin degradation by CAOSA. In these processes, hydroxyl radicals (HO·) were generated from molecular oxygen, and lignin with benzyl alcohol groups was oxidized into carbonyl groups by hydroxyl radicals. Moreover, due to the lower bond dissociation energy, the Cα-Cβ bond of the lignin side chain is broken during the CAOSA pretreatment. Specifically, the carbonyl or vinyl α-C at the lignin side chain is attacked by the active oxygen group to form a C-O-O-C ring structure. The further breakage of the four-atom ring causes the cleavage of Cα-Cβ, and many phenylpropane units are converted into benzoic acid (Ding et al., 2021; Yang et al., 2012). Chu et al. (2017) took alkaline oxygen soaking of the poplar wood before steam pretreatment. In this process, the carboxyl content of the pretreated lignin was increased from 33 mmol/kg in conventional stream pretreatment to 156 mmol/kg in the hybrid process, and the accessibility of cellulase was obviously enhanced. On this basis, a mild O_2-assisted alkaline treatment was developed (80°C and 1% NaOH solution), and the glucose and xylose yields after saccharification reached 89.7% and 90.2%, respectively (Song et al., 2021). In addition, some acidic hydrophilic compounds were also suggested to be used for lignocellulose pretreatment, in which the hydrophobic compounds were solubilized simultaneously with the biomass fractionation (Cai et al., 2023). For example, Su et al. (2021) highlighted the advantages of improving enzymatic digestibility by esterifying lignin through maleic acid (MA) pretreatment, where the retained carboxylated water-insoluble solids (WIS) exhibited higher enzymatic digestibility compared to the uncarboxylated WIS that were generated from the conventional process.

Another strategy is the chemical modification of lignin structure during the pretreatment process so that it decrease the toxicity of lignin to cellulase activity. For instance, alcohols as nucleophiles could attack carbon cations and convert lignin into more stable benzyl ether structures, and thus preventing the polycondensation of lignin (Cai et al., 2023). Dong et al. (2022) selectively enhanced the α-etherification of lignin under mild conditions (120°C). After diol pretreatment, the cellulose conversion reached >90% within 24 h in an experimental group that only loaded 7.5 FPU/g (glucan) of cellulase. In addition, aromatic compounds are also active in nucleophilic substitution, so lignin can be preserved from self-condensation based on this phenomenon. For example, 2-naphthol and its derivatives enhanced the lignin solubility in pretreatment process (Pielhop, 2023) (Fig. 10.2).

10.3.2 Passive Measures: The Reduction of Negative Effects of Lignin in Hydrolysis Process

In contrast to the proactive measures that aim to mitigate the lignin inhibition in the lignocellulose pretreatment stage, the passive measures are generally conducted during the enzymatic hydrolysis stage for monosugar production. In conventional processes, the phenolic inhibitors in hydrolysis can be separated by ether adsorption or extraction before or within the saccharification progress, thus avoided the chemical changes of the phenols structure (Zhai et al., 2022). For instance, Freitas et al. (2017; 2019) used sugarcane bagasse fly ash and coconut shell-based activated carbon as adsorbents, and used them in liquid hot water (LHW) pretreated pulp's saccharification. Results showed high affinity for phenolic substances. The sugarcane bagasse fly ash achieved 80% phenol removal at 5% (w/v) of dosage, and the glucose concentration of the treated hydrolysate was 52% higher than that of control group. The coconut-shell-activated carbon, however, exhibited high selectivity for adsorbing phenols without affecting the monosugars' production. The glucose concentration was increased by 30% after adding 1% (w/v) coconut shell activated carbon in buffer.

Another strategy is modification of phenolic compounds during pulp's saccharification. As aforementioned in **Section 10.2.3**, the inhibitory effects of phenolic compounds are highly dependent on their structure. Similar to the principle for lignin modification by proactive measures in pretreatment stage, the chemical modification of phenolics in buffer can also reduce the toxicity to cellulase (Soudham et al., 2011). For example, the catalysis reduction of soluble phenols can be realized by the addition of $NaBH_4$ in buffer, in which process the highly inhibitory phenolic aldehyde are transformed into the less inhibitory phenolic alcohols. In this process, although the overall mole content of the phenols is not changed in buffer, the inhibition of the reduced lignins that were rich in hydroxyl groups was eliminated (Cavka and Jönsson, 2013). However, despite the $NaBH_4$ detoxification is easy to operate, the low selectivity for inhibitors reduces the monosugars production. After comprehensively considering the cost and practicability, this process is unsatisfied with the large-scale operations (Ilanidis et al., 2021a). By contrast, sodium disulfate alleviates the inhibition of cellulase from phenolic compounds through the hydrophilization of inhibitors by sulfonation. It has industrial feasibility due to the low temperature and low doses of the agent (Ilanidis et al., 2021b). For instance, after adding 15 mM dithionite to a mixture of solution after SO_2 pretreatment and pretreated spruce wood, the monosugars production was increased by 33% in the enzymatic hydrolysis process (Soudham et al., 2011).

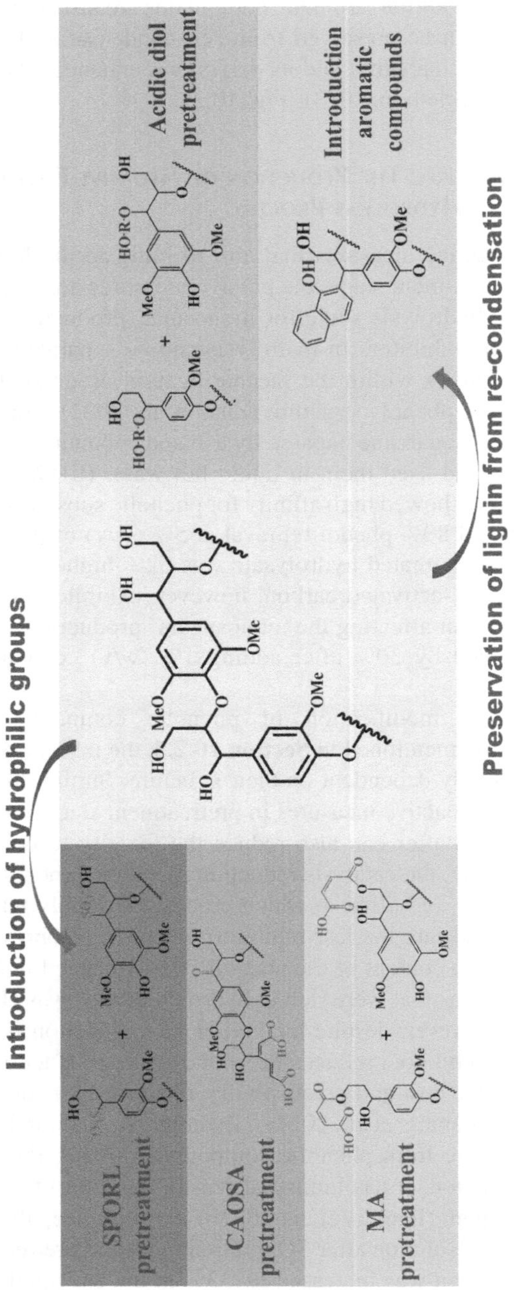

FIGURE 10.2 Proactive strategies to alleviate lignin inhibition to cellulase activity.

In addition to the biological detoxification by microorganisms, auxiliary enzymes are suggested to work together with the base cellulase in typical saccharification processes. For instance, laccases can be used to oxidize phenols and catalysis opening the benzene ring of lignin, and thus mitigate the inhibition of lignin to cellulase. Specifically, laccase could completely degrade coniferyl aldehyde, which has a high inhibition to cellulase's activity. As for the highly toxic vanillin, the molecule can be completely degraded in conjunction with the auxiliary oxidant 1-hydroxybenzotriazole (HOBt) (Saravanakumar et al., 2016). Meanwhile, the good stability in relatively harsh conditions allowed the wide applications of laccases in bioremediation as well as in the textile and pulping industries (Arregui et al., 2019).

Recently, there are primarily two directions for improving the catalytic detoxification and promoting the base cellulase performance by laccase. One is the screening of robust laccases with better catalysis oxidation performances. For instance, laccase (YlLac) that was isolated from the Yarrowia lipolytica exhibited superior catalytic oxidation efficiency on both phenolic and non-phenolic compounds in buffer (Kalyani et al., 2015). Another direction is construction of suitable immobilization carrier of laccase to achieve improved detoxification efficiency, stability, and reusability. For instance, recent research indicated that NiO nanoparticles, synthesized via green route (using phytochemicals from Salvadora persica leaf extract as reducing and capping agents), can effectively immobilize the metagenomic laccase PersiLac1. This immobilized enzyme retained 85.55% of its activity at 80°C and exhibited low enzyme leaching. After 15 biocatalytic cycles, it maintained 67.93% of the initial activity and showed significant efficiency in degrading phenolic compounds (Ariaeenejad et al., 2024). However, the optimized constitution of the as received enzyme cocktail is highly dependent on the pretreatment strategy, the conditions, the biomass types, and the chemicals dissolved in the buffer (Adsul et al., 2020; Pinheiro et al., 2021).

Non-enzymatic proteins are another group of promoters for enzymatic carbohydrate hydrolysis that have been extensively investigated in recent years. For example, bovine serum albumin (BSA) is often used as a model non-enzymatic protein for enzymatic hydrolysis of lignocellulose. Additional 5 g/L BSA in buffer allowed a 40% higher of glucose yield from dilute acid-treated corn stover (Brethauer et al., 2011). Many studies suggested that BSA blocks the non-productive binding of cellulase at lignin binding sites by adsorbing on lignin structure, thereby increasing the free enzyme amount in hydrolysate (Eriksson et al., 2002). However, Wang et al. (2013) found that when only using filter paper as substrate, BSA could significantly improve the enzymatic hydrolysis performance due to the increased cellulase stability.

Although BSA shows clear beneficial effect on pulp's saccharification, to pursue the maximum economic benefits of the lignocellulosic sugar platform, it is necessary to explore more cost-effective additives and replace expensive BSA. On this basis, the utilization of other low-cost protein-rich additives (such as corn infusions, yeast extracts, and peptones, etc.) in the enzyme cocktail has also been suggested in several reports (Wang et al., 2015). Among the proteins, the soybean protein is considered as a cost-effective candidate. When using the mixture secreted from *Aspergillus niger* and *Trichoderma reesei* as cellulase, after adding soybean protein in the buffer for enzymatic saccharification of the steam-exploded sugarcane bagasse, ~2 times higher glucose was received in the final hydrolysate, in contrast to the control group without protein addition (Florencio et al., 2016).

Generally, in the progress for pursuing cellulase tolerance improvement, the second class of hydrophobin, HFBII, has emerged as an important compound to be investigated (Linder et al., 2001). The HFBII exhibited the ability to form complexes with cellulase, and is capable of stabilizing the enzyme's structure through hydrogen bonding, van der Waals interactions, and de-solvation. The synergistic effect of the complexes not only enhances the flexibility and activity of the enzymes involved in carbohydrate hydrolysis, but also reduces the nonspecific adsorption of lignin residues, and thus improves the thermal stability of cellulase (Zong et al., 2016). By promoting the exposure of the hydrophobic regions, HFBII is able to compete for more hydrophobic adsorption sites in lignin structure and reduce the undesired interactions (Askolin, 2006). Literatures also indicated the recycled cellulase activity reached 73.1% of the initial activity after adding of 2 mg/g of HFBI, while there was only 65.3% in control group without the assistance of HFBI in saccharification. Meanwhile, HFBIIs also possessed high effectiveness in saccharification processes with high solid loading (Zong et al., 2016).

Another widely used passive measure is the addition of surfactants into the buffer for improved enzymatic carbohydrate hydrolysis. As aforementioned in **Section 10.2.2**, the dissolved phenols and lignin polymer interact with cellulase through charge, hydrophobic, and hydrogen bonding, leading to non-productive adsorption (Muñoz et al., 2022). In recent years, various types of surfactants are suggested as promoters of enzyme activity in carbohydrates' saccharification (Holmberg, 2018). According to the charge of the hydrophilic groups, the candidate surfactants can be divided into nonionic (e.g., Span, Tween, Polyethylene glycol), anionic (e.g., Sodium lauryl ether sulfate, Lignosulfonate, Alkylbenzene sulfonates), cationic (e.g., Benzalkonium chloride, Cetylpyridinium chloride) and amphoteric (e.g., Cocamidopropyl betaine) (Muñoz et al., 2022).

The promotion of commonly used surfactants on cellulase hydrolysis is summarized in Table 10.2. In general, the non-ionic surfactants are the most utilized ones in enzymatic hydrolysis of the pretreated pulps, as the high hydrophilic-lipophilic balance caused changes of the hydrophobicity and surface charge in lignin, thereby reducing the unproductive adsorption of cellulase to lignin. For instance, Xu et al. (2019) found that the addition of 0.4% (w/v) polyethylene glycol (PEG) in buffer could significantly increase glucan and xylan hydrolysis from alkali pretreated pulp. Meanwhile, PEG also proved to be a promoter for endoglucanase and xylanase activities. In another study, Chen et al. (2018) observed that the addition of Tween 20 greatly improved the cellulose hydrolysis by cellulase from the dilute acid pretreated wheat straw, and also found that the non-ionic surfactant changed the surface properties of lignin by hydrophobic adsorption. In addition, Tween 80 is another type of inexpensive non-ionic surfactant to promote the enzymatic hydrolysis of pulp. For the hydrolysis of the sugarcane bagasse obtained from $FeCl_3$-catalyzed organosolv pretreatment, the cellulase loading was decreased by half after adding Tween 80 (Zhang et al., 2018). In addition, some surfactants showed the ability to modify the cellulase structure and formed barriers through hydrogen bonds between the hydrophilic

TABLE 10.2
Effect of Different Types Surfactants on Enzymatic Hydrolysis of Pretreated Pulps

Type	Surfactant	Effect	Mechanisms	Reference
Anionic surfactant	Sodium dodecyl sulfate	Inhibition	Bound to cellulase through electrostatic interaction, resulted in partial loss of cellulase hydrolysis activity and a sharp decrease in α-helix content.	Xiang et al. (2006)
	Sodium polystyrene sulfonate	Promotion	Combined with the CBM of cellulase and forming a complex that containing more negative charge, resulted in electrostatic repulsion with lignin and reducing non-productive adsorption.	Wang et al. (2015)
Cationic surfactant	Cetyltrimethy-lammonium Bromide (CTAB)	Inhibition/ Promotion	At low dosages, the negative charge of the enzymatic solution is neutralized, which has a positive impact on enzymatic hydrolysis. while at high dosages, the hydrolysate will be positively charged, which has a negative impact on enzymatic hydrolysis.	Lin et al. (2017)
	PAD-X	Promotion	Precisely control the cation content to reduce the non-productive adsorption of lignin and cellulase.	Xu et al. (2023)
Nonionic surfactant	Tween	Promotion	Increase the polar surface energy component of lignin, making it more hydrophilic, thereby reducing the non-productive adsorption between enzyme and lignin.	Kaar and Holtzapple (1998)
	Triton			Menegol et al. (2014)
	PEG			Li et al. (2012a)
Bifunctional surfactant	Lignin-grafted quaternary ammonium phosphate	Promotion	Phosphobetaine reduced the ineffective adsorption of cellulases by forming a hydration film and electrostatic repulsion on lignin.	Li et al. (2023)

portion and water, thus strengthening the intermolecular forces and reducing the shear force under agitation, thereby reducing the cellulase loading in pulp's saccharification (Muñoz et al., 2022).

It is also suggested that there is a synergistic effect between the surfactant and other additives. For instance, the addition of L-cysteine and Tween 40 together in buffer ensured higher cellulose hydrolysis rate in contrast to the control group that added surfactant alone (Jiang et al., 2023). It was suggested the addition of L-cysteine and Tween 40 in 4-hydroxybenzaldehyde containing buffer led to an increase in the enzymatic hydrolysis efficiency of cellulose I, cellulose II, and cellulose III by 80.87%, 28.73%, and 3.77%, respectively. The reason was that L-cysteine and Tween 40 prevented irreversible/reversible binding of phenolic substances to cellulase by reacting with oxidative intermediates of phenolic compounds, and interacting with phenolic substances through hydrophobic interaction, electrostatic or van der Waals interactions. At the same time, the interaction between these additives and certain amino acids of cellulase could also promote the stabilization of the cellulase structure and maintaining the biological activity (Jiang et al., 2022).

For the function of ionic surfactants on carbohydrate-active enzymes, there is controversy over the interaction mechanism. For example, some anionic surfactants such as SDS caused conformational changes of cellulase through electrostatic attraction and hydrophobic interaction, which further resulted in enzyme activity reduction (Holmberg, 2018). Because of significant changes in the charge of the hydrolysis system on cellulase activity, the promotion of cationic surfactants on enzymatic hydrolysis performances are often insignificant. For example, Lin et al. (2017) study revealed the effect of cationic surfactants on Avicel hydrolysis and cellulase surface charge. When the content of cetyltrimethylammonium bromide (CnTAB), the cationic surfactant, exceeded 0.03 mM, the cellulase solution would be positively charged and negatively impacted on the enzymatic hydrolysis of Avicel. To address the inhibitory effect of high charge density, Xu et al. (2023) synthesized a surfactant namely PAD-X with adjustable cationic content by precisely controlling the ratio of the non-ionic unit Acrylamide (AM) and the cationic unit [2-(methacryloyloxy)ethyl] trimethylammonium chloride (DMC). In subsequent enzymatic hydrolysis process, a significant increase in the enzymatic hydrolysis efficiency of dilute acid-treated corn cob residue (CCR) was realized after adding the as prepared PAD-X, even with a low cellulase loading. Xu et al. (2023) also pointed out the enhanced saccharification performances after adding PAD-X was attributed to the elimination of the irreversible binding between lignin and cellulase.

In addition, there are also some bifunctional surfactants that can be used as the promoter for enhanced enzymatic saccharification performances. For instance, Li et al. (2023) grafted quaternary ammonium phosphate (QAP) onto enzymatic hydrolysis lignin (EHL) to obtain a temperature and pH responsive amphoteric lignin-grafted quaternary ammonium phosphate (LQAP). The positive effect of LQAP on enzymatic saccharification was as follows: On the one hand, the phosphobetaine in LQAP reduced the ineffective adsorption of cellulases by forming a hydration film and electrostatic repulsion on lignin surface. On the other hand, the salt-forming properties of positive and negative ions in phosphobetaine at low temperatures endow the LQAP with an upper critical solution temperature (UCST), thus promoting the co-precipitation of cellulase for recovery. Consequently, the enzymatic hydrolysis of CCR by LQAP was significantly improved.

10.3.3 Improving the Cellulase Tolerance by Modification

In addition to the proactive and passive measures, the continuous improvement of the lignin tolerances for carbohydrate-active enzymes is another eternal theme (Nordwald et al., 2014). In recent years, the emerging AI prediction strategies and the gradually matured metabolic engineering tools and methods for the proteins structure and expression opened a new window in the artificial construction of targeted enzymes that with specific performance, and thus largely prospered the related researches focused on highly lignin tolerance cellulases' designation and production (Pant et al., 2024). For instance, through rational design and random mutagenesis, 1.6 to 5.3 times higher cellulase activity was received after improving the thermal stability and pH stability (Contreras et al., 2020). This provides a direction to develop new anti-phenolic tolerant enzymes, although the research concentrated on the specific mitigation of lignin inhibitions are still rare at the current stage. In addition, through directed mutagenesis or engineering techniques, properties of cellulase can be enhanced (Akram et al., 2021). Charge engineering has been employed to simultaneously improve the tolerance of cellulase secreting from *T. reesei* toward the residual ionic liquids and lignin in saccharification process (Nordwald et al., 2014).

Another important direction to improve the cellulase tolerance of lignin is to chemical modifying the initial structure of enzymes. On this basis, polyethylene glycol derivatives (HO-PEG-COOH and HOOC-PEG-COOH) were used to modify cellulase, which tightened the configuration, altered the spatial distribution of peptide units and aromatic amino acid residues, and increased the α-helix content in order to increase cellulase activity and stability. (Chen et al., 2023). At the same time, the introduction of hydrophilic groups enabled cellulase to resist lignin adsorption. In addition, altering the microenvironment by immobilization is an effective method to minimize the negative interactions between enzymes and lignin. For example, to improve enzyme tolerance toward derived inhibitors from pretreatment, cellulase was immobilized on a magnetic core-shell metal-organic framework, resulting in increased tolerance to formic acid and vanillin. The reason was the inaccessibility of inhibitory phenols to the immobilized enzyme in the artificial nano-environment (Qi et al., 2018). However, in consideration of the accessibility of enzymes, enzyme immobilization requires properly avoiding the formation of new mass transfer resistance by the carrier, as the spatial structure of carbohydrates is difficult to be transformed into the inner species of the porous partials (Sigurdardóttir et al., 2018). This intrinsic limitation led to difficulties in realistic lignocellulosic monosugars production from the pretreated pulp using the immobilized cellulase (Eldin and Mita, 2014).

10.4 CONCLUSION AND PERSPECTIVES

For the lignocellulosic monosugars production, safer, cleaner, and more effective enzymatic hydrolysis of the carbohydrate-rich pulp is the primary demand. As an important issue to be concerned in pulp's saccharification, the proper mitigation of the negative influences from the lignin is critical to the reduction of enzyme's cost and maintaining the enzyme's activity. The inhibition of carbohydrate enzymatic hydrolysis is largely ascribed to the physical hindrance caused by steric effects of the condensed lignin polymer on the surface of pulp, the precipitation and inactivation of enzymes caused by the soluble phenols, and non-productive adsorption of cellulase via electrostatic, hydrogen bonding, and hydrophobic effects.

To date, various strategies have been suggested to mitigate the lignin inhibition toward the carbohydrate-active enzymes, and these strategies can be generally divided into proactive and passive technologies according to the detoxify mechanisms and the actions in different units. Generally, the passive measurements of the lignin inhibition in enzymatic saccharification process such as the addition of assistant enzymes, surfactant, and the non-enzymatic proteins is more practicable in most cases. The as-prepared enzyme cocktails could obviously decrease the cellulase dosage and promote the carbohydrates hydrolysis, thereby reducing the overall enzyme cost. In addition, changing the cellulase structure by chemical modification, and the stabilization of the enzyme structure by immobilization are also suggested in previous works, nonetheless there are still intrinsic weaknesses of these technologies such as the loss of enzyme activity after chemical modification, and the new steric hindrance caused by the immobilization carrier. With the breakthroughs in the AI and genetic technologies, the artificial designation of specific hyper-tolerant enzymes would become an important direction in future researches, whereas most of the current reports are focused on the improvement of the enzyme activity and the stability, rather than the tolerance of inhibitors in hydrolysate. Therefore, it is encouraged to design the high lignin-tolerant enzymes so that solve the inhibition once and for all.

AUTHOR'S CONTRIBUTION STATEMENT

D Cai provided necessary guidance to ensure the rigor of the content. YK Wang wrote the introduction section, and summarized all the sources and polish language of the Chapter. HG Zhang, HJ Zhang, and YX Gao wrote the draft manuscript of Sections 10.2, 10.3, and 10.4, respectively, and ZC Liao and CJ Chen prepared the figures in this chapter.

FUNDING

This chapter is supported by the National Natural Science Foundation of China (Grant Nos. 22478021).

CONFLICT OF INTEREST

There are no conflicts of interest to declare.

REFERENCES

Adsul, M., Sandhu, S.K., Singhania, R.R., Gupta, R., Puri, S.K., Mathur, A., 2020. Designing a cellulolytic enzyme cocktail for the efficient and economical conversion of lignocellulosic biomass to biofuels. *Enzyme Microb. Technol.* 133, 109442.

Akram, F., ul Haq, I., Aqeel, A., Ahmed, Z., Shah, F.I., 2021. Thermostable cellulases: structure, catalytic mechanisms, directed evolution and industrial implementations. *Renew. Sustain. Energy Rev.* 151, 111597.

Albornoz-Palma, G., Ching, D., Valerio, O., Mendonça, R.T., Pereira, M., 2020. Effect of lignin and hemicellulose on the properties of lignocellulose nanofibril suspensions. *Cellulose* 27, 10631–10647.

Ariaeenejad, S., Barani, M., Sarani, M., Lohrasbi-Nejad, A., Mohammadi-Nejad, G., Salekdeh, G.H., 2024. Green synthesis of NiO NPs for metagenome-derived laccase stabilization: detoxifying pollutants and wastes. *Int. J. Biol. Macromol.* 266, 130986.

Arregui, L., Ayala, M., Gómez-Gil, X., Gutiérrez-Soto, G., Hernández-Luna, C.E., Herrera De Los Santos, M., Levin, L., Rojo-Domínguez, A., Romero-Martínez, D., Saparrat, M.C., 2019. Laccases: structure, function, and potential application in water bioremediation. *Microb. Cell Fact.* 18, 1–33.

Askolin, S., 2006. *Characterization of the Trichoderma reesei hydrophobins HFBI and HFBII.* VTT Technical Research Centre of Finland.

Ayodele, B.V., Alsaffar, M.A., Mustapa, S.I., 2020. An overview of integration opportunities for sustainable bioethanol production from first-and second-generation sugar-based feedstocks. *J. Cleaner Prod.* 245, 118857.

Brethauer, S., Studer, M.H., Yang, B., Wyman, C.E., 2011. The effect of bovine serum albumin on batch and continuous enzymatic cellulose hydrolysis mixed by stirring or shaking. *Bioresour. Technol.* 102(10), 6295–6298.

Cai, C., Zhang, C., Li, N., Liu, H., Xie, J., Lou, H., Pan, X., Zhu, J., Wang, F., 2023. Changing the role of lignin in enzymatic hydrolysis for a sustainable and efficient sugar platform. *Renew. Sustain. Energy Rev.* 183, 113445.

Cai, D., Li, P., Luo, Z., Qin, P., Chen, C., Wang, Y., Wang, Z., Tan, T., 2016. Effect of dilute alkaline pretreatment on the conversion of different parts of corn stalk to fermentable sugars and its application in acetone–butanol–ethanol fermentation. *Bioresour. Technol.* 211, 117–124.

Cavka, A., Jönsson, L.J., 2013. Detoxification of lignocellulosic hydrolysates using sodium borohydride. *Bioresour. Technol.* 136, 368–376.

Chandra, R.P., Gourlay, K., Kim, C.-S., Saddler, J.N., 2015. Enhancing hemicellulose recovery and the enzymatic hydrolysis of cellulose by adding lignosulfonates during the two-stage steam pretreatment of poplar. *ACS Sustain. Chem. Eng.* 3(5), 986–991.

Chen, H., Fu, Y., Cui, C., Yu, S., Liu, S., Liu, Y., Liu, Y., Nie, G., Nie, S., Yao, S., 2023. Construction of novel cellulase with high activity, great stability and excellent lignin-resistant for efficient enzymatic hydrolysis by hydrophilic and negatively charged modification. *Ind. Crops Prod.* 191, 115910.

Chen, X., Zhai, R., Li, Y., Yuan, X., Liu, Z.-H., Jin, M., 2020. Understanding the structural characteristics of water-soluble phenolic compounds from four pretreatments of corn stover and their inhibitory effects on enzymatic hydrolysis and fermentation. *Biotechnol. Biofuels* 13, 1–13.

Chen, Y.-A., Zhou, Y., Qin, Y., Liu, D., Zhao, X., 2018. Evaluation of the action of Tween 20 non-ionic surfactant during enzymatic hydrolysis of lignocellulose: pretreatment, hydrolysis conditions and lignin structure. *Bioresour. Technol.* 269, 329–338.

Chu, Q., Chandra, R.P., Kim, C.-S., Saddler, J.N., 2017. Alkali-oxygen impregnation prior to steam pretreating poplar wood chips enhances selective lignin modification and removal while maximizing carbohydrate recovery, cellulose accessibility, and enzymatic hydrolysis. *ACS Sustain. Chem. Eng.* 5(5), 4011–4017.

Contreras, F., Pramanik, S., M. Rozhkova, A., N. Zorov, I., Korotkova, O., P. Sinitsyn, A., Schwaneberg, U., D. Davari, M., 2020. Engineering robust cellulases for tailored lignocellulosic degradation cocktails. *Int. J. Mol. Sci.* 21(5), 1589.

Dharmapriya, T.N., Huang, P.-J., 2024. Exploring recent significant catalytic systems for the conversion of glucose into bio-based chemicals: a concise review. *J. Taiwan Inst. Chem. Eng.* 162, 105585.

Ding, N., Liu, H., Sun, Y., Tang, X., Lei, T., Xu, F., Zeng, X., Lin, L., 2021. Lignin degradation in cooking with active oxygen and solid Alkali process: a mechanism study. *J. Cleaner Prod.* 278, 123984.

Duret, X., Fredon, E., Masson, E., Desharnais, L., Gérardin, P., 2013. Optimization of acid pretreatment in order to increase the Phenolic content of Picea abies bark by surface response methodology. *BioResources* 8(1), 13–27.

Eldin, M., Mita, D., 2014. Immobilized enzymes: strategies for overcoming the substrate diffusion-limitation problem. *Current Biotechnol.* 3(3), 207–217.

Eriksson, T., Börjesson, J., Tjerneld, F., 2002. Mechanism of surfactant effect in enzymatic hydrolysis of lignocellulose. *Enzyme Microb. Technol.* 31(3), 353–364.

Florencio, C., Badino, A.C., Farinas, C.S., 2016. Soybean protein as a cost-effective lignin-blocking additive for the saccharification of sugarcane bagasse. *Bioresour. Technol.* 221, 172–180.

Freitas, J. V., & Farinas, C. S. 2017. Sugarcane bagasse fly ash as a no-cost adsorbent for removal of phenolic inhibitors and improvement of biomass saccharification. *ACS Sustain. Chemis. Eng.* 5(12), 11727–11736.

Freitas, J. V., Nogueira, F. G., & Farinas, C. S. 2019. Coconut shell activated carbon as an alternative adsorbent of inhibitors from lignocellulosic biomass pretreatment. *Indust. Crops Prod.* 137, 16–23.

Haldar, D., Dey, P., Patel, A.K., Dong, C.-D., Singhania, R.R., 2022. A critical review on the effect of lignin redeposition on biomass in controlling the process of enzymatic hydrolysis. *Bioenergy Res.* 15, 1–12.

Hodgson, K.T., Berg, J.C., 1988. Dynamic wettability properties of single wood pulp fibers and their relationship to absorbency. *Wood Fiber Sci.* 20(1), 3–17.

Holmberg, K., 2018. Interactions between surfactants and hydrolytic enzymes. *Colloids Surf., B* 168, 169–177.

Hou, J., Zhang, Q., Tian, F., Liu, F., Jiang, J., Qin, J., Wang, H., Wang, J., Chang, S., Hu, X., 2024. Structure changes of lignin and their effects on enzymatic hydrolysis for bioethanol production: a focus on lignin modification. *J. Biotechnol*, 393, 61–73.

Hu, F., Jung, S., Ragauskas, A., 2013. Impact of pseudolignin versus dilute acid-pretreated lignin on enzymatic hydrolysis of cellulose. *ACS Sustain. Chem. Eng.* 1(1), 62–65.

Ilanidis, D., 2021. Biochemical conversion of biomass: hydrothermal pretreatment, by-product formation, conditioning, enzymatic saccharification, and fermentability, Umeå University.

Ilanidis, D., Stagge, S., Alriksson, B., Cavka, A., Jönsson, L.J., 2021a. Comparison of efficiency and cost of methods for conditioning of slurries of steam-pretreated softwood. *Front. Energy Res.* 9, 701980.

Ilanidis, D., Stagge, S., Alriksson, B., Jönsson, L.J., 2021b. Factors affecting detoxification of softwood enzymatic hydrolysates using sodium dithionite. *Processes* 9(5), 887.

Ilić, N., Milić, M., Beluhan, S., Dimitrijević-Branković, S., 2023. Cellulases: from lignocellulosic biomass to improved production. *Energies* 16(8), 3598.

Islam, M.K., Rehman, S., Guan, J., Lau, C.-Y., Tse, H.-Y., Yeung, C.S., Leu, S.-Y., 2021. Biphasic pretreatment for energy and carbon efficient conversion of lignocellulose into bioenergy and reactive lignin. *Appl. Energy* 303, 117653.

Jeoh, T., Wilson, D.B., Walker, L.P., 2006. Effect of cellulase mole fraction and cellulose recalcitrance on synergism in cellulose hydrolysis and binding. *Biotechnol. Progr.* 22(1), 270–277.

Jiang, X., Zhai, R., Leng, Y., Deng, Q., Jin, M., 2022. Understanding the toxicity of lignin-derived phenolics towards enzymatic saccharification of lignocellulose for rationally developing effective in-situ mitigation strategies to maximize sugar production from lignocellulosic biorefinery. *Bioresour. Technol.* 349, 126813.

Jiang, X., Zhai, R., Li, H., Li, C., Deng, Q., Wu, X., Jin, M., 2023. Binary additives for in-situ mitigating the inhibitory effect of lignin-derived phenolics on enzymatic hydrolysis of lignocellulose: enhanced performance and synergistic mechanism. *Energy* 282, 128062.

Kaar, W.E., Holtzapple, M.T., 1998. Benefits from Tween during enzymic hydrolysis of corn stover. *Biotechnol. Bioeng.* 59(4), 419–427.

Kalyani, D., Tiwari, M.K., Li, J., Kim, S.C., Kalia, V.C., Kang, Y.C., Lee, J.-K., 2015. A highly efficient recombinant laccase from the yeast Yarrowia lipolytica and its application in the hydrolysis of biomass. *PLoS One* 10(3), e0120156.

Kim, D., 2018. Physico-chemical conversion of lignocellulose: inhibitor effects and detoxification strategies: a mini review. *Molecules* 23(2), 309.

Kim, Y., Ximenes, E., Mosier, N.S., Ladisch, M.R., 2011. Soluble inhibitors/deactivators of cellulase enzymes from lignocellulosic biomass. *Enzyme Microb. Technol.* 48(4–5), 408–415.

Lan, T., Lou, H., Zhu, J., 2013. Enzymatic saccharification of lignocelluloses should be conducted at elevated pH 5.2–6.2. *Bioenergy Res.* 6, 476–485.

Lan, T., Wang, S., Li, H., Qin, Y., Yue, G., 2020. Effect of lignin isolated from p-toluenesulfonic acid pretreatment liquid of sugarcane bagasse on enzymatic hydrolysis of cellulose and cellulase adsorption. *Ind. Crops Prod.* 155, 112768.

Leu, S.-Y., Zhu, J., 2013. Substrate-related factors affecting enzymatic saccharification of lignocelluloses: our recent understanding. *Bioenergy Res.* 6, 405–415.

Li, F., Liang, H., Shan, J., Zhang, A., Lou, H., Tang, Y., 2023. Lignin-grafted quaternary ammonium phosphate with temperature and pH responsive behavior for improved enzymatic hydrolysis and cellulase recovery. *Int. J. Biol. Macromol.* 234, 123779.

Li, H., Pu, Y., Kumar, R., Ragauskas, A.J., Wyman, C.E., 2014. Investigation of lignin deposition on cellulose during hydrothermal pretreatment, its effect on cellulose hydrolysis, and underlying mechanisms. *Biotechnol. Bioeng.* 111(3), 485–492.

Li, J., Li, S., Fan, C., Yan, Z., 2012a. The mechanism of poly (ethylene glycol) 4000 effect on enzymatic hydrolysis of lignocellulose. *Colloids Surf., B* 89, 203–210.

Li, X., Luo, X., Li, K., Zhu, J., Fougere, J.D., Clarke, K., 2012b. Effects of SPORL and dilute acid pretreatment on substrate morphology, cell physical and chemical wall structures, and subsequent enzymatic hydrolysis of lodgepole pine. *Appl. Biochem. Biotechnol.* 168, 1556–1567.

Li, X., Zheng, Y., 2017. Lignin-enzyme interaction: mechanism, mitigation approach, modeling, and research prospects. *Biotechnol. Adv.* 35(4), 466–489.

Lin, W., Ning, D., Xu, D., Wu, S., Zou, Q., Mondal, A.K., Huang, F., 2024. Study on the effect of combined and free lignin to the cellulose enzymatic hydrolysis. *Ind. Crops Prod.* 222, 119516.

Lin, X., Cai, C., Lou, H., Qiu, X., Pang, Y., Yang, D., 2017. Effect of cationic surfactant cetyltrimethylammonium bromide on the enzymatic hydrolysis of cellulose. *Cellulose* 24, 61–68.

Linder, M., Selber, K., Nakari-Setälä, T., Qiao, M., Kula, M.-R., Penttilä, M., 2001. The hydrophobins HFBI and HFBII from Trichoderma r eesei showing efficient interactions with Nonionic Surfactants in aqueous two-phase systems. *Biomacromolecules* 2(2), 511–517.

Liu, H., Sun, J., Leu, S.Y., Chen, S., 2016. Toward a fundamental understanding of cellulase-lignin interactions in the whole slurry enzymatic saccharification process. *Biofuels, Bioprod. Biorefin.* 10(5), 648–663.

Liu, T., Yang, L., Liu, B., Tan, L., 2019. Hydroxycinnamic acids release during bioconversion of corn stover and their effects on lignocellulolytic enzymes. *Bioresour. Technol.* 294, 122116.

Lou, H., Zhu, J., Lan, T.Q., Lai, H., Qiu, X., 2013. pH-Induced lignin surface modification to reduce nonspecific cellulase binding and enhance enzymatic saccharification of lignocelluloses. *ChemSusChem* 6(5), 919–927.

Luo, Z., Li, P., Cai, D., Chen, Q., Qin, P., Tan, T., Cao, H., 2017. Comparison of performances of corn fiber plastic composites made from different parts of corn stalk. *Ind. Crops Prod.* 95, 521–527.

Ma, J., Shi, S., Jia, X., Xia, F., Ma, H., Gao, J., Xu, J., 2019. Advances in catalytic conversion of lignocellulose to chemicals and liquid fuels. *J. Energy Chem.* 36, 74–86.

Maitra, S., Singh, V., 2021. Balancing sugar recovery and inhibitor generation during energycane processing: coupling cryogenic grinding with hydrothermal pretreatment at low temperatures. *Bioresour. Technol.* 321, 124424.

Menegol, D., Scholl, A.L., Fontana, R.C., Dillon, A.J.P., Camassola, M., 2014. Increased release of fermentable sugars from elephant grass by enzymatic hydrolysis in the presence of surfactants. *Energy Convers. Manage.* 88, 1252–1256.

Michelin, M., Ximenes, E., de Moraes, M.D.L.T., Ladisch, M.R., 2016. Effect of phenolic compounds from pretreated sugarcane bagasse on cellulolytic and hemicellulolytic activities. *Bioresour. Technol.* 199, 275–278.

Mohanram, S., Amat, D., Choudhary, J., Arora, A., Nain, L., 2013. Novel perspectives for evolving enzyme cocktails for lignocellulose hydrolysis in biorefineries. *Sustain. Chem. Process.* 1, 1–12.

Moxley, G., Gaspar, A.R., Higgins, D., Xu, H., 2012. Structural changes of corn stover lignin during acid pretreatment. *J. Ind. Microbiol. Biotechnol.* 39(9), 1289–1299.

Muñoz, S.S., Balbino, T.R., Alba, E.M., Barbosa, F.G., de Pier, F.T., de Almeida, A.L.M., Zilla, A.H.B., Antunes, F.A.F., Hilares, R.T., Balagurusamy, N., 2022. Surfactants in biorefineries: role, challenges & perspectives. *Bioresour. Technol.* 345, 126477.

Nakagame, S., Chandra, R.P., Kadla, J.F., Saddler, J.N., 2011a. Enhancing the enzymatic hydrolysis of lignocellulosic biomass by increasing the carboxylic acid content of the associated lignin. *Biotechnol. Bioeng.* 108(3), 538–548.

Nakagame, S., Chandra, R.P., Saddler, J.N., 2011b. The influence of lignin on the enzymatic hydrolysis of pretreated biomass substrates. In J.Y. Zhu, X. Zhang, & X. Pan (Eds.), *Sustainable production of fuels, chemicals, and fibers from forest biomass*, 145–167. American Chemical Society.

Nordwald, E.M., Brunecky, R., Himmel, M.E., Beckham, G.T., Kaar, J.L., 2014. Charge engineering of cellulases improves ionic liquid tolerance and reduces lignin inhibition. *Biotechnol. Bioeng.* 111(8), 1541–1549.

Palonen, H., Tjerneld, F., Zacchi, G., Tenkanen, M., 2004. Adsorption of Trichoderma reesei CBH I and EG II and their catalytic domains on steam pretreated softwood and isolated lignin. *J. Biotechnol.* 107(1), 65–72.

Pan, X., 2008. Role of functional groups in lignin inhibition of enzymatic hydrolysis of cellulose to glucose. *J. Biobased Mater. Bioenergy* 2(1), 25–32.

Pant, S., Prakash, A., Vundavilli, P.R., Khadanga, K.C., Kuila, A., Aminabhavi, T.M., Garlapati, V.K., 2024. Concomitant inhibitor-tolerant cellulase and xylanase production towards sustainable bioethanol production by Zasmidiumcellare CBS 146.36. *Fuel* 375, 132593.

Pei, Z., Liu, X., Chen, J., Wang, H., Li, H., 2024. Research progress on lignin depolymerization strategies: a review. *Polymers* 16(17), 2388.

Pielhop, T., 2023. Suppression of lignin repolymerisation to enhance cellulose bioconversion and lignin valorisation–A review. *Chimia* 77(6), 403–416.

Pinheiro, V.E., Horváth, I.S., Lundin, M., Polizeli, M.D.L.T.D.M., 2021. Screening and cocktail optimization using experimental mixture design: enzymatic saccharification as a biological pretreatment strategy. *Biofuels, Bioprod. Biorefin.* 15(5), 1447–1460.

Qaseem, M.F., Shaheen, H., Wu, A.-M., 2021. Cell wall hemicellulose for sustainable industrial utilization. *Renew. Sustain. Energy Rev.* 144, 110996.

Qi, B., Luo, J., Wan, Y., 2018. Immobilization of cellulase on a core-shell structured metal-organic framework composites: better inhibitors tolerance and easier recycling. *Bioresour. Technol.* 268, 577–582.

Qin, C., Clarke, K., Li, K., 2014. Interactive forces between lignin and cellulase as determined by atomic force microscopy. *Biotechnol. Biofuels* 7, 1–10.

Qin, L., Li, W.-C., Liu, L., Zhu, J.-Q., Li, X., Li, B.-Z., Yuan, Y.-J., 2016. Inhibition of lignin-derived phenolic compounds to cellulase. *Biotechnol. Biofuels* 9, 1–10.

Rahikainen, J.L., Evans, J.D., Mikander, S., Kalliola, A., Puranen, T., Tamminen, T., Marjamaa, K., Kruus, K., 2013. Cellulase–lignin interactions—the role of carbohydrate-binding module and pH in non-productive binding. *Enzyme Microb. Technol.* 53(5), 315–321.

Rakotoarivonina, H., Hermant, B., Aubry, N., Rémond, C., 2015. Engineering the hydrophobic residues of a GH11 xylanase impacts its adsorption onto lignin and its thermostability. *Enzyme Microb. Technol.* 81, 47–55.

Roy, R., Rahman, M.S., Raynie, D.E., 2020. Recent advances of greener pretreatment technologies of lignocellulose. *Curr. Res. Green Sustain. Chem.* 3, 100035.

Saini, J.K., Patel, A.K., Adsul, M., Singhania, R.R., 2016. Cellulase adsorption on lignin: a roadblock for economic hydrolysis of biomass. *Renew. Energy* 98, 29–42.

Saravanakumar, T., Park, H.-S., Mo, A.-Y., Choi, M.-S., Kim, D.-H., Park, S.-M., 2016. Detoxification of furanic and phenolic lignocellulose derived inhibitors of yeast using laccase immobilized on bacterial cellulosic nanofibers. *J. Mol. Catal. B: Enzym.* 134, 196–205.

Selig, M.J., Viamajala, S., Decker, S.R., Tucker, M.P., Himmel, M.E., Vinzant, T.B., 2007. Deposition of lignin droplets produced during dilute acid pretreatment of maize stems retards enzymatic hydrolysis of cellulose. *Biotechnol. Progr.* 23(6), 1333–1339.

Sigurdardóttir, S.B., Lehmann, J., Ovtar, S., Grivel, J.C., Negra, M.D., Kaiser, A., Pinelo, M., 2018. Enzyme immobilization on inorganic surfaces for membrane reactor applications: mass transfer challenges, enzyme leakage and reuse of materials. *Adv. Synth. Catal.* 360(14), 2578–2607.

Song, W., Peng, L., Bakhshyar, D., He, L., Zhang, J., 2021. Mild O2-aided alkaline pretreatment effectively improves fractionated efficiency and enzymatic digestibility of Napier grass stem towards a sustainable biorefinery. *Bioresour. Technol.* 319, 124162.

Soudham, V.P., Alriksson, B., Jönsson, L.J., 2011. Reducing agents improve enzymatic hydrolysis of cellulosic substrates in the presence of pretreatment liquid. *J. Biotechnol.* 155(2), 244–250.

Stamogiannou, I., Van Camp, J., Smagghe, G., Van de Walle, D., Dewettinck, K., Raes, K., 2021. Impact of phenolic compound as activators or inhibitors on the enzymatic hydrolysis of cellulose. *Int. J. Biol. Macromol.* 186, 174–180.

Su, C., Hirth, K., Liu, Z., Cao, Y., & Zhu, J. Y. 2021. Acid hydrotropic fractionation of switchgrass at atmospheric pressure using maleic acid in comparison with p-TsOH: Advantages of lignin esterification. *Indust. Crops Prod.* 159, 113017.

Sun, S., Huang, Y., Sun, R., Tu, M., 2016. The strong association of condensed phenolic moieties in isolated lignins with their inhibition of enzymatic hydrolysis. *Green Chem.* 18(15), 4276–4286.

Tian, Y., Jiang, Y., Ou, S., 2013. Interaction of cellulase with three phenolic acids. *Food Chem.* 138(2–3), 1022–1027.

Trajano, H.L., Engle, N.L., Foston, M., Ragauskas, A.J., Tschaplinski, T.J., Wyman, C.E., 2013. The fate of lignin during hydrothermal pretreatment. *Biotechnol. Biofuels* 6, 1–16.

Vohra, R., Shirkot, C., Dhawan, S., Gupta, K., 1980. Effect of lignin and some of its components on the production and activity of cellulase (s) by Trichoderma reesei. *Biotechnol. Bioeng.* 22(7), 1497–1500.

Wang, H., Kobayashi, S., Mochidzuki, K., 2015. Effect of non-enzymatic proteins on enzymatic hydrolysis and simultaneous saccharification and fermentation of different lignocellulosic materials. *Bioresour. Technol.* 190, 373–380.

Wang, H., Mochidzuki, K., Kobayashi, S., Hiraide, H., Wang, X., Cui, Z., 2013. Effect of bovine serum albumin (BSA) on enzymatic cellulose hydrolysis. *Appl. Biochem. Biotechnol.* 170, 541–551.

Wang, K., Guo, H., You, S., Yang, H., Li, N., Liang, X., Qing, Y., Huang, Y., 2024a. Interpretive analysis of the relationship between pseudolignin structure and enzymatic hydrolysis. *Ind. Crops Prod.* 214, 118519.

Wang, Y., Zhang, Y., Cui, Q., Feng, Y., Xuan, J., 2024b. Composition of lignocellulose hydrolysate in different biorefinery strategies: nutrients and inhibitors. *Molecules* 29(10), 2275.

Wu, Y., Su, C., Liao, Z., Zhang, G., Jiang, Y., Wang, Y., Zhang, C., Cai, D., Qin, P., Tan, T., 2024. Sequential catalytic lignin valorization and bioethanol production: an integrated biorefinery strategy. *Biotechnol. Biofuels Bioprod.* 17(1), 8.

Wu, Y., Wen, J., Su, C., Jiang, C., Zhang, C., Wang, Y., Jiang, Y., Ren, W., Qin, P., Cai, D., 2023. Inhibitions of microbial fermentation by residual reductive lignin oil: concerns on the bioconversion of reductive catalytic fractionated carbohydrate pulp. *Chem. Eng. J.* 452, 139267.

Xiang, J., Fan, J.-B., Chen, N., Chen, J., Liang, Y., 2006. Interaction of cellulase with sodium dodecyl sulfate at critical micelle concentration level. *Colloids Surf., B* 49(2), 175–180.

Ximenes, E., Kim, Y., Ladisch, M.R., 2013. Biological conversion of plants to fuels and chemicals and the effects of inhibitors. In C.E. Wyman (Ed.), *Aqueous pretreatment of plant biomass for biological and chemical conversion to fuels and chemicals*, 39–60. Wiley.

Ximenes, E., Kim, Y., Mosier, N., Dien, B., & Ladisch, M. 2010. Inhibition of cellulases by phenols. *Enzyme Microb. Technol.* 46(3–4), 170–176.

Ximenes, E., Kim, Y., Mosier, N., Dien, B., Ladisch, M., 2011. Deactivation of cellulases by phenols. *Enzyme Microb. Technol.* 48(1), 54–60.

Xu, L., Zhang, A., Liao, R., Pang, Y., Yang, D., Lou, H., Qiu, X., 2023. Copolymer with tunable cation content as an efficient promoter in enzymatic saccharification of lignocellulose. *ACS Appl. Polym. Mater.* 5(11), 8997–9006.

Xu, C., Zhang, J., Zhang, Y., Guo, Y., Xu, H., Xu, J., & Wang, Z. 2019. Enhancement of high-solids enzymatic hydrolysis efficiency of alkali pretreated sugarcane bagasse at low cellulase dosage by fed-batch strategy based on optimized accessory enzymes and additives. *Biores. Technol.* 292, 121993.

Yang, Q., Pan, X., 2016. Correlation between lignin physicochemical properties and inhibition to enzymatic hydrolysis of cellulose. *Biotechnol. Bioeng.* 113(6), 1213–1224.

Yang, Q., Shi, J., Lin, L., 2012. Characterization of structural changes of lignin in the process of cooking of bagasse with solid alkali and active oxygen as a pretreatment for lignin conversion. *Energy Fuels* 26(11), 6999–7004.

Yao, L., Yang, H., Meng, X., Ragauskas, A.J., 2022. Toward a fundamental understanding of the role of lignin in the biorefinery process. *Front. Energy Res.* 9, 804086.

Ying, W., Shi, Z., Yang, H., Xu, G., Zheng, Z., Yang, J., 2018. Effect of alkaline lignin modification on cellulase–lignin interactions and enzymatic saccharification yield. *Biotechnol. Biofuels* 11, 1–13.

Yu, H., Li, X., Zhang, W., Sun, D., Jiang, J., Liu, Z., 2015. Hydrophilic pretreatment of furfural residues to improve enzymatic hydrolysis. *Cellulose* 22, 1675–1686.

Yuan, Y., Jiang, B., Chen, H., Wu, W., Wu, S., Jin, Y., Xiao, H., 2021. Recent advances in understanding the effects of lignin structural characteristics on enzymatic hydrolysis. *Biotechnol. Biofuels* 14, 1–20.

Zhai, R., Hu, J., Jin, M., 2022. Towards efficient enzymatic saccharification of pretreated lignocellulose: enzyme inhibition by lignin-derived phenolics and recent trends in mitigation strategies. *Biotechnol. Adv.* 61, 108044.

Zhang, H., Zhang, S., Yuan, H., Lyu, G., Xie, J., 2018. $FeCl_3$-catalyzed ethanol pretreatment of sugarcane bagasse boosts sugar yields with low enzyme loadings and short hydrolysis time. *Bioresour. Technol.* 249, 395–401.

Zhang, X., Yang, W., Blasiak, W., 2011a. Modeling study of woody biomass: interactions of cellulose, hemicellulose, and lignin. *Energy Fuels* 25(10), 4786–4795.

Zhang, Y., Jiang, X., Wan, S., Wu, W., Wu, S., Jin, Y., 2020. Adsorption behavior of two glucanases on three lignins and the effect by adding sulfonated lignin. *J. Biotechnol.* 323, 1–8.

Zhang, Y., Vitry, A., Caughey, G., Roughead, E.E., Ryan, P., Gilbert, A., Shakib, S., McDermott, R., 2011b. The association between co-morbidity and the use of antidiabetics or adjunctive cardiovascular medicines in Australian veterans with diabetes. *Diabetes Res. Clin. Pract.* 91(1), 115–120.

Zhao, X., Meng, X., Ragauskas, A.J., Lai, C., Ling, Z., Huang, C., Yong, Q., 2022. Unlocking the secret of lignin-enzyme interactions: recent advances in developing state-of-the-art analytical techniques. *Biotechnol. Adv.* 54, 107830.

Zheng, W., Qin, Y., Zhuang, X., Xiao, H., Liu, C., Fu, X., Lin, Q., Lan, T., 2024. Recycling enzymatic hydrolysis lignin residues saved cellulase in enzymatic hydrolysis of lignocellulose: an insight from cellulase adsorption mechanism. *Ind. Crops Prod.* 208, 117884.

Zhou, M., Tian, X., 2022. Development of different pretreatments and related technologies for efficient biomass conversion of lignocellulose. *Int. J. Biol. Macromol.* 202, 256–268.

Zhu, J., Pan, X., Wang, G., Gleisner, R., 2009. Sulfite pretreatment (SPORL) for robust enzymatic saccharification of spruce and red pine. *Bioresour. Technol.* 100(8), 2411–2418.

Zong, Z., He, R., Fu, H., Zhao, T., Chen, S., Shao, X., Zhang, D., Cai, W., 2016. Pretreating cellulases with hydrophobins for improving bioconversion of cellulose: an experimental and computational study. *Green Chem.* 18(24), 6666–6674.

11 Carbohydrate-Active Enzymes

Role in Food and Agriculture

Pallavi S. Shere, Nihar N. Jog, and
**Shadab Ahmed*

ABBREVIATIONS

AOS	Alginate oligosaccharides
CAGR	Compound annual growth rate
CAZymes	Carbohydrate-active enzymes
E1	Emulsification index
GH	Glycosyl hydrolase
GTs	Glycosyltransferases
HFCS	High fructose corn syrup
HMCS	High maltose corn syrup
ICT	Intramolecular charge transfer
PELs	Pectate lyases
PLs	Polysaccharide lyases
UOS	Ulvan oligosaccharides
XOS	Xylo-oligosaccharides

11.1 INTRODUCTION

According to the article from Grandview research (www.grandviewresearch.com/industry-analysis/food-enzymes-market.), the worldwide market for industrial enzymes was valued at US$ 7.42 billion during the year 2023 and is projected to see an increase at a compound annual growth rate (CAGR) of 6.3% from 2024 to 2030. This growth is driven by increasing demand in different industries, particularly food and beverages, detergents, animal feed, and more. Enzymes can substitute synthetic catalysts in various processes, resulting in lower energy consumption and reduced environmental impact. A complex enzymatic process recycles a vast array of polysaccharides into simple sugars, which serve as one among the available vital energy sources for microbes. The reprocessing of various organic compounds relies mostly on carbohydrate-active enzymes (CAZymes), which degrade polysaccharides using a wide range of sophisticated mechanisms. CAZymes are modular enzymes that are attached to carbohydrate-binding modules, which are important in polysaccharide recognition, leading to the promotion of effective catalysis (Gavande et al., 2023). Based on their primary sequence homology, CAZymes have

DOI: 10.1201/9781003561682-11

been categorized into various families, each with its own functional significance; these include glycoside hydrolases, glycosyltransferase (GT), polysaccharide lyase (PL), carbohydrate esterase (CE), and auxiliary activities family. Various CAZymes are in demand in the food and agriculture industry.

The utilization of different enzymes has been able to help increase the food supply to attain the ever-increasing demand of the food industry. Enzymes are known to be safely incorporated directly into food to enhance product features, like texture and appearance, enhance nutritional quality, extend storage periods, improve food safety, and generate new products of commercial interest. Enzymes are employed in an extended range of applications, like bakery, dairy manufacturing, starch processing, and the production of fruit juice, wine, beer, and other beverages (de Souza et al., 2021). Despite the speedy enhancement of the global food enzyme industry, driven by the implementation of diverse enzymes for different types of food processing applications, the available biocatalysts are still insufficient to meet all the demands. Consumers' emphasis on preventive health and the increasing requirement for functional foods have created a requirement for multifunctional biocatalysts with enhanced catalytic efficiency and stability, essential for transforming into bioactive compounds (Sathya et al., 2014).

CAZymes hold significant importance in agriculture due to various applications. CAZymes such as cellulases and hemicellulases break down organic matter in the soil, improving nutrient availability for plants and promoting healthier growth. These enzymes facilitate the decomposition of plant residues and organic matter, providing essential nutrients for plant development and enhancing crop resistance to environmental stress (Salam et al., 2018). CAZymes are incorporated into animal feed to enhance digestion and nutrient absorption, leading to healthier livestock and more efficient use of feed resources (Plouhinec et al., 2023). Enzymes aid in the disintegration of organic waste, supporting sustainable agricultural practices and minimizing environmental impact. Certain CAZymes guard crops from pests and diseases, minimizing dependence on chemical pesticides and fostering organic farming practices (Salam et al., 2018). These applications highlight the significant role CAZymes play in enhancing sustainable agriculture and food production. Thus, with this chapter, we briefly elaborate on the classification of different enzymes of the CAZymes family such as glycoside hydrolase, including amylase, xylanase, pectinases, β-galactosidase, cellulases, glycosyl transferase, PL, and CE and their implementations in food and agriculture industry.

11.2 GLYCOSYL HYDROLASE

Glycoside hydrolases (GHs) (EC 3.2.1.), or glycosidases, are group of enzymes known to break glycosidic bonds in carbohydrates or oligosaccharides. These are crucial for the processing and metabolism of carbohydrates, helping to convert complex sugars into simpler forms that the body can absorb and use for energy (Sathya et al., 2014). GHs indeed have a remarkable presence across all domains of life—prokaryotic, eukaryotic, and archaea. GHs are categorized into 189 different families considering their modes of action and amino acid sequences according to the CAZy database. Glycosyl hydrolase families include enzymes such as amylases, xylanases, chitinases, pectinases, invertases, and β-galactosidases. Some of these are discussed below for their applications in food and agriculture.

11.2.1 Amylase

Amylases are enzymes that cleave starch into amylose and amylopectin. These enzymes have various sources like plants, microorganisms, and higher organisms. Among these, glucoamylase, α-amylase, and β-amylase are divided into glycoside hydrolase families 13, 14, and 15, respectively, as given in the CAZy database (Divya Balakrishnan et al., 2019).

11.2.1.1 Role in Food Industry

As the most abundant carbohydrate source in most foods, amylases, with their ability to break down starch and its derivatives, such as maltodextrin, cyclodextrin, and various modified starch products, have various utilizations in the food industry (Divya Balakrishnan et al., 2019).

1. Baking
   In bread preparation, amylases hydrolyze starch in the dough, and during baking, these enzymes aid in starch liquefaction. Additionally, amylases act as anti-staling agents, helping to retain softness and extend the storage life of baked products. However, excessive amylase use can cause gumminess in bread, so it is used alongside pullulanase to mitigate this effect (Muhammad Adeel Farooq et al., 2020).
2. Sweetening Agents
   Low-cost starch sources can be converted into sugar syrups using various amylase enzymes. The following table lists different types of sweeteners developed with the aid of these enzymes.
3. Fat Mimics
   Low-fat foods are increasingly preferred for their contribution to a healthy lifestyle. Various starch-hydrolyzing amylases are utilized to create carbohydrate-based fat mimics from cornstarch. Mingyuan Li et al. (2024) employed

TABLE 11.1
Classification of Amylase in GH Family (According to CAZy)

Glycosyl Hydrolase Family	3D Structure	Catalytic Nucleophile Base	Catalytic Proton Donor
GH-13, GH-14, GH-31, GH-97	$(\beta/\alpha)_8$	Asp/Glu	Glu/Asp
GH-57, GH-119	$(\beta/\alpha)_7$	Asp/Glu	Glu/Asp
GH-15, GH-126, GH-144, GH-144, GH-176	$(\alpha/\alpha)_6$	Glu	Glu

TABLE 11.2
Different Types of Sweeteners Developed Using Amylase

Sr.no	Type of Sugar Syrup	Enzymes Used	Types of Food in Which They Are Used as Sweeteners	References
1.	HMCS (High Maltose Corn Syrup)	β-Amylases, fungal α-amylases	Frozen desserts, Hard candy	Balakrishnan et al. (2019)
		α-Amylase (Spezyme® Powerliq), Secura enzyme (Novozymes) thermostable β-amylase >80 % maltose obtained		Mahmoud M. Abdrabuo et al. (2024)
		α-Amylase, β-amylase and pullulanase 60–62% maltose obtained		Nataliia Sabadash et al. (2021)
	High maltose syrup (from brown rice, corn, and buckwheat)	Extrusion puffing, amylases		Mingyuan Li et al. (2024)
2.	HFCS (High Fructose Corn Syrup)	α-Amylases, glucoamylases, glucose isomerase	Soft drinks, jellies, jams, candies, yogurts, and breads	Divya Balakrishnan et al. (2019)
		α-Amylases, glucose isomerase (Both from *Streptomyces* species) Saccharification 90.9% with 40.8% glucose to fructose conversion		Tanim Arpit Singh et al. (2020)
		Amyloglycosidase (from *Aspergillus niger*), Glucose isomerase (from *Streptomyces* species) Saccharification 95% with 48% glucose to fructose conversion		Murilo Amaral-Fonseca et al. (2021)

(continued)

TABLE 11.2 (Continued)
Different Types of Sweeteners Developed Using Amylase

Sr.no	Type of Sugar Syrup	Enzymes Used	Types of Food in Which They Are Used as Sweeteners	References
3.	High fructose syrup (from lesser Yam tubers)	Inulinase, α-amylases, glucoamylases		Sri Winartiet al. (2022)
3.	Chocolate syrup	Amylases	Cocoa-flavored frozen confections	Ritu Saini et al. (2017)
4.	Malt oligosaccharides mixture	α-Amylases, β-amylases, pullulanase	Frozen foods	Divya Balakrishnan et al. (2019)
5.	Isomalt oligosaccharides	α-Amylases, α, β-glucosidase, dextran sucrase, and 4, 6–α–Glucotransferase from the GH familyies, including GH31, GH13, GH70, and GH57	Jam-jelly, cakes, infant foods formula, dairy products, fat mimics	Vidhya K. Shinde (2021)
6.	Maltotetraose	α-Amylase	Baking, frozen foods, chocolate, beverage	Xinyu Wang et al. (2024)
7.	Alo mixture	β-Amylase and trans glucosidase	Frozen foods	Prasanna V. Aiyer (2005)

ultra-high-pressure microfluidization technology combined with enzymatic hydrolysis to enhance the oil-holding capacity, emulsification index (E1), and emulsion stability index of potato starch-based fat mimics.

4. Brewing
amylases are used in the beer industry in mashing for starch hydrolysis. The activity of these enzymes is affected due to the temperature progression step, which is applied to enhance gelatinization. Thus, a kinetic study was carried out by C.F. De Schepper et al. (2022), related to heat-induced deactivation of barley malt α-amylases and β-amylases in the mashing process, which showed that β-amylase exhibits considerably higher thermostability in wort than in an aqueous buffer.

11.2.1.2 Role in Agriculture

Amylase-producing bacteria in soil can impact soil quality and nutrient availability, thereby affecting plant development and overall ecosystem nutrition dynamics (Oyenado et al., 2024). Amylase inhibitors are crucial in protecting plants against insects like seed weevils, which rely on starch for nutrients (Octa Âvio L. Franco et al., 2002). Amylase helps in the decomposition of organic matter, speeding up the composting process and producing nutrient-rich compost. It is also used to treat seeds, improving germination rates and seedling vigor by breaking down starch reserves in the seeds. It is produced using agricultural waste materials, such as rice husk and wheat bran, making the process cost-effective and sustainable. In intensive poultry production, α-amylase is utilized as a feed additive to improve the digestibility of starches present in grains. Research indicates that adding amylase to chicken feed can lead to increased corn consumption while reducing reliance on soybean meal, thus optimizing feed formulations for better nutrient absorption and growth rates in poultry. The environmental impact assessments of such practices show that the use of amylase contributes positively by reducing greenhouse gas emissions associated with feed production, highlighting its potential as a sustainable agricultural practice (Britta Gernaey et al., 2018). Genetic engineering techniques are being employed to produce more efficient strains of amylase-producing microbes, which can be used in biofertilizers or as part of integrated pest management strategies. These engineered microbes can help break down organic matter more effectively, releasing nutrients that are otherwise locked away in complex organic compounds (Subash C B Gopinath et al., 2017).

11.2.2 Xylanase

Xylanase is an enzyme that enhances the hydrolysis of xylan, a principal component of plant cell walls and hemicellulose, into simpler sugars like xylose. This process is facilitated with the disintegration of β-1,4-glycosidic bonds within xylan. This enzyme is known to be present in different types of microorganisms, such as bacteria, fungi, and some plants and animals, though microbial sources are preferred for industrial applications due to their high production efficiency and robustness in diverse conditions (Nishant Shahi et al., 2016; Kedar Sharma et al., 2019). Xylanase's ability to decompose hemicellulose has led to its widespread utilization in industries such as biofuel production, textiles, animal feed, food, and pharmaceuticals. This enzyme is particularly valuable because of its eco-friendly, efficient catalytic abilities under mild reaction conditions, which align well with industrial goals to reduce energy consumption and minimize environmental impact (Priyanka Bajaj et al., 2019).

11.2.2.1 Role of Xylanase in Food

Xylanase has multiple critical applications in the food industry, contributing to improved product quality, enhanced nutritional value, and optimized production processes. This enzyme's versatility and eco-friendly characteristics make it highly desirable in various food processing sectors.

1. Baking Industry
    - Improving Dough Quality and Bread Texture: In baking, xylanase enhances the quality of dough by modifying the hemicellulose structure, which strengthens gluten networks and improves dough elasticity. This results in better dough handling, higher loaf volumes, and finer crumb textures in bread and other baked goods. By breaking down arabinoxylans in wheat flour, xylanase reduces the dough's viscosity and stickiness, making it easier to process in industrial settings where high-speed machinery is used (Kedar Sharma et al., 2019; Handray Fernandes de Souza et al., 2022).
    - Increased Freshness and Shelf-Life: Xylanase can help retain moisture in baked products, contributing to prolonged freshness and a softer texture over time. By enhancing the structural properties of bread, xylanase reduces staling, which is a common challenge in the baking industry. This application is particularly valuable for large-scale bakeries aiming to reduce waste and extend product shelf-life (Nishant Shahi et al., 2016; Handray Fernandes de Souza et al., 2022).
    - Reducing Additives in Recipes: Xylanase acts as a natural dough improver, which can minimize the need for chemical additives typically used to enhance bread quality. This aligns with consumer demands for cleaner labels and more natural ingredients in food products, as xylanase allows for improved quality without synthetic enhancers (Kedar Sharma et al., 2019).
2. Fruit Juice Clarification
    - Enhancing Clarity and Stability of Fruit Juices: In fruit juice production, xylanase is used to break down hemicellulose components that cause turbidity in juices. By degrading the xylan in plant cell walls, the enzyme facilitates the release of pectin and other polysaccharides, leading to a clearer, more visually appealing juice. This process is especially popular in the making of apple and grape juices, where clarity is a key quality factor (Handray Fernandes de Souza et al., 2022; Priyanka Bajaj et al., 2019).
    - Improving Filtration Efficiency: Xylanase assists in reducing the viscosity of juice, making it easier to filter and process. This results in improved efficiency during filtration, reducing processing times and operational costs for manufacturers. It also enables the production of high-quality juice

TABLE 11.3
Classification of Xylanase in the GH family

Family	Mechanism	Activity	Clan	3D Structure	Catalytic Nucleophile/Base	Catalytic Proton Donor
GH 11	Retaining e	Endo-β-1,4-xylosidase	GH-C	β-Jelly roll	Glu	Glu
GH 26	Retaining e	Endo-β-1,3-xylosidase, endo-β-1,4-glucosidase and exo-β-1,4-xylosidase	GH-A	$(\beta/\alpha)_8$ Barrel	Glu	Glu
GH 43	Inverting e	Endo-β-1,4-xylosidase and exo-β-1,4-xylosidase	GH-F	5-Fold β-propeller	Asp	Glu
GH 48	Inverting e	Cellulose endo-β-1,4-glucosidase and cellulose exo-β-1,4-glucobiosidase	GH-M	$(\alpha/\alpha)_6$ Barrel	Unknown	Glu
GH 51	Retaining e	Endo-β-1,4-xylosidase and exo-β-1,4-xylosidase	GH-A	$(\beta/\alpha)_8$ Barrel	Glu	Glu
GH 62	Inverting e	Endo-β-1,4-xylosidase	GH-F	5-Fold β-propeller	Glu	Asp
GH 98	Inverting e	Xylan endo-β-1,4-xylosidase	Unknown	$(\beta/\alpha)_8$ Barrel	Asp + Glu dyad	Glu
GH 141	Unknown e	Xylan endo-β-1,4-xylosidase	Unknown	Parallel β-helix	Asp	Asp

with fewer impurities and better stability, ensuring a consistent product appearance (Nishant Shahi et al., 2016).
- Reducing the Need for Harsh Processing Techniques: Traditionally, fruit juice clarification required extensive mechanical and chemical treatments to achieve the desired clarity. The use of xylanase simplifies the process, reducing the dependency on physical filtering techniques and chemical agents, which aligns with the industry's movement toward sustainable, natural processing methods (Priyanka Bajaj et al., 2019).

3. Production of Prebiotics and Functional Foods
   - Production of Xylo-oligosaccharides (XOS): Xylanase is used to produce XOS, which are prebiotic fibers that aid in the growth of beneficial gut bacteria. XOS is increasingly added to foods as a functional ingredient due to its positive effects on gut health. By converting xylan into XOS, xylanase enables the development of functional foods aimed at improving digestive health and supporting immunity (Handray Fernandes de Souza et al., 2022).
   - Prebiotic Supplementation: XOS derived from xylanase activity is used as a dietary supplement, especially in products marketed for their prebiotic content. These supplements assist the growth of various probiotic bacteria present in the gut, which are beneficial to health as they lead to improved digestion and enhanced immune function. As consumer demand for gut-health-promoting products grows, the use of xylanase in producing XOS is becoming a valuable addition to functional food formulations (Kedar Sharma et al., 2019).
4. Brewing Industry
   - Improving Filtration and Reducing Haze: In the brewing industry, xylanase is employed to improve the filtration of beer and reduce haze formation. During beer production, hemicellulose in grains like barley can cause haze and filtration challenges. Xylanase breaks down these polysaccharides, leading to improved clarity and stability in the final product, which is essential for achieving a high-quality appearance in commercial beer (Kedar Sharma et al., 2019; Priyanka Bajaj et al., 2019).
   - Enhancing Flavor and Consistency: By assisting in the breakdown of plant cell walls in brewing ingredients, xylanase helps release fermentable sugars and flavor compounds. This contributes to a more consistent flavor profile in beer, allowing brewers to maintain high standards in taste and quality across batches. The enzyme's use can lead to a more efficient brewing process and a smoother, more appealing beer (Kumar et al., 2018).
5. Dairy and Plant-Based Milk Processing
   - Improving Texture in Plant-Based Milk Alternatives: Xylanase can be used to elevate the texture and mouthfeel of plant-based milk products derived from oats, almonds, or soy. By breaking down hemicellulose, xylanase reduces viscosity and enhances the smoothness of these beverages, creating a texture more like traditional dairy milk. This is particularly beneficial in

the production of oat milk, where hemicellulose content can affect the final product's viscosity (Handray Fernandes de Souza et al., 2022).

- Enhanced Yield in Dairy Alternatives: By hydrolyzing xylan in plant materials, xylanase increases the release of soluble fibers and other nutrients, enhancing the yield and nutritional profile of plant-based milks. This is essential in a growing industry focused on meeting consumer demands for high-quality, dairy-free alternatives (Handray Fernandes de Souza et al., 2022).

11.2.2.2 Role of Xylanase in Agriculture

1. Animal Feed Enhancement:
   - Improving Digestibility and Nutrient Absorption: It is commonly added to animal feed, especially for poultry and swine, which consume feed high in plant materials containing hemicellulose. The xylan in hemicellulose can form a viscous gel in the animal's gut, which hinders nutrient absorption and digestibility. It breaks down the xylan structure, reducing the viscosity in the animal's digestive system, thereby enhancing nutrient absorption and gut health (Kedar Sharma et al., 2019; Priyanka Bajaj et al., 2019). This breakdown releases simple sugars and other accessible nutrients, enabling better energy extraction from the feed.
   - Supporting Gut Health and Growth Performance: Its use in feed has been shown to contribute to better growth rates and feed conversion ratios in livestock. By reducing anti-nutritional factors in the feed, xylanase minimizes digestive issues and improves feed efficiency, leading to enhanced growth and overall animal health. In poultry, for instance, supplementation of xylanase leads to increased egg production and healthier weight gain in broilers (Nishant Shahi et al., 2016).
   - Cost-Efficiency and Environmental Impact: By improving nutrient uptake and feed efficiency, xylanase allows producers to use less feed to achieve desired growth, which reduces costs and environmental impact associated with feed production. Additionally, the breakdown of plant-based feed components minimizes waste, as animals excrete fewer undigested nutrients, thus reducing nutrient run-off and pollution in agricultural settings (Handray Fernandes de Souza et al., 2022).
2. Biofuel Production from Agricultural Residues
   - Breaking Down Lignocellulosic Biomass: Lignocellulosic residues such as straw, corn stover, and rice husks are rich in lignocellulosic material, with hemicellulose being a major component. Xylanase is crucial for the hydrolysis of xylan within hemicellulose, converting it into fermentable sugars such as xylose, which can then be transformed into bioethanol and other biofuels. This enzymatic breakdown is essential for efficient biofuel production, as it reduces the need for harsh chemical pretreatment and optimizes the saccharification process, making the conversion more environmentally friendly and affordable (Kumar et al., 2018; Priyanka Bajaj et al., 2019).

- Synergistic Use with Cellulase in Bioethanol Production: In bioethanol production, xylanase often works in conjunction with cellulase. While xylanase breaks down the hemicellulose, cellulase breaks down cellulose. This dual action not only produces more fermentable sugars but also improves the accessibility of cellulose, leading to higher bioethanol yields. Studies have shown that adding xylanase increases ethanol production by boosting sugar release during saccharification, highlighting its synergistic role in enhancing the efficiency of biofuel production from agricultural waste (Priyanka Bajaj et al., 2019).
- Environmental Sustainability: Utilizing xylanase in biofuel production promotes a circular economy by converting agricultural waste into valuable energy resources. This leads to a reduced requirement of fossil fuels and lowered greenhouse gas emissions associated with conventional fuel production. Moreover, enzymatic processes, such as those facilitated by xylanase, generate fewer harmful byproducts than chemical methods, aligning with the growing emphasis on eco-friendly biotechnological solutions in agriculture (Nishant Shahi et al., 2016).

3. Soil Fertility and Crop Residue Management
    - Enhancing Composting Processes: Xylanase can be applied in composting agricultural waste, helping accelerate the decomposition of plant materials high in hemicellulose. By breaking down complex plant polymers into simpler sugars, it speeds up microbial activity, leading to faster composting and the production of nutrient-rich organic matter. This compost can then be used to enhance soil fertility, supporting sustainable agricultural practices (Handray Fernandes de Souza et al., 2022; Priyanka Bajaj et al., 2019).
    - Improving Soil Health: Applying xylanase-treated compost to soil not only recycles nutrients but also supports soil microbial biodiversity. The breakdown of complex carbohydrates through xylanase generates substrates that promote the development of soil microbes, known to execute a vital role in the cycling of nutrients and plant health. This contributes to long-term soil sustainability and better crop yields (Kumar et al., 2018).

TABLE 11.4
Classification of Pectinases in the GH Family (According to CAZy)

Family	Mechanism	Activity	Clan	3D structure	Catalytic Nucleophile/Base	Catalytic Proton Donor
GH 28	Inverting a	Endo-α-1,4-galacturonidase and exo-α-1,4-galacturonidase	GH-N	Parallel β-helix	Asp	Asp

11.2.3 Pectinases (EC 3.2.1.15)

Pectinase refers to a category of enzymes known to hydrolyze pectin, a complex polysaccharide in plant cell walls, particularly in the middle lamella where cells are joined. This enzyme group is essential for degrading pectin into simpler molecules, facilitating various applications in agriculture and the food industry. Pectinases are categorized into depolymerizing enzymes, which break down the main pectin chains, and de-esterifying enzymes, which modify side chains, aiding in the full degradation of pectin (Jebin Ahmed et al., 2021; Rolf A. Prade et al., 1999).

11.2.3.1 Role in Food Industry

TABLE 11.5
Applications of Pectinases in Food

Industry	Application	Function	References
Food	Juice Clarification and Extraction	Clarification: Breaks down pectin in fruit juice, reducing turbidity and enhancing clarity for products like apple, grape, and citrus juices	Jebin Ahmed et al. (2021), Rolf A. Prade et al. (1999)
		Increased Yield: Breaks down cell walls, allowing for more juice extraction, which benefits production efficiency and reduces waste	González-Carranza et al. (2012)
	Winemaking	Maceration and Clarification: Enhances juice extraction in winemaking, releasing pigments and flavors from grape skins, and improving color and flavor in wines, particularly red wines	Jebin Ahmed et al. (2021), González-Carranza et al. (2012)
		Reducing Sediments: Breaks down the pectin, thus reducing sediments and haze in wine, facilitating filtration and improving product appearance and quality	Rolf A. Prade et al. (1999)
	Enhancing Quality of Vegetable Purees	Improves texture and consistency in vegetable purees by breaking down cell walls, making products like tomato paste, mashed potatoes, and pumpkin purees smoother and easier to process	Jebin Ahmed et al. (2021)
	Oil Extraction from Fruits and Seeds	Increases oil yield from oil-rich fruits and seeds (e.g., olives, avocados) by breaking down pectin, aiding in oil release, and reducing the need for mechanical or chemical extraction methods	González-Carranza et al. (2012)
	Production of Value-Added Products	Prebiotics and Dietary Fibers: Converts pectin into prebiotic compounds like galacturonic acid, promoting the growth of bacteria for improved health of gut	Jebin Ahmed et al. (2021)
		Low-Calorie Thickeners and Stabilizers: Used to create low-calorie thickeners and stabilizers for food products where consumers desire a better texture with fewer calories	Rolf A. Prade et al. (1999)

11.2.3.1 Role in Agriculture

TABLE 11.6
Applications of Pectinases in Agriculture

Application Area	Application	Description	References
Agricultural Sector	Improving Soil Fertility and Crop Residue Management	Composting and Decomposition: Enhances composting by breaking down pectic substances in plant waste, speeding up decomposition and producing nutrient-rich compost	Jebin Ahmed et al. (2021), Rolf A. Prade et al. (1999)
		Bioremediation and Waste Management: Treats agricultural waste, turning pectin-containing residues into biodegradable compounds, contributing to waste management and eco-friendly bioremediation	Jebin Ahmed et al. (2021)
	Plant Pathogen Control and Disease Management	Protection Against Fungal Pathogens: Alters plant cell wall structure, making it harder for pathogens to adhere and penetrate tissues, which helps in developing disease-resistant crops	Jeremy A. González-Carranza et al. (2012) Rolf A. Prade et al. (1999)
		Development of Biofungicides: Utilizes pectinase-producing microorganisms to create biofungicides that degrade fungal cell walls, providing an eco-friendly alternative to chemical fungicides	Rolf A. Prade et al. (1999)
	Fiber Retting for Natural Fibers	Accelerates retting, a process that separates fibers from plants by breaking down the pectin, yielding cleaner, higher quality fibers used in textiles and biocomposites, and is more sustainable than chemical retting	Jebin Ahmed et al. (2021), González-Carranza et al. (2012)

11.2.4 Galactosidases

Galactosidases (EC 3.2.1.23) are known as the important glycosyl hydrolases in higher plants, microbes, and animals. They selectively breakdown the β-(1-3), (1-4) or (1-6) galactoside bonds present in oligo and disaccharides (Liu et al., 2023). In addition to their hydrolytic function, these enzymes are indeed capable of catalyzing the transfer of sugar moieties, like galactose, to other molecules. Beta-galactosidases catalyze the transfer of galactose to a sugar acceptor, leading to the formation of galacto-oligosaccharides. Galactosidases have been categorized into various glycosyl hydrolase families; specific characteristics of each are listed in Table 11.7.

11.2.4.1 Role in Food

In the food industry, α-galactosidases serve as gelling agents. Meanwhile, α-galactosidases are used in animal feeds to reduce viscosity and enhance metabolizable energy (Menon et al., 2024). Transgalactosylation activity of α-galactosidase is utilized in the production of prebiotic oligosaccharides. Its hydrolytic activity when used on lactose makes a sweetening agent that is readily soluble.

These enzymes are also beneficial for people with lactose intolerance, as these are essential for the generation of lactose-free products. They are also required for making frozen yogurt, leading to increased creaminess and taste (Kumar et al., 2024). Lactulose is a synthetic, nondigestible disaccharide that does not occur naturally, which is produced from lactose by the thermal treatment of milk. In food, lactulose serves as an effective functional ingredient because it promotes the development of beneficial bacteria, such as *bifidobacteria* and *lactobacilli*, within the intestine.

Galactosyl sugar alcohols make excellent food sugar substitutes because they offer multiple functionalities in food products. These include providing sweetness, reducing bulk calories, and offering multiple physiological benefits, such as the prevention of dental caries (Liu et al., 2023).

11.2.4.2 Role in Agriculture

α-Galactosidase is a vital enzyme due to its significant applications in agriculture. It plays a defined role in converting agricultural waste into high-value products, such as ethanol.

TABLE 11.7
Classification of Galactosidase in the GH Family (According to CAZy)

GH Family	Enzyme Activity	3D Structure	Catalytic Nucleophile Base	Catalytic Proton Donor
GH-4, GH-27, GH-31, GH-36, GH-57, GH-97	α-Galactosidase	$(\beta/\alpha)_8$	Glu/Asp	Glu/His
GH-1, GH-2, GH-3, GH-5, GH-35, GH-59	β-Galactosidase	$(\beta/\alpha)_8$	Glu	Glu
GH-154		$(\alpha/\alpha)_6$ Barrel		
GH-165		5-Bladed β-propeller		

β-Galactosidase plays a vital function in the fruit ripening process by disintegration of pectin in the plant cell walls. In addition to its role in fruit softening, β-galactosidase is involved in various other vital processes within plants. It contributes to:

- Reshaping cell walls.
- Floral organ development.
- Formation of secondary cell walls.
- Secretion of mucus.
- Elongation of hypocotyls.

This enzyme's multifunctional nature makes it an essential component in plant maturation (Ranwala et al., 1992).

The expression levels of β-galactosidase in plants increase in tandem with their growth, development, and maturation phases. This enzyme's activity is finely tuned to support various physiological processes as the plant progresses through its life cycle. β-Galactosidase also plays a crucial role in the metabolism of glycoside inside seeds, particularly when the seed germination and initial sprouting phases are present. It plays a critical role in hydrolyzing complex sugars and cycling soil carbon, which directly impacts soil quality. By facilitating the disintegration of organic matter, this enzyme contributes to the discharge of nutrients and the improvement of soil structure (Nthebere et al., 2024).

β-Galactosidase activity is highlighted in the presence of environmental stresses, such as heavy metal stresses, ascertaining it as a marker of plant growth events. β-Galactosidase activity is critical in responding to various environmental pressures, including heavy metal stresses. This enzyme's function of degrading complex sugars and cycling soil carbon becomes even more crucial under stressful conditions. As a result, the level of β-galactosidase activity can serve as a marker of plant growth dynamics and overall plant health in challenging environments.

This role of β-galactosidase was exploited in various studies to design probes to monitor β-gal-related events in plants (Chen et al., 2023; Li et al., 2024). One probe specifically investigates engagement with β-galactosidase activity and ripening of fruits; another one has a role for in vitro detection and qualitative measurement, thus emphasizing the connection involving β-galactosidase activity with processing and preservation. An engageable fluorescent probe, developed because of intramolecular charge transfer (ICT) and incorporating benzothiazole coumarin-bearing β-gal (BC-βgal), has been reported. This probe enables distinct in situ imaging of β-gal in plant cells under heavy metal stress (Zhao et al., 2024).

Various immobilization measures have been employed that limit operational costs and enhance batch hydrolysis efficiency in enzymatic processes.

1. Covalent bonding
2. Cross-linking
3. Entrapment/Encapsulation
4. Adsorption

Genetic engineering of beta-galactosidases:

1. Site-directed mutagenesis
2. Site-saturation mutagenesis
3. Truncation or deletion mutagenesis
4. DNA shuffling
5. Monobody modifications

Through various techniques, β-galactosidases have been endowed with superior properties, such as:

- Improved Efficiency of Transglycosylation: Enhanced capability to transfer glycosyl groups, leading to more efficient biochemical reactions.
- Changed Glycosyl Donor Specificity: Altered specificity for glycosyl donors, allowing for more versatile applications.
- Tailor-Made Product Structures: Customization of product structures to meet specific needs and applications.
- Increased Thermal Stability: Greater resilience to higher temperatures, making the enzyme more robust in diverse conditions.
- Reduced Product Inhibition: Lower sensitivity to inhibition by products, ensuring sustained enzyme activity and efficiency (Liu et al., 2023).

11.2.5 Cellulases

Cellulases are indeed modular enzymes composed of distinct units, each with specific structural and functional roles. These units, known as domains or modules, enable the enzyme to perform its functions efficiently by allowing different parts of the enzyme to operate independently yet cohesively (Kuhad et al., 2011). Cellulase comes under the glycosyl hydrolase (GH) family, which comprises three major enzymes: endoglucanase, which breaks any intrinsic glycosidic bonds that generate glucan chains; exoglucanase, which targets the ends of cellulose chain, releasing cellobiose as the end product; and β-glycosidase, which selectively targets cellobiose disaccharides, converting them into glucose (Sutaoney et al., 2024). In the CAZy database, cellulases are grouped into various families; their specific characteristics are provided in Table 11.8.

11.2.5.1 Role in Food

Cellulases prove useful in the fruit and vegetable juice industry, as they are used to break down the cell walls of the plant material, thus enhancing the extraction process and leading to higher juice yields. They are used as macerating enzymes in combination with pectinases in food processing. These enzymes are also shown to increase the sensory qualities such as texture and flavor in fruits and vegetables. They are also beneficial in enhancing texture and quality of bakery products (Sutaoney et al., 2024).

Cellulases are also useful in extraction of carotenoids, olive oil, and phenolic compounds (Ejaz et al., 2021).

TABLE 11.8
Classification of Cellulase in the GH Family(According to CAZy)

Glycosyl Hydrolase Family	Cellulase Family	Clan	3D Structure	Catalytic Nucleophile Base	Catalytic Proton Donor
GH-5	A	GH-A	$(\beta/\alpha)_8$ Barrel	Glu	Glu
GH-7	C	GH-B	β-Jelly roll	Glu	Glu
GH-9	E		$(\alpha/\alpha)_6$ Barrel	Asp	Glu
GH-10	F	GH-A	$(\beta/\alpha)_8$ Barrel	Glu	Glu
GH-12	H	GH-C	β-Jelly roll	Glu	Glu
GH-16		GH-B	β-Jelly roll	Glu	Glu
GH-45	K		6 Stranded β barrel	Asp	Asp
GH-48	L	GH-M	$(\alpha/\alpha)_6$ Barrel	-	Glu

Cellulases are used to increase nutritional quality of animal feed. Thus, they are shown to increase weight gain in animals. It is also used in fodder preparation as it helps to increase digestibility.

Cellulases are useful in the beer industry. By breaking down cell walls and reducing viscosity, they improve the extraction of fermentable sugars from the grains. This enhances both the quality and rate of production of the beer, leading to a smoother brewing process and a more consistent final product. In wine production, these enzymes, along with other biocatalysts, hydrolyze plant cell wall polysaccharides, significantly refining the maceration process, color, value, aroma, stability, and clarity of the wine. β-Glucosidases play a crucial role in enhancing wine fragrance by modifying glycosylated precursors (Sutaoney et al., 2024).

11.2.5.2 Role in Agriculture

Multi-enzyme blends of cellulolytic and xylanolytic enzymes are known to comprise important functions in agriculture. These enzymes are known to enhance plant development and prevent plant diseases. β-1,3-Glucanase, isolated from *Trichoderma harzianum* CECT 2413, is known to produce morphological changes in plants, such as hyphal tip expansion, cytoplasm leakage, septae formation, and control of plant diseases by inhibiting the development of detrimental *Rhizopus solani* and *Fusarium* species (Sutaoney et al., 2024). Cellulases and other related enzymes from specific types of fungi have the potential to degrade the cell wall of plant pathogens, leading to control diseases in plants (Kuhad et al., 2011).

Cellulases have also been shown to improve the soil quality (Sutaoney et al., 2024). Application of cellulolytic bacteria to soil enhanced the straw breakdown and improved soil fertility. Thus, using these enzymes speeds up the straw breakdown to improve soil quality (Kuhad et al., 2011).

Wheat straw, which is obtained as a crop byproduct, is used in the production of superabsorbent resin. This helps in the controlled release of fertilizers and increases the water-holding capacity of soil.

11.3 GLYCOSYLTRANSFERASES (EC NO 2.4)

GTs are found commonly in most living organisms, including plants, animals, and microorganisms. These enzymes transport sugar molecules from nucleotide sugars to a varied molecule, such as hormones or secondary metabolites. The incorporation of a sugar moiety by GTs changes the hydrophilicity of the molecule, thereby altering its chemical properties (Humara Naz Majeed et al., 2024). GTs are a large group of enzymes that lead to the generation of glycosidic bonds by transferring sugar moieties to specific acceptor molecules, such as proteins, lipids, or other sugars. This process forms glycosidic bonds and is critical in various biological processes, which include glycosylation, cell signaling, energy storage, and metabolism. There are about 138 families of GTs. The most common ones are as follows:

Along with these families, the GTs are broadly classified into two types based on their mechanism of glycosidic bond formation.

1. Retaining GTs: Retain the anomeric configuration of the donor sugar (e.g., α remains α, β remains β).
2. Inverting GTs: Invert the anomeric configuration during the reaction (e.g., α to β or vice versa).

11.3.1 Role in Food

Glycosylation is used to maintain the stability of natural compounds such as Vitamin C (Humara Naz Majeed et al., 2024). It is also used as sweeteners and additives in the food industries (J. Wang et al., 2024). It is also used in food applications as a gelling agent and emulsifying agent (J. Wang et al., 2024).

11.3.2 Role in Agriculture

GTs are known to increase stress tolerance in plants (J. Wang et al., 2024). Different types of bioactive compounds have been formed by glycosylation, which are useful in plant protection (Humara Naz Majeed et al., 2024). These can modify secondary metabolites to enhance plant defence mechanisms against pathogens and pests (Jun Wang et al., 2024). These participate in the biosynthesis of secondary metabolites, phytohormones, and aroma and flavor compounds, which are essential for plant growth and metabolism (J. Wang et al., 2024).

11.4 POLYSACCHARIDE LYASES

PLs (EC 4.2.2.-) are enzymes responsible for breaking uronic acid-containing polysaccharide chains through a β-elimination mechanism, leading to the development of an unsaturated hexenuronic acid residue and a new reducing end (Chakraborty et al., 2017). As of this writing, PLs have been categorized into 44 different families as per the CAZy database (January 2025). The characterized PLs include endopectate lyase (EC 4.2.2.2), exopectate lyase (EC 4.2.2.9), pectin lyase (EC 4.2.2.11), rhamnogalacturonan lyase (EC 4.2.2.23 and 4.2.2.24), alginate lyase (EC

TABLE 11.9
Classification of Glycosyl Transferases in the CAZy Database

Family	Key Functions
GT1	Involved in the synthesis of glycosides. Found in diverse organisms, including bacteria and plants
GT2	One of the largest families. Includes cellulose synthases, chitin synthases, and mannosyltransferases. Functions in synthesizing structural polysaccharides
GT3	Contains enzymes involved in the biosynthesis of lipopolysaccharides and glycolipids. Uses unusual donor molecules like polyprenyl phosphates
GT4	Encompasses a wide range of enzymes, including those responsible for N-glycan biosynthesis. Plays a crucial role in protein glycosylation
GT8	Includes enzymes like galactosyltransferases. Involved in the biosynthesis of glycoproteins and glycosaminoglycans
GT20	Involved in sialylation (attachment of sialic acids). Important in cell–cell communication and immune response
GT43 and GT47	Play roles in plant cell wall biosynthesis, including xylan and cellulose production
GT64	Involved in complex carbohydrate biosynthesis and modification

4.2.2.3), ulvan lyase, chondroitin AC lyase (EC 4.2.2.5), chondroitin ABC lyase (EC 4.2.2.20 and 4.2.2.21), heparin lyase (EC 4.2.2.7), heparan lyase (EC 4.2.2.8), and hyaluronate lyase (EC 4.2.2.1). These enzymes cleave a variety of substrates, including alginate, chondroitin, heparin, gellan, hyaluronan, pectin, xanthan, and several other bacterial exopolysaccharides. They are obtained from a diverse range of organisms across various areas of life, including bacteria, fungi, and marine animals (Chakraborty et al., 2017). Applications of alginate lyase, ulvan lyase, and pectate lyase are discussed briefly in this chapter.

Alginate lyases, a vital class within PL (EC 4.2.2.-), is known for the biodegradation of alginate. These enzymes carry out the breakdown of 1,4-glycosidic bonds in alginate, resulting in the formation of double bonds between C4 and C5 at the non-reducing ends of products. According to their mode of action, alginate lyases are divided into endo-type and exo-type lyases (EC 4.2.2.26). Endo-type alginate lyases degrade alginate polymers to generate alginate oligosaccharides (AOS), and exo-type alginate lyases lead to generate unsaturated monomers or dimers from ends of alginate. Alginate lyases are commonly found in algae, molluscs, bacteria, fungi, and viruses (Jiang et al., 2024). As per the CAZy database, alginate lyases are classified into various PL families; their characteristic features are listed in Table 11.10.

Ulvan lyases consist of enzymes that particularly degrade ulvan; these are primarily sourced and purified from marine bacteria (Wang et al., 2024). As per the CAZymes database, they are classified under five PL families: PL-24, PL-25, PL-28, PL-37, and PL-40.

Pectate lyases (PELs) are key enzymes produced by pathogenic organisms to break down pectin, a crucial component of plant cell walls. Pectate lyases (PELs) randomly break the β (1–4) linkages among galacturonosyl residues in pectin through

TABLE 11.10
Classification of Alginate Lyase in Polysaccharide Lyase Family (According to CAZy)

Polysaccharide Lyase Family	3D Structure
PL-5	$(\alpha/\alpha)_6$
PL-6, PL-31	Parallel β-helix
PL-7, PL-14, PL-18, PL-36	β-Jelly roll
PL-8, PL-15, PL-17, PL-39	$(\alpha/\alpha)_6$ Toroid + anti-parallel β-sheet

a process known as β-elimination. In the CAZy database, these enzymes are divided into six PELs.

11.4.1 Role in Food

Alginate can be modified by alginate lyase, which catalyzes the beta elimination of glycosidic bonds to form oligosaccharides with unsaturated double bonds. AOS have antioxidant activity as well as antimicrobial activity and thus can be used in food preservation (Wang et al., 2024). Alginate lyases depolymerize alginate into uronic acid–containing alginate oligosaccharides, which can bind various cations to form complexes and are used as sources of metal ions in food additives (Murata et al., 1993).

Ulvan lyase enzymes particularly degrade ulvan, and are mostly primarily sourced and purified from marine bacteria. Ulvan oligosaccharides (UOS) are seen to be useful as prebiotics to maintain lipid metabolism in high-fat diet mice, thus leading to reduced blood lipid levels and liver lipid accumulation. They are also used in the preparation of probiotics. These enzymes have also been shown to retard obesity (Wang et al., 2024).

Pectate lyases are known to clarify fruit juices by hydrolyzing pectin, resulting in a clearer and more appealing product. These enzymes prove beneficial in the extraction and processing of fruits and vegetables by breaking down pectin, which improves yield and quality (de Souza et al., 2021). These enzymes also enhance the extraction of vegetable oils by breaking down pectin, which helps in releasing the oil from the plant material (Ling Zheng et al., 2021). These are also involved in the fermentation process of wine, aiding in the breakdown of pectin to enhance the clarity and stability of the wine.

11.4.2 Role in Agriculture

Oligo alginates have been shown to act as elicitors for plant metabolite production. They have also been shown to increase the growth of *Bifidobacteria*. They could also be used as biofertilizers in the agriculture industry (Michaud et al., 2003).

UOS have been found to trigger natural defences, such as oxidative bursts and hypersensitive reactions. UOS treatment entirely stopped the growth of blue and grey mold on fruit, and hence, could be used as a fungicide (Wang et al., 2024).

TABLE 11.11
Classification of Ulvan Lyase in Polysaccharide Lyase Family (According to CAZy)

Polysaccharide Lyase Family	3D Structure
PL-24, PL-25	7-Bladed β-propeller
PL-28	β-Jelly roll

TABLE 11.12
Classification of Pectate Lyase in Polysaccharide Lyase Family (According to CAZy)

Polysaccharide Lyase Family	3D Structure
PL-1, PL-3, PL-9	Parallel β-helix
PL-2	$(\alpha/\alpha)_7$ Barrel
PL-10	$(\alpha/\alpha)_3$ Barrel

Pectate lyases (PELs) play a crucial part in the softening of fruits and seem to be an essential component of the enzyme cocktail responsible for ripening-related cell wall remodeling (Selman Uluisika et al., 2020; Yuanxiu Lin et al., 2023). These also can help plants adapt to environmental stresses such as drought, cold temperature, and mineral toxicity by regulating accumulation of glycosides. These enzymes can modify secondary metabolites to enhance plant defence mechanisms against pathogens and pests (de Souza et al., 2021).

11.5 CARBOHYDRATE ESTERASES (EC NO: 3.1.1 TO 3.1.31)

CEs are a category of enzymes utilized in the modification of carbohydrates by catalyzing the hydrolysis of ester bonds. They typically act on polysaccharides or oligosaccharides, removing acetyl or feruloyl groups, and thereby modulating the physicochemical properties of the carbohydrate substrates.

Functions of CEs

1. Deacetylation: Removal of acetyl groups from polysaccharides, improving accessibility for other enzymes like GHs.
2. Defucosylation: Hydrolysis of ferulic acid esters, aiding in the breakdown of plant cell walls.
3. Modification of Carbohydrate Structures: Alters solubility, stability, and reactivity of carbohydrate substrates.
4. Role in Biomass Degradation: Facilitates the degradation of plant biomass by removing ester-linked modifications that hinder enzymatic access to carbohydrates.

TABLE 11.13
Classification of Carbohydrate Esterase Family in the CAZy Database

CE Family	Enzyme Types	Substrates	Functions and Applications
CE1	Acetyl xylan esterases, Feruloyl esterase	Acetylated xylans, Feruloylated arabinoxylans	Deacetylation of xylans; involved in plant biomass degradation and lignocellulose processing
CE2	Acetyl xylan esterases, Cutinases	Acetylated polysaccharides	Degrades plant cell wall components; aids in textile and biofuel production
CE3	Esterases	Pectin and related compounds	Degrades pectic polysaccharides, particularly in plant cell wall remodeling
CE4	Chitin deacetylases, Acetyl xylan esterases	Chitin, Acetylated polysaccharides	Deacetylation of chitin and xylan; important in fungal cell wall degradation and biomass utilization
CE5	Cutinases, Feruloyl esterases	Plant cuticle polymers, Feruloylated substrates	Breakdown of plant cuticle and ferulic acid esters; aids in agriculture and biorefining
CE6	Acetyl xylan esterases	Acetylated xylans	Removal of acetyl groups; facilitates enzymatic hydrolysis of xylans for biofuel production
CE7	Acetyl esterases	Various carbohydrate esters	Hydrolysis of acetylated carbohydrates; aids in lignocellulosic biomass conversion
CE8	Pectin methylesterases	Pectin (methylated)	Demethylates pectins; essential in fruit ripening, juice clarification, and biomass breakdown
CE9	Acetyl esterases	Acetylated glycans	Modifies carbohydrate esters for various industrial and biotechnological applications
CE10	Esterases	Diverse carbohydrate esters	Broad substrate specificity; used in enzymatic processing of plant biomass and glycan remodeling
CE11	Deacetylases	Acetylated polysaccharides	Modifies the structure of polysaccharides, aiding in biomass breakdown
CE12	Feruloyl esterases	Feruloylated polysaccharides	Hydrolyzes ester linkages of ferulic acid; used in food industry for flavor enhancement and biofuel production
CE13	Esterases	Pectin derivatives	Breakdown of complex carbohydrates; involved in fruit and vegetable processing

(*continued*)

TABLE 11.13 (Continued)
Classification of Carbohydrate Esterase Family in the CAZy Database

CE Family	Enzyme Types	Substrates	Functions and Applications
CE14	Deacetylases	Chitin, Acetylated xylans	Hydrolysis of acetyl groups; assists in fungal degradation and chitin modification
CE15	Glucuronoyl esterases	Lignin–carbohydrate complexes	Breaks lignin–polysaccharide ester bonds; important for lignocellulosic biomass degradation
CE16	Esterases	Plant-derived esters	Participates in plant biomass deconstruction; used in food and textile industries
CE17	Esterases	Plant esters and polysaccharides	Broad-spectrum activity on carbohydrate esters; biotechnological applications
CE18	Deacetylases	Chitin and acetylated carbohydrates	Modifies structural carbohydrates; applications in agricultural waste processing
CE19	Esterases	Polysaccharides and their derivatives	Degradation of esterified carbohydrates; relevant for industrial enzymatic processes
CE20	Esterases	Plant polysaccharides	Hydrolysis of ester bonds in plant-derived carbohydrates; used in food and biomass industries

11.6 COMMERCIALLY AVAILABLE CARBOHYDRATE-ACTIVE ENZYMES WITH APPLICATION IN FOOD AND AGRICULTURE

11.7 MARKET ANALYSIS OF CARBOHYDRATE-ACTIVE ENZYMES

The global market for food enzymes has witnessed significant growth in recent years, driven by the increasing demand for processed and functional foods, as well as advancements in enzyme technology. According to early 2020 studies, the food enzyme market was valued at US$1,906.7 billion and is projected to grow at a compound annual growth rate (CAGR) of 6.2% between 2020 and 2031. By the end of 2031, the market is expected to reach an estimated valuation of US$4,124.7 billion (according to www.alliedmarketresearch.com/food-enzyme-market).

Consumers are increasingly looking for food products that are healthier, more digestible, and have improved texture and flavor. Enzymes play a crucial role in achieving these characteristics, making them highly desirable in the food industry. Among the diverse types of food enzymes (proteases, lipases, and carbohydrate enzymes), carbohydrate enzymes are projected to see the highest rise in usage. These enzymes aid in breaking down complex carbohydrates into simpler sugars, enhancing food texture, taste, and nutritional value. With continuous advancements in enzyme

TABLE 11.14
Commercially Available Carbohydrate-Active Enzymes

Megazyme products			
Enzyme	**Temperature Optima**	**Source**	**Applications**
Endo-1,4-β-xylanase	50°C	*Neocallismastix patriciarum*	Research in carbohydrate and biofuels and in food and feeds and paper pulping industries
α-Glucosidase	40°C	Yeast	Diagnostics and research within the food and feed, carbohydrate, and biofuel industries
β-Glucosidase	80°C	*Aspergillus* sp.	
β-Glucosidase	50°C	*Agrobacterium* sp.	
β-Galactosidase	60°C	*Aspergillus* sp.	
α-Glucosidase	60°C	Recombinant from *Bacillus stearothermophilus*	Carbohydrate and biofuels research
Pectate lyase	55°C	*Aspergillus* sp.	
Feruloyl esterase	40°C	Rumen microorganism	Biofuels, paper and pulp, medical, food nutrition, and pharmacological industries

Megazyme International Ireland Ltd. (n.d.). *Home*. www.megazyme.com/"https://www.megazyme.com (Accessed January 27, 2026).

Novozymes Products			
Industry Used	**Product Name**	**Enzyme Used**	**Benefit**
Baking	Pentopan®	Xylanase	Improved dough stability and quality
	AMG®	Glucoamylase	Enhanced fermentation
	BAN®	Amylase	Improved crust
	Celluclast®	Cellulase	Improved crumb texture
	Fungamyl®	Amylase	Improved crust, Enhanced fermentation
	Goldcrust®	Glucoamylase	
Clarification	Amylase®AG 300L	Amylase	Fruit juice clarification
	Amylase®AGXXL		
	Pectinex® Ultra clear juice clarification	Multiblend enzyme	Lemon juice clarification
Dairy	Lactozyme® Pure 6500L	β-Galactosidase	Lactose-free products
	Saphera®		UHT lactose-free milk
	Saphera®900L		Low lactose infant formula
	Saphera®2600L		Lactose-free milk and yogurt products
Ethanol	Biofuels Fiberex® F1	Cellulase	Increased corn oil yields, Higher fiber to ethanol yields
Beer	AMG®300L BrewQ	Glucoamylase	Achieved desired attenuation, Accelerated fermentation

Novozymes A/S. (n.d.). *Home*. www.novozymes.com/en" www.novozymes.com/en (Accessed January 27, 2026).

engineering, new and more efficient enzyme solutions are being developed, further fueling market expansion. Enzymes are natural biocatalysts that reduce the need for artificial additives, making them an attractive choice for consumers seeking clean-label and sustainable food products.

With increasing research, evolving consumer preferences, and technological developments, the food enzyme market—particularly carbohydrate enzymes—is set for steady and robust growth over the coming decade. As industries continue to adopt enzymatic processes to enhance food quality and production efficiency, the demand for food enzymes will remain on the rise.

11.8 CONCLUSION AND FUTURE PROSPECTS

Among the various CAZymes, amylase has been useful in various food applications, such as baking and sweetening agent, and also in fat mimics and the brewing industry. It also has applications in agriculture, as it can be helpful to improve soil quality. Xylanases are utilized in the food industry as well as in agriculture. Galactosidases also have a specific role in the production of bioactive compounds in the use of functional foods, probiotics, and prebiotics. These are also useful in agriculture, as their levels in plants are useful in monitoring stress conditions and improving soil quality. Also these are responsible for overall plant development. Pectinases are also responsible for increasing stress tolerance in plants as well as plant defence mechanisms. In the food and wine industries, they are beneficial in the clarification of fruit juices and as sweeteners and emulsification agents, respectively. Cellulases also have many applications in food production, as they are helpful in the extraction of various components and clarification of fruit juices and are also an important part of macerating enzymes. These are also applied to overall enhance the texture, flavor, and stability of various foods. GTs are also beneficial in the food industry, and they are also responsible for increasing stress tolerance in plants as well as in plant defence mechanisms. Various PL enzymes are required for extraction in fruit juices, food preservation, and production of prebiotics. These are important in plants, as they have been shown to be helpful in mechanisms to adapt to stresses as well as in defence. CEs are utilized for flavor enhancement and fruit and vegetable processing in the food industry; they also enhance biomass degradation.

For further enhanced use of these carbohydrate enzymes, genetic engineering approaches could be applied to develop enzymes with improved stability with a broad pH range, high temperature, and solvent stability. Genetic engineering can also help to develop functional foods. Also, we can focus on discovering new microbial sources for these enzymes to explore various properties of these enzymes to be incorporated in the food and agriculture industry.

ACKNOWLEDGMENT

The authors thank the Department of Biotechnology (jointly merged with Institute of Bioinformatics and Biotechnology), Savitribai Phule Pune University, Pune for providing the computational facility, research journal access, software plagiarism software access and other necessary infrastructure for writing the manuscript.

AUTHOR CONTRIBUTION

Pallavi Shere, Nihar Jog: original writing and first draft preparation and editing. Shadab Ahmed: manuscript outline, reviewing and editing.

FUNDING

SA is thankful to DST-ECRA (File No ECR/2015/000529) for providing funds for the personal computer which was used to write, review and edit the manuscript.

CONFLICT OF INTEREST STATEMENT

The authors declare that the research was done without any commercial or financial relationship that could be constructed as potential conflict of interest.

REFERENCES

Abdrabuo, M.M., Hefnawy, T.H., Al-Khatib, A.Y. and El-Maghraby, L.M., 2024. Comparison of some commercial enzymes used in the production of high maltose syrup from corn starch. *Egyptian Journal of Chemistry*, *67*(3), pp.483–494.

Ahmed, J., Thakur, A. and Goyal, A., 2021. Emerging trends on the role of recombinant pectinolytic enzymes in industries–An overview. *Biocatalysis and Agricultural Biotechnology, 38*, p.102200.

Aiyer, P.V., 2005. Amylases and their applications. *African Journal of Biotechnology*, *4*(13), pp. 1525–1529.

Allied Market Research. 2025. *Food Enzymes Market Size, Share, Trends, and Growth Report 2031*. Portland, OR: Allied Market Research www.alliedmarketresearch.com/food-enzyme-market

Amaral-Fonseca, M., Morellon-Sterling, R., Fernandez-Lafuente, R. and Tardioli, P.W., 2021. Optimization of simultaneous saccharification and isomerization of dextrin to high fructose syrup using a mixture of immobilized amyloglucosidase and glucose isomerase. *Catalysis Today*, *362*, pp.175–183.

Bajaj, P. and Mahajan, R., 2019. Cellulase and xylanase synergism in industrial biotechnology. *Applied Microbiology and Biotechnology*, *103*(21–22), pp.8711–8724.

Balakrishnan, D., Kumar, S.S. and Sugathan, S., 2019. Amylases for food applications—Updated information. In Binod Parameswaran, Sunita Varjani, Sindhu Raveendran (Eds.), *Green Bio-processes: Enzymes in Industrial Food Processing* (pp. 199–227). Springer.

CAZy Database. (n.d.). *Glycoside Hydrolases*. www.cazy.org/Glycoside-Hydrolases.html (Accessed January 27, 2026)

CAZy Database. (n.d.). *GlycosylTransferases*. www.cazy.org/GlycosylTransferase-family (Accessed January 27, 2026).

CAZy Database. (n.d.). *Polysaccharide Lyases*. www.cazy.org/Polysaccharide-Lyases.html (Accessed January 27, 2026)

Chakraborty, S., Rani, A., Dhillon, A. and Goyal, A., 2017. Polysaccharide lyases. In Ashok Pandey, Sangeeta Negi, and Carlos Ricardo Soccol (Eds.), *Current Developments in Biotechnology and Bioengineering* (pp. 527–539). Elsevier.

De Schepper, C.F., Buvé, C., Van Loey, A.M. and Courtin, C.M., 2022. A kinetic study on the thermal inactivation of barley malt α-amylase and β-amylase during the mashing process. *Food Research International, 157*, p.111201.

de Souza, T.S. and Kawaguti, H.Y., 2021. Cellulases, hemicellulases, and pectinases: Applications in the food and beverage industry. *Food and Bioprocess Technology, 14*(8), pp.1446–1477.

Ejaz, U., Sohail, M. and Ghanemi, A., 2021. Cellulases: From bioactivity to a variety of industrial applications. *Biomimetics, 6*(3), p.44.

Farooq, M.A., Ali, S., Hassan, A., Tahir, H.M., Mumtaz, S. and Mumtaz, S., 2021. Biosynthesis and industrial applications of α-amylase: A review. *Archives of Microbiology, 203*, pp.1281–1292.

Fernandes de Souza, H., Aguiar Borges, L., Dédalo Di Próspero Gonçalves, V., Vitor Dos Santos, J., Sousa Bessa, M., Fronja Carosia, M., Vieira de Carvalho, M., Viana Brandi, I. and Setsuko Kamimura, E. (2022). Recent advances in the application of xylanases in the food industry and production by actinobacteria: A review. *Food Research International (Ottawa, Ont.), 162*(Pt B), p.112103.

Franco, O.L., Rigden, D.J., Melo, F.R. and Grossi-de-Sá, M.F., 2002. Plant α-amylase inhibitors and their interaction with insect α-amylases: Structure, function and potential for crop protection. *European Journal of Biochemistry, 269*(2), pp.397–412.

Gavande, P.V., Goyal, A. and Fontes, C.M., 2023. Carbohydrates and carbohydrate-active enZymes (CAZyme): An overview. *Glycoside Hydrolases, 2,* pp.1–23.

Gernaey, B., Sorbara, J.O.B. and Nielsen, P.H., 2018. Environmental assessment of amylase used as digestibility improvement factor for intensive chicken production in Brazil. *Sustainability, 10*(8), p.2735. https://doi.org/10.3390/su10082735

González-Carranza, Z.H., Shahid, A.A., Zhang, L., Liu, Y., Ninsuwan, U. and Roberts, J.A., 2012. A novel approach to dissect the abscission process in Arabidopsis. *Plant Physiology, 160*(3), pp.1342–1356.

Gopinath, S.C., Anbu, P., Arshad, M.K., Lakshmipriya, T., Voon, C.H., Hashim, U. and Chinni, S.V., 2017. Biotechnological processes in microbial amylase production. *BioMed Research International, 2017*, p.1272193. https://doi.org/10.1155/2017/1272193

Grand View Research. 2024. *Food Enzymes Market Size, Share & Trends Analysis Report.* San Francisco, CA: Grand View Research. www.grandviewresearch.com

Jiang, J., Jiang, Z., Yan, Q., Han, S. and Yang, S., 2024. Biochemical characterization of a novel bifunctional alginate lyase from Microbulbifer arenaceous. *Protein Expression and Purification, 213*, p.106372.

Kumar, A., Dhiman, S., Krishan, B., Samtiya, M., Kumari, A., Pathak, N., Kumari, A., Aluko, R.E. and Dhewa, T., 2024. Microbial enzymes and major applications in the food industry: A concise review. *Food Production, Processing and Nutrition, 6*(1), p.85.

Kuhad, R. C., Gupta, R., & Singh, A. (2011). Microbial cellulases and their industrial applications. Enzyme research, 2011, 280696. https://doi.org/10.4061/2011/280696

Kumar, V., Dangi, A.K. and Shukla, P., 2018. Engineering thermostable microbial Xylanases toward its industrial applications. *Molecular Biotechnology, 60*(3), pp.226–235.

Li, M., Yong, Y., Zheng, Y., Hao, S., Wu, J. and Yang, X., 2024. Effect of dynamic ultra-high pressure microfluidization on the properties of Potato starch fat mimetics. *Starch-Stärke, 77,* p.2300249.

Lin, Y., He, H., Wen, Y., Cao, S., Wang, Z., Sun, Z., Zhang, Y., Wang, Y., He, W., Li, M. and Chen, Q., 2023. Comprehensive analysis of the pectate lyase gene family and the role of FaPL1 in strawberry softening. *International Journal of Molecular Sciences, 24*(17), p.13217.

Liu, P., Chen, Y., Ma, C., Ouyang, J. and Zheng, Z., 2023. β-Galactosidase: A traditional enzyme given multiple roles through protein engineering. *Critical Reviews in Food Science and Nutrition, 65*, pp.1–20.

Lombard, V., Golaconda Ramulu, H., Drula, E., Coutinho, P. M. and Henrissat, B., 2021. The carbohydrate-active enzyme database: functions and literature. *Nucleic Acids Research,*

49(D1), pp. D571–D577. https://doi.org/10.1093/nar/gkab1046 (Accessed January 27, 2026)

Majeed, H.N., Shaheen, S. and Kashif, M., 2024. Glycosyltransferases: Unraveling molecular insights and biotechnological implications. *Science Reviews. Biology*, *3*(1), p.16.

Menon, A., Pandurangan Maragatham, V., Samuel, M. and Arunraj, R., 2024. Properties and applications of α-galactosidase in agricultural waste processing and secondary agricultural process industries. *Journal of the Science of Food and Agriculture*, *104*(1), pp.21–31.

Michaud, P., Da Costa, A., Courtois, B. and Courtois, J., 2003. Polysaccharide lyases: Recent developments as biotechnological tools. *Critical Reviews in Biotechnology*, *23*(4), pp.233–266.

Murata, K., Inose, T., Hisano, T., Abe, S., Yonemoto, Y., Yamashita, T., Takagi, M., Sakaguchi, K., Kimura, A. and Imanaka, T., 1993. Bacterial alginate lyase: Enzymology, genetics and application. *Journal of Fermentation and Bioengineering*, *76*(5), pp.427–437.

Nthebere, K., Prakash, T.R., Bhimireddy, P., Chandran, L.P., Gudapati, J., Admala, M. and Prasad, K., 2024. Conservation agricultural practices promoted arbuscular mycorrhizal fungal colonization and glomalin content on sandy clay loam of southern India. *Heliyon*, *11*(1).

Oyenado, O.A. and Omoruyi, I.M., 2024. Review of amylase production by microorganisms and their industrial application. *Ife Journal of Science*, *26*(2), pp.537–553.

Plouhinec, L., Neugnot, V., Lafond, M. and Berrin, J.G., 2023. Carbohydrate-active enzymes in animal feed. *Biotechnology Advances*, *65*, p.108145.

Prade, R.A., Zhan, D., Ayoubi, P. and Mort, A.J., 1999. Pectins, pectinases and plant-microbe interactions. *Biotechnology & Genetic Engineering Reviews*, *16*, pp.361–391.

Ranwala, Anil P., Chiyuki Suematsu, and Hiroshi Masuda. 1992. "The Role of β-Galactosidases in the Modification of Cell Wall Components during Muskmelon Fruit Ripening." *Plant Physiology* 100 (3): 1318–25. **https://doi.org/10.1104/pp.100.3.1318.**

Sabadash, N., Hrabovska, O., Fesych, I., Avramenko, A. and Serhiienko, A., 2021. Effect of the combined use of β-amylase and pullulanase on the carbohydrate composition of maltose syrups. *Ukrainian Food Journal*, 10(4), 673–684.

Saini, R., Saini, H.S. and Dahiya, A., 2017. Amylases: Characteristics and industrial applications. *Journal of Pharmacognosy and Phytochemistry*, *6*(4), pp.1865–1871.

Salam, L.B., 2018. Detection of carbohydrate-active enzymes and genes in a spent engine oil-perturbed agricultural soil. *Bulletin of the National Research Centre*, *42*, pp.1–18.

Sathya, T.A. and Khan, M., 2014. Diversity of glycosyl hydrolase enzymes from metagenome and their application in food industry. *Journal of Food Science*, *79*(11), pp.R2149–R2156.

Shahi, N., Hasan, A., Akhtar, S., Siddiqui, M., Sayeed, U. and Khan, M., 2016. Xylanase: A promising enzyme. *Journal of Chemical and Pharmaceutical Research*, *8*, pp.334–339.

Sharma, K., Thakur, A. and Goyal, A., 2019. Xylanases for food applications. In Binod Parameswaran, Sunita Varjani, and Sindhu Raveendran (Eds.), *Green Bioprocesses: Enzymes in Industrial Food Processing*. Springer. https://doi.org/10.1007/978-981-13-3263-0_7.

Shinde, V.K. and Vamkudoth, K.R., 2022. Malt oligosaccharide forming amylases and their applications in food and pharma industry. *Journal of Food Science and Technology*, *59*, pp.3733–3744.

Singh, T.A., Jajoo, A. and Bhasin, S., 2020. Production and application of glucose isomerase from *Streptomyces enissocaesilis* and amylase from Streptomyces sp. for the synthesis of high fructose corn syrup. *SN Applied Sciences*, *2*(12), p.1968.

Sutaoney, P., Rai, S.N., Sinha, S., Choudhary, R., Gupta, A.K., Singh, S.K. and Banerjee, P., 2024. Current perspective in research and industrial applications of microbial cellulases. *International Journal of Biological Macromolecules*, *264*, p.130639.

Uluisik, S. and Seymour, G.B., 2020. Pectate lyases: Their role in plants and importance in fruit ripening. *Food Chemistry*, *309*, p.125559.

Wang, J., Hu, Y., Guo, D., Gao, T., Liu, T., Jin, J., Zhao, M., Yu, K., Tong, W., Ge, H. and Pan, Y., 2024. Evolution and functional divergence of glycosyltransferase genes shaped the quality and cold tolerance of tea plants. *The Plant Cell*, *37*, p.268.

Wang, X., Xu, M., Ren, X., Li, M., Wang, C., Yang, F. and Li, X., 2024. High-level expression and biochemical characterization of a Maltotetraose Amylase in Pichia pastoris X-33 for Maltotetraose production. *Applied Biochemistry and Biotechnology*, *196*, pp.6745–6758.

Winarti, S. and Anggreini, R.A., 2022. Development of liquid sugar HFS (High Fructose Syrup) from Lesser Yam Tubers using enzyme-mix Inulinase and Amylase. In *MATEC Web of Conferences* (Vol. 372, p. 03004). EDP Sciences.

www.cazy.org/Glycoside-Hydrolases.html

www.cazy.org/GlycosylTransferase-family

www.cazy.org/Polysaccharide-Lyases.html

Zhao, K., Tan, H., Fang, C., Zhou, Z., Wu, C., Zhu, X., Liu, F., Zhang, Y. and Li, H., 2024. An activatable fluorescence probe for rapid detection and in situ imaging of β-galactosidase activity in cabbage roots under heavy metal stress. *Food Chemistry*, *452*, p.139557.

Zheng, L., Xu, Y., Li, Q. and Zhu, B., 2021. Pectinolytic lyases: A comprehensive review of sources, category, property, structure, and catalytic mechanism of pectate lyases and pectin lyases. *Bioresources and Bioprocessing*, *8*, pp.1–13.

Index

H

I

L

M